**Advanced Modular Sciences**

Chemistry

A2

Series Editor: Colin Chambers

Peter Harwood

Mike Hughes

This book has been designed to support AQA Chemistry specification B. It contains some material which has been added in order to clarify the specification. The examination will be limited to material set out in the specification document.

Published by HarperCollins*Publishers* Limited
77-85 Fulham Palace Road
Hammersmith
London W6 8JB

www.CollinsEducation.com
Online support for schools and colleges

© Peter Harwood and Mike Hughes
First published 2001
Reprinted 2002
ISBN 0 00 327754 2

Peter Harwood and Mike Hughes assert the moral right to be identified as the authors of this work.

British Library Cataloguing in Publication Data
A catalogue record for this publication is available from the British Library

Cover design by Chi Leung
Design by Ken Vail Graphic Design and Caroline Grimshaw
Illustrations by Barking Dog Art, Peter Harper and Illustrated Arts
Picture research by Caroline Thompson
Commissioned photos by Andrew Lambert
Index by Julie Rimington
Production by Kathryn Botterill
Edited by Nick Allen, Mary Sanders and Pat Winter
Commissioned by Martin Davies

Printed and bound by Scotprint

The publisher wishes to thank the Assessment and Qualifications Alliance for permission to reproduce examination questions.

You might also like to visit
www.**fire**and**water**.com
The book lover's website

Further Physical and Organic Chemistry

Thermodynamics and Further Inorganic Chemistry

Acknowledgements

Every effort has been made to contact the holders of copyright material, but if any have been inadvertently overlooked the publishers will be pleased to make the necessary arrangements at the first opportunity.

The publishers would like to thank the following for permission to reproduce photographs (T = Top, B = Bottom, C = Centre, L= Left, R = Right):

Ace Photoagency/Adams, 80BR, A Hughes, 68;
Adams Picture Library, 172, 238CR;
Heather Angel, 116TC;
Photos from www.John Birdsall.co.uk, 32, 80CR, 220;
Chris Bonington Picture Library, 111CL;
Rijksmuseum Kroller-Muller, Otterlo, Netherlands/Bridgeman Art Library, 213;
Camera Press/McInery, 74, Paynter, 95CL;
J Allan Cash Ltd, 71;
Cephas/M Rock, 62T, S Boreham, 62C;
Colorsport, 35;
Corbis/P Ward, 14CL, A Nachoum, 14CR, H Diltz, 166TC, D Jones, 166BC;
The Craft Council Photo Library/J Malone, E Lewenstein, P Lane, 206T;
Sue Cunningham Photographic, 86TR;
John Feltwell, 63L;
Food Features Ltd, 89;
Geophotos/Tony Waltham, 82;
GettyOne Stone, 50;
Glenmorangie plc, 21;
Robert Harding Picture Library, 28, 67, 154R, Westlight/Morgan, 111TR;
Hutchison Library, 78BL;
ICCE/Boulton, 133CL;
ICI, 63R;
ICI Chemicals & Polymers, 60;
Chris Kapolka, 25B;
Koolpack, 173;

Andrew Lambert, 2CL, 5, 6, 11, 14B, 15, 20, 29, 34, 39, 65, 80BL, 98, 106, 108C, 110, 117, 118, 119, 120, 154L&B, 163, 167, 179, 180, 181, 185, 188, 192, 194, 198, 214, 215, 216, 221, 223, 224, 232R, 239, 240, 243, 250;
BIOPOL: Trademark and Property of Monsanto plc, 112;
NASA Dryden Flight Center, 204;
Natural History Museum, 140;
ëPAí News, 2C, 233;
Vinegar and Oil Set by Simone ten Homple, 1991, reproduced with permission of the P & O Makower Trust, 232L;
Phototake NYC/B Masini, 108L;
Gareth Price, 154TR;
Quadrant Picture Library/Auto Express, 133(inset);
Rex Features Ltd, 25TL, 80CL, 88, 95CR;
By permission of the President and Council of the Royal Society, 116TR;
Science Photo Library, 2T, 78BR, 86TL, 113, 126, 135, 143, 176, 196, 206B, 207, 211, 217, 235, 238T;
Still Pictures/M Edwards, 23, 212;
Topham Picturepoint, 70B.

Front cover:
Images supplied by: Science Photo Library (top left)
Robert Harding Picture Library/Westlight/Morgan (Centre)
GettyOne Stone (top right)

To the student

This book covers the content needed for the AQA Specification in Chemistry at A2-level. It aims to give you the information you need to get a good grade and to make your study of advanced chemistry successful and interesting. Chemistry is constantly evolving and, wherever possible, modern issues and problems have been used to make your study stimulating and to encourage you to continue studying chemistry after you complete your current course.

Using the book

Don't try to achieve too much in one reading session. Chemistry is complex and some demanding ideas need to be supported with a lot of facts. Trying to take in too much at one time can make you lose sight of the most important ideas – all you see is a mass of information.

Each chapter starts by showing how the chemistry you will learn is applied somewhere in the world. At other points in the chapter you may find more examples of the way the chemistry you are covering is used. These detailed contexts are not needed for your examination but should help to strengthen your understanding of the subject.

The numbered questions in the main text allow you to check that you have understood what is being explained. These are all short and straightforward in style – there are no trick questions. Don't be tempted to pass over these questions, they will give you new insights into the work. Answers are given in the back of the book.

The Key Facts summarise the information you will need in your examination. However, the examination will test your ability to apply these facts rather than simply to remember them. The main text in the book explains these facts. The application boxes encourage you to apply them in new situations.

Past paper questions are included at the end of each chapter. These will help you to test yourself against the sorts of questions that will come up in your examination. There is also a section that covers synoptic assessment questions to help you prepare for the synoptic element of your final examination.

Good luck!

1 Kinetics

Being relatively unreactive at ground level, CFCs find their way to the stratosphere and cause ozone depletion there.

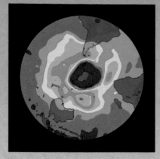

It takes a relatively small, well placed explosion – a reaction that generates energy very rapidly – to demolish old flats. Magnesium reacts with hydrochloric acid much more slowly.

By 2000 the ozone hole in the stratosphere over the southern hemisphere had extended as far as the southern parts of Africa, South America and Australia. Ozone, normally present in the upper atmosphere, absorbs harmful high-energy ultraviolet-B rays from the Sun. Without ozone, this radiation reaches the Earth's surface and can be the cause of skin cancers and eye cataracts. This is why schoolchildren in Australia must now wear hats outdoors.

In 1987, many countries agreed to reduce the use of chlorofluorocarbons (CFCs), chemicals that help to deplete the ozone layer. The reductions are starting to take effect: levels of CFCs in the lower atmosphere are beginning to decline, and so less of the CFCs will make their way to the upper atmosphere. Some scientists expect the ozone hole to start shrinking within ten years, and that it might even vanish by 2050.

Because CFCs are so unreactive at ground level, and therefore safe to health, and because they are easily liquefied under pressure and at moderately low temperatures, they were judged ideal for use as refrigerants and as propellants in aerosols. But the energy in ultraviolet light in the stratosphere breaks them down, producing chlorine free radicals. These particles react with ozone, O_3, which is broken down to oxygen, O_2, a gas which does not shield the Earth from UV-B.

It is the slow rate at which CFCs themselves react that is the problem. If they were reactive, they would do so before reaching the upper atmosphere. Other reactions we are familiar with take place at different rates.

Rusting is quite slow, and its rate can be difficult to measure; so is the rate of an explosion, which is a very fast reaction; while the rate at which magnesium reacts with acid is much easier to follow.

1.1 Reaction kinetics

Reaction kinetics is the study of the rate of chemical reactions, and in this chapter we look at the kinetics of reactions whose rates are measurable.

The rate of a chemical reaction is the change in molar concentration of a substance in unit time.

In the laboratory, we calculate the rate of a reaction by measuring the change in concentration over time, for example either as a decrease in concentration of reactant per unit time, or as an increase in concentration of product per unit time. The rate is often measured as change in moles per dm^3 per second, $mol\ dm^{-3}\ s^{-1}$, but slower reactions may be measured in moles per dm^3 per minute or per hour.

Chemists who design industrial processes need to know the rate of reactions to answer questions such as: Will the reaction give a good yield of product per hour at normal temperatures? What conditions will give optimum reaction rate? Questions like these are answered by studying reaction kinetics. The answers can be found only by experiment, and cannot be predicted merely by looking at an equation for the reaction, as we shall see later.

Following Chapter 6 in *AS Chemistry*, this section examines further aspects of reaction kinetics. We will be looking at how reaction rates change, why they change, and what this tells us about how the species are reacting.

Collision theory
In any gas, liquid or solution, the particles are constantly moving and colliding with each other. At each collision, energy is transferred between the particles, so the particles are

continually changing their energies. The distribution of energies between particles in a system is called the **Maxwell–Boltzmann distribution** of energies. The curve in Fig. 1 shows the numbers of particles at each energy value at temperature T_1. The area under the curve represents the total number of particles, so the number, and therefore the area below the curve, does not change when the temperature changes.

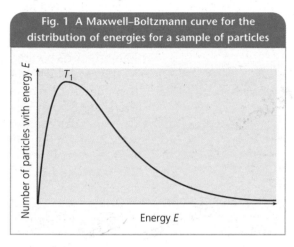

Fig. 1 A Maxwell–Boltzmann curve for the distribution of energies for a sample of particles

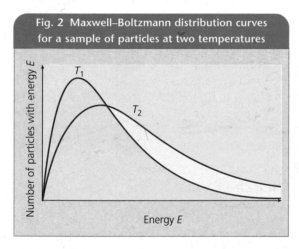

Fig. 2 Maxwell–Boltzmann distribution curves for a sample of particles at two temperatures

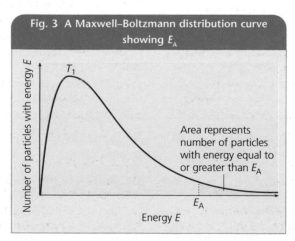

Fig. 3 A Maxwell–Boltzmann distribution curve showing E_A

Area represents number of particles with energy equal to or greater than E_A

Increasing the temperature adds energy to the system and changes the distribution of energies of the particles. At a higher temperature (T_2 in Fig. 2), more particles have higher energies.

Collision theory tells us that chemical reactions occur when particles have enough energy from these collisions to break existing chemical bonds and form new ones. The amount of energy needed is called the **activation energy**, E_A, see Fig. 3.

1 Look at the Maxwell–Boltzmann distribution curves.

a What does the peak of the curve tell you?

b The curves start at the origin. What does this tell us about the energies of the particles?

c Would the curve ever touch the energy axis? Explain your reasoning.

Any factor which affects the *number* of collisions or the *energy* of the collisions will affect the rate of reaction, because it changes the number of collisions possessing the activation energy. Such factors are described in detail in *AS Chemistry*, pages 94–97, and are summarised below.

The effect of temperature

Increasing the temperature increases the average kinetic energy of the particles, they move faster and hence there are more collisions. But far more significant in predicting the number of reacting particles is *the number of collisions between particles that possess energies equal to or greater than the activation energy*. As a rough guide, a 10 K rise in temperature will double the number of collisions having energies equal to or greater than the activation energy, so twice as many particles can react at any one time.

You may be wondering: Since higher temperatures cause particles to move faster, giving more collisions, doesn't the increased overall number of collisions have a large effect on the reaction rate as well as particle energies? The answer is that over a 10 K rise, there are only about 3% to 4% more collisions, so this effect is quite small.

Consider a set of particles at 300 K: if we heat the particles to 310 K, then twice as many particles will have energies equal to or greater than the activation energy, so twice as many particles can react at any instant. Hence the rate at the higher temperature will be approximately twice that at the lower temperature:

An increase in temperature of 10 K approximately doubles the reaction rate.

Most reactions follow this simple rule, so the Maxwell–Boltzmann curve is a similar shape for their particles. Fig. 4 shows that increasing the temperature by 10 K doubles the number of particles possessing the activation energy.

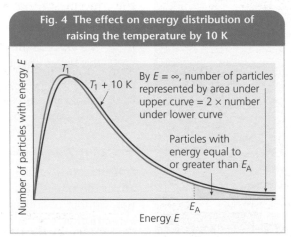

Fig. 4 The effect on energy distribution of raising the temperature by 10 K

The higher temperature gives a greater spread of energies of particles, taking the maximum to a higher energy value, so the peak of the curve moves lower and to the right. Remember: the area under the curve stays the same, while the area of the curve beyond the activation energy value is greater – it is twice the area for 300 K.

The effect of surface area
When we increase the surface area of a solid, such as by grinding it to a powder, it reacts faster than if it were still one solid lump. This is because collisions occur only at the surface of a solid, and if a greater surface area is exposed, there are more locations for collisions to occur, and therefore the reaction happens faster. For example a 1 cm cube has an area of 6 cm², but if this is ground to a powder of 10^{-7} cm particles, then the surface area is about the same as 30 tennis courts.

The effect of concentration
As the concentration of a reactant increases, there are more particles in a given volume, so there will be a greater number of particles with energies equal to or greater than the activation energy. This produces a greater number of collisions that result in a reaction. The same applies to a gas when pressure is increased.

In the next section, we look at the effect of concentration in more detail.

2

a If you want to make a jelly by dissolving jelly cubes in water, how much quicker will it dissolve if you slice each of the 2 cm cubes up into eight 1 cm cubes?

b If we increase the temperature, what effect does this have on the rate of dissolving? Explain your answer in terms of collisions that result in a reaction.

1.2 Deriving the rate equation

Chemists define the rate of a reaction as the rate at which a reactant is used up, or the rate at which a product is formed. We state the amount of reactant or product in terms of its concentration, and this is measured as mol dm^{-3}. By measuring concentrations at time intervals, *at a fixed temperature*, we can measure the rate of change of concentration.

The students are measuring a rate of reaction using a colorimeter. They are monitoring the reaction of a haloalkane. As it reacts it releases bromide ions which form a silver bromide precipitate with silver nitrate solution. The students measure the rate at which the solution goes cloudy. This allows them to calculate the concentrations of products and hence the concentrations of the reactants

Fig. 5 Rate at which a reactant is used up in a reaction

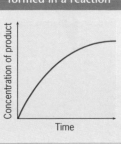

In Fig. 5, the line is a curve because the reactant is being used up and its concentration is decreasing more slowly with each successive time interval. This is because, as the reaction proceeds, fewer reactant particles per unit volume are available for collisions, so the time taken for a given number of particles to react increases. The rate of reaction is decreasing and eventually it stops.

Fig. 6 Rate at which a product is formed in a reaction

3

a At what stage in a reaction is the reactant used up fastest?

b At what stage is the formation of the product fastest?

At a fixed temperature, increasing the concentration of the reactants usually increases the rate of reaction. In some reactions, but by no means all, doubling the concentration of a reactant (call it A) doubles the number of particles of A with energies equal to or greater than the activation energy, and the reaction rate doubles. When you halve the concentration, the number of particles of A with the activation energy will halve, and the rate will be halved. In other words, the rate of reaction is proportional to the concentration of A, given by the expression:

rate $\propto$ concentration of A

By including a **rate constant**, k, we can replace the proportional sign by an equals sign, to give the **rate equation**:

$$\text{rate} = k \times \text{concentration of A}$$

Using the symbol [A] for concentration of A, gives:

$$\text{rate} = k[A]$$

In this example, the rate of reaction is determined by the concentration of A, which is $[A]^1$. In other reactions, the concentrations of two reactants may affect the rate in a similar way:

$$\text{rate} = k[A]^1 [B]^1$$

The powers in the equation are usually small numbers, or zero, and are called the **orders of reaction**.

So, in this example the rate is first order with respect to A and first order with respect to B, and overall the reaction is second order: the powers are added to give the overall order.

A general rate equation for two reactants is given as:

$$\text{rate} = k[A]^m [B]^n$$

Orders of reaction are found from experimental data and cannot be assumed from the stoichiometric equation.

Zero-order reactions

There are reactions whose rate does not change when the concentration of a reactant changes. If this applies to reactant A whose rate equation is:

$$\text{rate} = k[A]^x$$

then the expression for $[A]^x$ must always equal 1. This is achieved by using the power zero, hence the reaction is a **zero-order reaction**:

$$\text{rate} = k[A]^0 \quad \text{which gives} \quad \text{rate} = k$$

An example of a reaction which is zero order is the decomposition of ammonia to nitrogen on a hot tungsten wire:

$$2NH_3 \rightarrow N_2 + 3H_2$$

For a zero-order reaction, the reaction rate is always constant and always equal to the rate constant k, over the concentration range of A being considered. Fig. 7 shows graphs of rate against time and concentration against time for such a zero-order reaction.

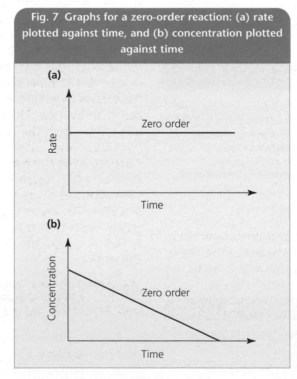

Fig. 7 Graphs for a zero-order reaction: (a) rate plotted against time, and (b) concentration plotted against time

(a)

Rate | Time — Zero order

(b)

Concentration | Time — Zero order

First-order reactions

We have seen that the rate of a reaction may be directly proportional to the concentration of one species, and that, taking this species to be A, the rate equation for reactant A is given as:

$$\text{rate} = k[A]^1 \quad \text{or} \quad \text{rate} = k[A]$$

Any such reaction is called a **first-order**

reaction with respect to A, and the figure for [A] inserted into the rate equation is the value for the concentration of A. If the concentration doubles, the rate of the reaction will double also.

A reaction which behaves in this way is the hydrolysis of the ester methyl ethanoate:

$$CH_3COOCH_3 + H_2O \rightarrow CH_3COOH + CH_3OH$$

The order for the reaction of the ester is calculated by finding its concentration at different stages in the reaction. An easy way is to determine how much has been converted to the acid, and then to calculate the concentration of ester left in the mixture. You do this by titrating a *small*, measured volume of the reaction mixture with alkali. This tells you how much acid has been produced, and so you can calculate the concentration of ester remaining in the reaction mixture. The results from a typical reaction are shown in Table 1.

Table 1 Results for the reaction of methyl ethanoate	
Time/min	Ester concentration/mol dm^{-3}
0	0.200
5	0.134 = A
10	0.095
15	0.068 = B
20	0.046
25	0.034 = C

The students are titrating the reaction mixture. The small sample they use is mixed with some ice to stop the reaction while they do the titration

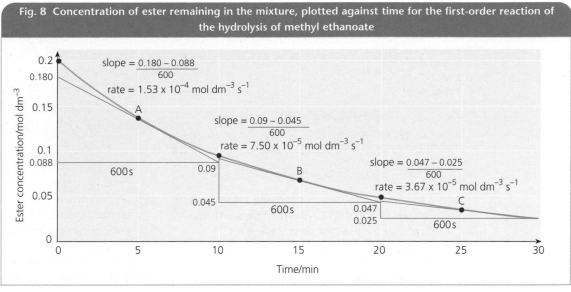

Fig. 8 Concentration of ester remaining in the mixture, plotted against time for the first-order reaction of the hydrolysis of methyl ethanoate

These results are plotted on the graph of Fig. 8 which shows the typical declining curve for a first-order reaction. You can find the rate of reaction at a particular time by drawing a tangent to the curve. The rate is the slope of the curve at the tangent.

$$\text{rate} = \frac{\text{change in concentration}}{\text{time taken}} = \text{slope of curve}$$

The rates of ester hydrolysis in mol dm^{-3} s^{-1} at points A, B and C are 0.000153, 0.0000750 and 0.0000367.

Plotting the rates against the concentrations shown in Table 1 gives a straight line (Fig. 9). The graph shows that for this reaction:

$$\begin{matrix}\text{rate of hydrolysis} \\ \text{of ester}\end{matrix} \propto \begin{matrix}\text{concentration} \\ \text{of ester}\end{matrix}$$

and: $\text{rate} = k[CH_3COOCH_3]$

The rate of the reaction is proportional to the concentration of the reagent to the power one, namely rate = $k[CH_3COOCH_3]^1$, so the reaction is first order.

In some reactions it may not be easy to find the reactant concentration, or there may be a more convenient way of following the rate of reaction. For example, the decomposition of hydrogen peroxide in aqueous solution can easily be followed by measuring the volume of oxygen gas evolved (Fig. 10). A catalyst such as powdered metal or manganese(IV) oxide is required to speed up the reaction and is introduced into the flask in a small test-tube. The reaction starts when the catalyst is submerged in the hydrogen peroxide solution. The oxygen evolved is monitored using the gas syringe.

You cannot take samples in this type of reaction, but you know the concentration at the start of the reaction. Plotting a graph of the volume of oxygen evolved against time will allow a tangent to be drawn at time = 0, giving the *initial* rate of reaction when we did know the concentration of the hydrogen peroxide.

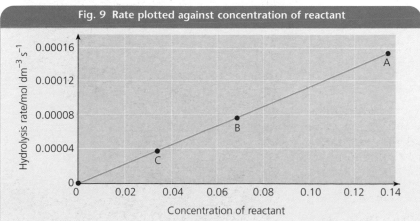

Fig. 9 Rate plotted against concentration of reactant

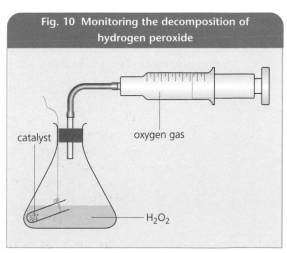

Fig. 10 Monitoring the decomposition of hydrogen peroxide

Table 2 Volumes of oxygen produced in the decomposition of hydrogen peroxide			
Initial concn. of H₂O₂: $A: 0.1 \text{ mol dm}^{-3}$ $B: 0.2 \text{ mol dm}^{-3}$ $C: 0.3 \text{ mol dm}^{-3}$			
Time/s	Volume of O_2/cm^3		
	A	B	C
0	0	0	0
10	3	6	9
20	6	11	17
30	8	16	23
40	10	19	28
50	11	22	32
60	12	24	36
70	13	26	38
80	14	27	40
90	14	28	41.5
100	14	29	42
110	15	29	43
120	15	30	43.5
130	15	30	43.5
140	15	30	44

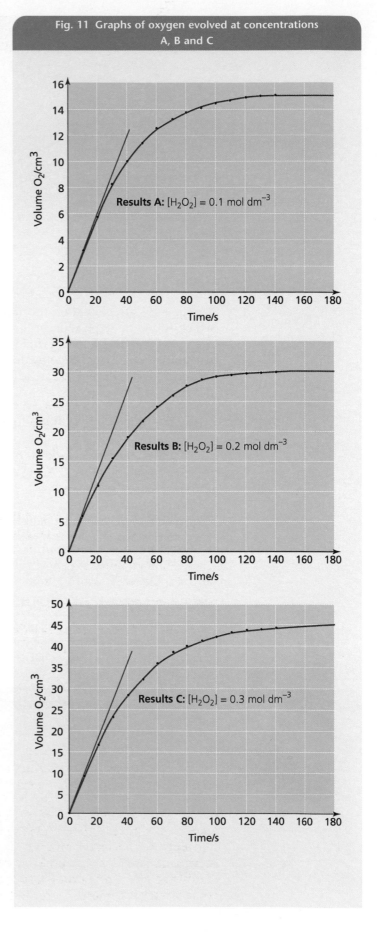

Fig. 11 Graphs of oxygen evolved at concentrations A, B and C

Table 2 shows a set of results for the decomposition of hydrogen peroxide using different initial concentrations. The total volume of O_2 produced is given at 10-second intervals. The graphs of the results and the initial rate tangents are shown in Figs. 11A–C. By drawing a tangent to each curve at 0 seconds, the initial rates can be calculated as volume of oxygen gas given off per second. The rates are given in this table:

	Initial concn H₂O₂/mol dm⁻³	Rate O₂ prodn./cm³ s⁻¹
A	0.1	0.3
B	0.2	0.6
C	0.3	0.9

Using these results and plotting initial rate against initial concentration gives Fig. 12.

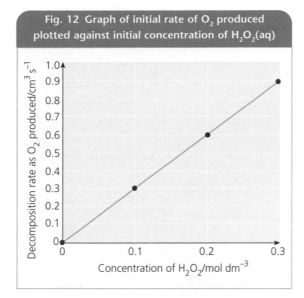

Fig. 12 Graph of initial rate of O_2 produced plotted against initial concentration of H_2O_2(aq)

You can see from the graph that:

$$\text{rate} \propto \text{concentration}$$

so:
$$\text{rate of decomposition of hydrogen peroxide} = k[H_2O_2]^1$$

Hence, the reaction for oxygen produced from hydrogen peroxide solution is also first order.

The reaction between the phosphinate ion and the hydroxide ion produces hydrogen gas:

$$H_2PO_2^- + OH^- \rightarrow HPO_3^- + H_2$$

You can find the overall order of this reaction by looking at the effect of changing the initial concentration of each reactant in turn. Table 3 gives data for a series of experiments in which phosphinate reacts with hydroxide ions.

Table 3 Results for six experiments

Expt.	[$H_2PO_2^-$]	[OH^-]	Initial rate (mol dm^{-3} s^{-1})
1	0.1	6.0	14.4
2	0.2	6.0	28.8
3	0.3	6.0	43.2
4	0.6	1.0	2.4
5	0.6	2.0	9.6
6	0.6	3.0	21.5

If the hydroxide ion concentration is kept constant and phosphinate ion concentration is doubled from 0.1 to 0.2 mol dm^{-3}, the initial reaction rate also doubles (from 14.4 in Expt. 1 to 28.8 in Expt. 2). Trebling the phosphinate concentration causes a threefold increase in rate (Expts.1 and 3). In other words, the rate is proportional to the concentration of phosphinate, and we say that the reaction is first order with respect to phosphinate.

Second- and third-order reactions

If in the above reaction we then keep the concentration of phosphinate constant and double the concentration of hydroxide (from 1.0 to 2.0 mol dm^{-3}), the rate increases by a factor of 4 (from 2.4 in Expt. 4 to 9.6 in Expt. 5). Trebling the hydroxide concentration causes the rate to increase by a factor of 9 (Expts. 4 and 6).

Consider Experiments 4 and 5 where [$H_2PO_2^-$] remains constant and [OH^-] doubles. We can write the following equation and cancel:

$$\frac{\text{rate 5}}{\text{rate 4}} = \frac{9.6}{2.4} = \frac{k[2.0]^m \, \cancel{[0.6]^1}}{k[1.0]^m \, \cancel{[0.6]^1}} = 4$$

to give:
$$\frac{2.0^m}{1.0^m} = 4$$

1 to any power is 1, so $2.0^m = 4$ or 2^2. Hence, the small, whole-number value of m that satisfies this expression is 2, and the rate of the reaction is proportional to [OH^-]^2. We describe the reaction as **second order** with respect to hydroxide ions.

We have therefore found that the reaction is first order with respect to phosphinate, and second order with respect to hydroxide, so that the overall rate equation for the reaction is given by the equation:

$$\text{rate} = k[H_2PO_2^-]^1 [OH^-]^2 \quad \text{or} \quad \text{rate} = k[H_2PO_2^-] [OH^-]^2$$

We now refer again to the general rate equation (page 5):

$$\text{rate} = k[A]^m [B]^n$$

and define the order for the whole reaction:

The overall order of a reaction is the sum of the powers of the concentration in terms of the rate equation.

In the rate equation for the phosphinate/hydroxide reaction, $m = 1$ and $n = 2$, the sum of the powers is $1 + 2 = 3$, so the overall order of reaction is three. This makes the reaction between phosphinate and hydroxide ions a **third-order reaction**.

If a particular reagent is present in large excess, its concentration will change negligibly as the reaction proceeds and will not appear to affect the rate. The hydrolysis of an ester (page 6) is an example of this: the whole reaction is carried out in water, so the amount of water used up is negligible.

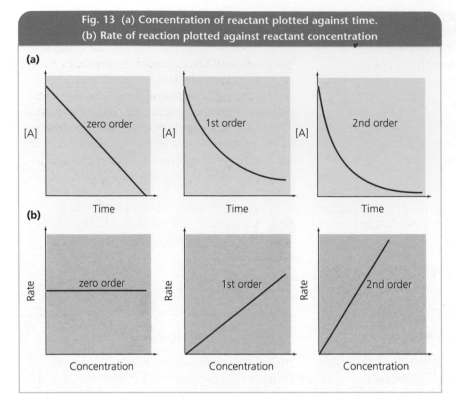

Fig. 13 (a) Concentration of reactant plotted against time.
(b) Rate of reaction plotted against reactant concentration

Summarising orders of reaction

The graphs of Fig. 13(a) show how the concentration of reactant A changes with time for reactions of orders 0, 1 and 2, and the graphs of Fig. 13(b) show how the reaction rate varies with concentration for reactions of orders 0, 1 and 2.

Units for rate constant *k* for different orders of reaction

We have seen that rate = change in concentration/time taken, and that rate is measured as $mol\ dm^{-3}\ s^{-1}$.

We have also seen that (1st order)
rate = k × concentration.

For a zero-order reaction, where rate = $k[A]^0$, and $[A]^0 = 1$, rate constant: $k = \dfrac{rate}{1}$

So the units for a zero-order reaction are those for rate, namely $mol\ dm^{-3}\ s^{-1}$.

For a first-order reaction, in which rate = $k[A]$, we can write

$$k = \frac{rate}{[A]} \quad or \quad k = \frac{rate}{concentration}$$

We can insert the units of rate and concentration and cancel:

$$\frac{\cancel{mol\ dm}^{-3}\ s^{-1}}{\cancel{mol\ dm}^{-3}} = s^{-1}$$

Therefore, for a first-order reaction, the units for *k* are s^{-1}.

For a second-order reaction, we can write

$$rate = k[A]^2, \text{ giving } k = \frac{rate}{[A]^2}$$

Alternatively, the rate can be $k[A][B]$, giving

$$k = \frac{rate}{[A][B]}$$

In both cases, we can say that

$$k = \frac{rate}{[concentration]^2}$$

Inserting units and cancelling:

$$\frac{\cancel{mol\ dm}^{-3}\ s^{-1}}{(\cancel{mol\ dm}^{-3})(mol\ dm^{-3})} = \frac{s^{-1}}{(mol\ dm^{-3})} = mol^{-1}\ dm^3\ s^{-1}$$

Therefore, for a second-order reaction, the units for *k* are $mol^{-1}\ dm^3\ s^{-1}$.

Similar rules apply for higher-order reactions: the rate constant *k* does not always have the same units since these depend on the order of reaction (see Table 4).

4 For the reaction $2NO + Cl_2 \rightarrow 2NOCl$, the rate equation is rate = $k[NO]^2[Cl_2]$.

a Give the order of this reaction with respect to the reagent NO.

b Give the order of this reaction with respect to the reagent Cl_2.

c Give the overall order of reaction.

d Work out the units for *k* in the rate equation.

Effect of temperature on the rate constant

When carrying out kinetics investigations and reporting your data to others, you must make a note of the conditions. If you are monitoring the volume of a gas produced, then the atmospheric pressure can affect the volume, so this should be quoted. You should also quote the temperature at which you carried out your experiments because temperature will affect the rate of reaction and hence the value for the rate constant.

Table 4 Summary of units for rate constants		
Order of reaction = n + m	**Rate equation rate = k x [reactants]$^{n+m}$**	**Units of the rate constant**
0	rate = k x conc^0	$mol\ dm^{-3}\ s^{-1}$
1	rate = k x conc^1	s^{-1}
2	rate = k x conc^2	$mol^{-1}\ dm^3\ s^{-1}$
3	rate = k x conc^3	$mol^{-2}\ dm^6\ s^{-1}$
4	rate = k x conc^4	$mol^{-3}\ dm^9\ s^{-1}$

We have seen that the rate depends upon temperature, since increasing the temperature increases the number of molecules having energies equal to or greater than the activation energy.

$$rate = k[A]x$$

If the temperature is increased without altering the concentration of A, k must increase because the rate increases when the temperature increases.

Remember, a 10 K rise in temperature will approximately double the rate of some simple reactions so the equation shows it must double the rate constant.

When acid is added to a solution containing thiosulphate ions, free sulphur is gradually formed, and this turns the solution a cloudy yellow. For the same volumes and concentrations of reactants, the time at different temperatures can be measured for a cross to be obscured.

5 An investigation which you might have done in the laboratory is on the reaction between thiosulphate ions and hydrochloric acid. The reaction gives sulphur as a product: the solution goes cloudy and you can measure the time taken to obscure a cross at the bottom of the flask, as shown below left. Some results for this reaction are:

Temperature/°C	20	30	40	50	60
Time to obscure cross/s	140	71	34	17	8

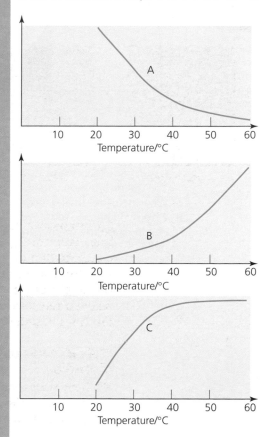

a Which graph represents the results of time plotted against temperature?

b Which graph represents the results of rate plotted against temperature?

c State the relationship between the temperature in this reaction and the rate constant.

KEY FACTS

- The order of a reaction is the sum of the powers of the concentration terms.
- The general form of the rate equation is:
  $rate = k[A]^m[B]^n$
- The units for k in the rate equation depend upon the order of reaction.

11

1 The diagram below shows the distribution curve for the energies of the molecules in a mixture of two gases at a given temperature.
E_a is the activation energy for the reaction between the two gases.

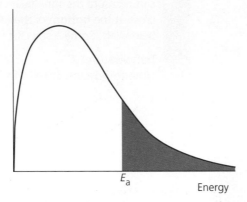

a Label the vertical axis.
b i) How do the molecules in a gas exchange energy?
ii) What effect, if any, does the exchange have on the distribution shown above?
c What does the shaded area on the graph represent?
d Give two changes to this reaction mixture which would cause the shaded area to increase.
NEAB CH03 Feb 1997 Q1

2
a The rate of the reaction between the gases nitrogen(II) oxide, NO, and oxygen was studied at a given temperature. Some results obtained are shown in the table below.

Experiment	Initial concentration of NO/mol dm^{-3}	Initial concentration of O$_2$/mol dm^{-3}	Initial rate /mol dm^{-3} s^{-1}
1	0.2	0.3	1.2×10^{-4}
2	0.2	0.6	2.4×10^{-4}
3	0.4	0.3	4.8×10^{-4}

Use the data in the table to deduce:
the order of reaction with respect to O$_2$, (1)
and the order of reaction with respect to NO. (1)
Hence, write a rate equation for the reaction. (1)
Calculate the value of the rate constant for the reaction at this temperature and state its units. (3)
b State how the value of the rate constant for a reaction changes with an increase in temperature. (1)
NEAB CH03 Feb 1997 Q2

3
a The curve below shows the distribution of energies at a temperature, T_1, for the molecules in a mixture of gases which react together.

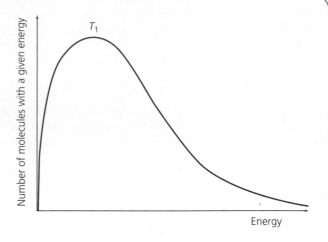

i) On the same axes, add a second curve for the distribution of energies in the same sample at a higher temperature, T_2, (1)
and label the curve T_2. (1)
ii) At which temperature is the rate of reaction between the gases in the mixture higher? (1)
Give a reason for your choice. (1)
iii) Why do collisions between molecules of gaseous reactants not always lead to a reaction? (1)
NEAB CH03 June 1997 Q1(part)

4
a The rate of the reaction between reactants P and Q can be represented by the equation
$$\text{Rate} = k[\text{P}][\text{Q}]^2$$
Without giving practical details, outline the experiments you would perform to show that the reaction is first order with respect to P and second order with respect to Q. Given that the units of rate are mol dm^{-3} s^{-1}, deduce the units of the rate constant, k. (7)
b Give three ways in which you could increase the rate of reaction between a solid and a substance in solution other than by the addition of a catalyst. For each of your methods, explain why the rate of reaction increases. (9)
c Explain why an increase in pressure increases the rate of any reaction involving gases but has little effect on the rate of a reaction involving solids and solutions. (4)
NEAB CH03 Mar 1998 Q8

5
a The diagram below shows the Maxwell–Boltzmann energy distribution curves for molecules of a gas under two sets of conditions A and B. The total area under curve B is the same as the total area under curve A.

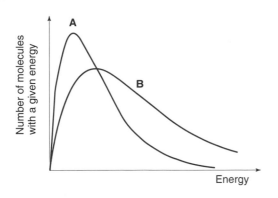

i) What change of condition is needed to produce curve **B** from curve **A**? (1)

ii) What is represented by the total area under curve **A**? (1)

iii) Why is the total area under curve **B** the same as that under curve **A**? (1)

b i) Explain the meaning of the term *activation energy.* (1)

ii) In a reaction involving gas molecules, if all other conditions are kept constant, state the effect, if any, on the value of the activation energy when:
a catalyst is added, (1)
the volume of the vessel is decreased. (1)

c Explain why reactions between solids usually occur very slowly, if at all. (2)
NEAB CH03 June 1998 Q1

6 The rate equation for a reaction between substances A, B and C is of the form:

$$\text{rate} = k[A]^x[B]^y[C]^z \text{ where } x + y + z = 4$$

The following data were obtained in a series of experiments at a constant temperature.

Experiment	Initial concentration of **A**/mol dm^{-3}	Initial concentration of **B**/mol dm^{-3}	Initial concentration of **C**/mol dm^{-3}	Initial rate /mol dm^{-3} s^{-1}
1	0.10	0.20	0.20	8.0×10^{-5}
2	0.10	0.05	0.20	2.0×10^{-5}
3	0.05	0.10	0.20	2.0×10^{-5}
4	0.10	0.10	0.10	to be calculated

a Use the data in the table to deduce:
the order of reaction with respect to A (1)
and the order of reaction with respect to B. (1)
Hence deduce the order of reaction with respect to C. (1)

b Calculate the value of the rate constant, k, (1) stating its units (1)
and also the value of the initial rate in experiment 4. (1)

c How does the value of k change when the temperature of the reaction is increased? (1)
NEAB June 99, CH03(A), Q3

7
a The diagram below shows the Maxwell–Boltzmann energy distribution curves for molecules of a sample of a gas at two different temperatures.

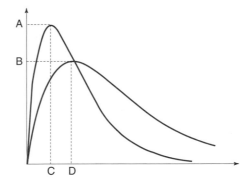

i) Label both axes. (2)

ii) Choose one of the letters on the axes above which represents the most probable energy of the molecules at the lower temperature. (1)

b Give two requirements for a reaction to occur between molecules in the gas phase. (2)

c Which has the greater effect on the initial rate of a first-order reaction:
either increasing the temperature of the reaction mixture from 300 K to 309 K (a 3% increase)
or increasing the concentration of the reagent from 1.00 M to 1.03 M (also a 3% increase)? (1)
Briefly explain your answer.
NEAB March 2000 CH03(A), Q1

8 Esters can be hydrolysed by heating with dilute hydrochloric acid to form an alcohol and a carboxylic acid.

a i) Name the ester CH_3COOCH_3 (1)

ii) Write an equation for the hydrolysis of this ester. (1)

iii) Suggest the role of hydrochloric acid in this hydrolysis. (1)

b The rate equation for the hydrolysis of this ester is
$$\text{rate} = k[CH_3COOCH_3][H^+]$$
When the initial concentration of the ester is 0.50 M and that of hydrochloric acid is 1.0 M, the initial rate of the reaction is 2.4×10^{-3} mol dm^{-3} s^{-1} at 320 K.

i) Calculate the value of the rate constant at this temperature and give its units. (3)

ii) Calculate the initial rate of the reaction if the concentration of hydrochloric acid is increased to 1.5 M but all other conditions remain unchanged. (1)

iii) Calculate the initial rate of the reaction at 320 K if more solvent is added to the original mixture so that the total volume is doubled. (1)
NEAB CH03 March 2000 Q2(part)

2 Equilibria

We require oxygen to break down glucose and provide energy to our body tissues, especially our muscles when we exercise. Carbon dioxide is produced as a waste product. The haemoglobin in our blood can reversibly bind either to oxygen or to carbon dioxide, depending on the concentration of these gases in the tissue and blood fluids.

The concentration of blood oxygen is high in the arteries from our lungs. There, haemoglobin bonds preferentially to oxygen and carries this gas to our cells. If the cells are metabolising, they are using up oxygen, so blood and tissue oxygen concentration is low, and oxygen dissociates from haemoglobin.

When the concentration of carbon dioxide is high, as it is when leaving active muscles, haemoglobin bonds to this gas and carries it back to our lungs where, with blood at a lower carbon dioxide level, the gas dissociates, diffuses into the lungs, and we breathe it out.

$$Hb + O_2 \underset{\text{in tissues}}{\overset{\text{in lungs}}{\rightleftharpoons}} HbO_2 \qquad Hb + CO_2 \underset{\text{in lungs}}{\overset{\text{in tissues}}{\rightleftharpoons}} HbCO_2$$

The athlete has used a lot of oxygen, the equilibrium has shifted, so an increase in oxygen concentration is needed to make up for the oxygen deficiency.

Free divers train their bodies to adapt to low blood oxygen levels.

These are *dynamic equilibrium* reactions, and the position of equilibrium shifts to respond to our body's needs.

If we hold our breath swimming under water, the level of carbon dioxide in our blood increases, and this stimulates the need to breathe. An inexperienced diver is very sensitive to carbon dioxide levels and a high concentration of carbon dioxide in the lungs will trigger that 'burning' sensation you feel when you hold your breath.

Free divers have trained their bodies so that they can hold their breath for long periods – up to 2 minutes. The human body is capable of other remarkable adaptations to the underwater environment. Trained divers can lower their heart rate to an incredible 20 beats per minute. Blood vessels in the skin contract under conditions of low oxygen in order to leave more blood available for important organs, namely the heart, brain and muscles. Changes in blood chemistry allow the body to carry and use oxygen more efficiently. These changes, in effect, squeeze the last molecule of available oxygen from non-essential organs.

The reversible gas reactions are vital for our body's needs, but reversible reactions are also extremely important in economic terms, as this chapter explains.

We have seen in Chapter 7 of *AS Chemistry* that some reactions go more or less to completion. Other reactions do not: when the reaction seems to have stopped there are still significant amounts of reactants as well as products in the reaction mixture. Such reactions are **reversible**, and when the reaction appears to have stopped, it is in fact at **dynamic equilibrium**. As an analogy, imagine a runner on a treadmill at the sports centre: to an observer, the runner isn't moving in any direction, yet he or she is going forward at the same rate as the treadmill is moving backwards.

The reaction between hydrogen and iodine is an example of a reversible reaction. At room temperature, hydrogen is a gas and iodine is a solid, but at temperatures over 457 K and at 101 kPa pressure (1 atm.), iodine is a gas.

Nitrogen dioxide gas is brown and exists in equilibrium with the colourless dimer, N_2O_4: $2NO_2(g) \rightleftharpoons N_2O_4(g)$. Low pressure favours the formation of nitrogen dioxide. If you apply more pressure, the equilibrium position moves to the right. This is observed as a change in colour as the brown fades.

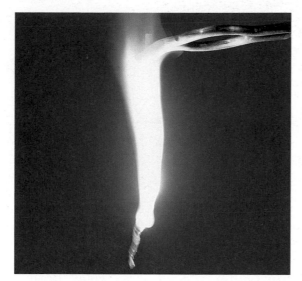

The reaction between magnesium and oxygen is not reversible to any significant extent.

If equimolar quantities of the reactants are heated to 600 K and the two gases are sealed in a container, they combine and form hydrogen iodide:

$$H_2(g) + I_2(g) \rightarrow 2HI(g)$$

However, no matter how long you keep the temperature at 600 K, there is always some hydrogen and iodine present in the container – the reaction does not go to completion. Under these conditions the final reaction mixture contains the three substances in the proportions of 0.2 moles of hydrogen and iodine to 1.6 moles of hydrogen iodide.

As the product HI(g) is formed in the reaction, it decomposes to form gaseous hydrogen and gaseous iodine again, so the forward and reverse reactions are happening at the same time. At equilibrium the

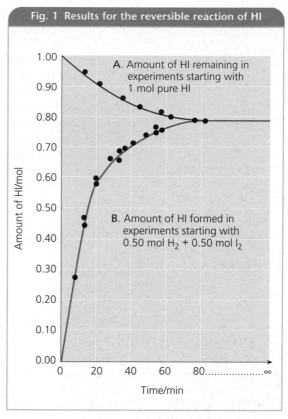

Fig. 1 Results for the reversible reaction of HI

A. Amount of HI remaining in experiments starting with 1 mol pure HI

B. Amount of HI formed in experiments starting with 0.50 mol H_2 + 0.50 mol I_2

concentrations of reactants and products do not change because the rates of the forward and reverse reactions are equal: this is referred to as a **dynamic equilibrium**.

If all the reactants and products are in the same phase the reaction is called **homogeneous**; if there is a mixture of phases, then the reaction is **heterogeneous**. In this example, the reactants and products are all gases at 600 K, and as they are all in the same phase, we refer to the reaction as **homogeneous**.

2.1 Equilibrium constant K_c

In *AS Chemistry* we looked at the equilibrium constant only qualitatively. At equilibrium the ratio of the concentrations stays constant and, knowing the concentrations of the various substances, it is easy to calculate the value of the constant. This is very useful since you can predict how a reaction will behave when the concentrations are changed. We call the ratio of the concentrations of the

products and reactants the **equilibrium constant**, K_c (where the c stands for molar concentration). Its value is constant only so long as the temperature is constant, and changes if the temperature changes.

For the reaction $H_2(g) + I_2(g) \rightleftharpoons 2HI(g)$ the equilibrium constant is written as:

$$K_c = \frac{[HI]^2}{[H_2][I_2]}$$

Expressions for product concentrations are above the line, while those for reactants are below the line. Note also that the concentration terms for a reaction are raised to the power that is stated in the stoichiometric equation. For example, in the reaction above, 2 moles of HI in the reaction equation gives $[HI]^2$ in the equilibrium constant equation.

In the reaction between ethanol, C_2H_5OH, and ethanoic acid, CH_3COOH, the ester ethyl ethanoate, $CH_3COOC_2H_5$, is formed. The reaction equation is:

$$C_2H_5OH(l) + CH_3COOH(l)$$
$$\rightleftharpoons CH_3COOC_2H_5(l) + H_2O(l)$$

and the equilibrium constant for this reaction, K_c, is expressed as:

$$K_c = \frac{[CH_3COOC_2H_5][H_2O]}{[C_2H_5OH][CH_3COOH]}$$

We can write a general equation for any homogeneous reaction at equilibrium:

$$aA + bB \rightleftharpoons cC + dD$$

where a, b, c and d are the numbers of moles of substances A, B, C and D in the balanced

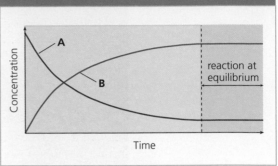

Fig. 2 Changes in concentration with time for the reversible reaction of HI(g)

chemical equation for the equilibrium. The equilibrium constant, K_c, is then expressed as:

$$K_c = \frac{[C]^c[D]^d}{[A]^a[B]^b}$$

1

a In Fig. 2, which line, A or B, represents the H_2/I_2 concentration, and which line represents the HI concentration? Explain your answer.

b What will be the effect on the rates of these two reactions of increasing the temperature?

2.2 Units of K_c

The units of K_c depend upon the number of moles in the chemical equation. For any equilibrium the units of K_c can be calculated by substituting the units of concentration $(mol\ dm^{-3})$ into the equation for K_c.

For the final reaction in the industrial synthesis of ammonia,
$N_2(g) + 3H_2(g) \rightleftharpoons 2NH_3(g)$:

$$K_c = \frac{[NH_3]^2}{[N_2][H_2]^3}$$

$$K_c = \frac{(\text{mol dm}^{-3})^2}{(\text{mol dm}^{-3})\ (\text{mol dm}^{-3})^3}$$

and after cancelling the units are:

$$\frac{1}{(\text{mol dm}^{-3})^2} \quad \text{i.e. mol}^{-2}\ dm^6$$

Similarly for the oxidation of sulphur dioxide, $2SO_2(g) + O_2(g) \rightleftharpoons 2SO_3(g)$:

$$K_c = \frac{[SO_3]^2}{[O_2][SO_2]^2}$$

$$K_c = \frac{(\text{mol dm}^{-3})^2}{(\text{mol dm}^{-3})(\text{mol dm}^{-3})^2}$$

and the units are:

$$\frac{1}{(\text{mol dm}^{-3})} \quad \text{i.e. mol}^{-1}\ dm^3$$

For the esterification equilibrium above, the units for K_c are given by:

$$K_c = \frac{[CH_3COOC_2H_5][H_2O]}{[C_2H_5OH][CH_3COOH]}$$

$$= \frac{(\text{mol dm}^{-3})(\text{mol dm}^{-3})}{(\text{mol dm}^{-3})(\text{mol dm}^{-3})}$$

And in this case cancelling gives no units.

2 Write an expression for K_c for the reactions and calculate the units of K_c.

a $CO_2(g) + NO(g) \rightleftharpoons CO(g) + NO_2(g)$

b $C_2H_6(g) \rightleftharpoons C_2H_4(g) + H_2(g)$

c $2NO(g) + O_2(g) \rightleftharpoons 2NO_2(g)$

2.3 Calculating equilibrium constants

For the esterification reaction, let us look at how we calculate the value of the equilibrium constant at a given temperature. Table 1 gives the concentrations in mol dm^{-3} of the substances present at equilibrium.

Table 1		
Reagent	Concentration at the start	Concentration at equilibrium
C_2H_5OH	1.0	0.33
CH_3COOH	1.0	0.33
$CH_3COOC_2H_5$	0	0.67
H_2O	0	0.67

$$K_c = \frac{[CH_3COOC_2H_5][H_2O]}{[C_2H_5OH][CH_3COOH]}$$

$$= \frac{0.67 \times 0.67}{0.33 \times 0.33} = 4.12$$

We have seen that there are no units for this value because all the concentration terms cancel.

In general, if the position of equilibrium of a reaction lies well over to the product side, then the equilibrium constant will be relatively large. A small equilibrium constant indicates that the position of equilibrium lies well over to the side of the reactants and that not much conversion takes place. For example, the NH_3 equilibrium constant at 600 K is 3.0×10^{-2} mol^{-2} dm^6. This indicates a low equilibrium yield of NH_3: the gas that comes out of the converter is mostly nitrogen and hydrogen, which is why the unconverted gases are recycled in the Haber process.

3 Dioximes are compounds that can be used to recover copper ions from water because the two substances form strong complexes together. The equilibrium can be represented by a simplified equation, where H_2A is the dioxime:

$Cu^{2+}(aq) + H_2A(aq) \rightleftharpoons CuA(aq) + 2H^+(aq)$

A solution containing 0.005 mol dm^{-3} of copper ions was mixed with a solution containing 0.005 mol dm^{-3} of dioxime and at equilibrium the mixture contained 0.0049 mol dm^{-3} of complex CuA(aq).

a Write an expression for the equilibrium constant for the reaction.

b Would you expect K_c to be small or large?

c Calculate the concentrations of copper ions, dioxime molecules and hydrogen ions at equilibrium.

d Calculate the value for the equilibrium constant and quote the units (if any).

e What will happen to the position of equilibrium if acid is added to the solution?

f What will happen to the value for the equilibrium constant if acid is added to the solution?

KEY FACTS

■ The general expression for an equilibrium constant in terms of concentrations is given by:

$$K_c = \frac{[C]^c[D]^d}{[A]^a[B]^b}$$

where a, b, c and d are the numbers of moles of each substance A, B, C and D in the balanced chemical equation:

aA + bB $\rightleftharpoons$ cC + dD

■ The units for K_c depend upon the particular equilibrium.

■ A high value for K_c indicates a high equilibrium yield of product(s).

■ The value of the equilibrium constant for a reaction is constant at a fixed temperature.

2.4 Factors that affect equilibrium

The effect of making changes to the conditions of a reaction at equilibrium can be predicted by using Le Chatelier's principle:

The position of the equilibrium of a system changes to minimise the effect of any imposed change in conditions, and eventually to restore equilibrium.

The changes imposed can be changes of concentration, pressure or temperature.

Changing concentration

Consider the esterification reaction again. Le Chatelier's principle predicts that if the concentration of the reactant ethanol is increased, the position of equilibrium shifts to minimise the change. The reaction moves to the right to remove ethanol, when ethanol reacts with more of the ethanoic acid to produce more ethyl ethanoate and water. At the same time the concentration in the reaction mixture of ethanoic acid is reduced.

So, if we know the value of the equilibrium constant for a reaction in the first place, we can use it to calculate the effect of changing any of the concentrations:

When different proportions of reactants in a reversible reaction are mixed at a fixed temperature, the concentrations of the reactants and products will adjust until the value of the equilibrium constant is achieved.

Consider the following data for the reaction of ethanol with ethanoic acid at 25 °C:

$$C_2H_5OH(l) + CH_3COOH(l) \rightleftharpoons CH_3COOC_2H_5(l) + H_2O(l)$$
initial concentrations :
$$a \qquad b \qquad 0 \qquad 0$$

The concentrations in the equilibrium mixture can be measured by analysing each of the reaction mixtures. If the original amounts in moles of reactants are a and b, and at equilibrium the amount of ester and water are both x, then the amounts of ethanol and acid remaining at equilibrium are $(a - x)$ and $(b - x)$ respectively (i.e. the initial amount minus the amount reacted).

$$C_2H_5OH(l) + CH_3COOH(l) \rightleftharpoons CH_3COOC_2H_5(l) + H_2O(l)$$
concentrations at equilibrium:
$$(a - x) \qquad (b - x) \qquad x \qquad x$$

$$K_c = \frac{[CH_3COOC_2H_5][H_2O]}{[C_2H_5OH][CH_3COOH]}$$

Where V is the total volume of the reaction mixture:

$$K_c = \frac{\left(\dfrac{x}{V}\right)\left(\dfrac{x}{V}\right)}{\left(\dfrac{a - x}{V}\right)\left(\dfrac{b - x}{V}\right)}$$

$$= \frac{x^2}{(a - x)(b - x)}$$

Table 2 gives data for different initial concentrations of ethanol and the resulting equilibrium concentrations. Note that the equilibrium remains constant (given variations in concentration measurements) even though the initial concentrations of ethanol were changed.

Table 2 Data for the esterification reaction at 298 K		
Initial moles of ethanol (a)	Initial moles of ethanoic acid (b)	Equilibrium moles of ethanol (a – x)
0.500	1.000	0.075
1.000	1.000	0.330
2.000	1.000	1.153
4.000	1.000	3.067
Equilibrium moles of ethanoic acid	Equilibrium (b – x) moles of ethyl ethanoate (x)	Equilibrium constant
0.575	0.425	4.188
0.330	0.670	4.122
0.153	0.847	4.067
0.067	0.933	4.236

Table 2 shows that when the concentration of ethanol increases, the number of moles of ethyl ethanoate at equilibrium also increases. This is predicted from Le Chatelier's principle – the imposed change is to increase the concentration of ethanol and the system minimises that effect by removing ethanol. In so doing, the concentration of ethyl ethanoate increases and the concentrations of ethanol and ethanoic acid decrease. The value of the equilibrium constant does not change.

Changing pressure

In many industrial processes both temperature and pressure are set to obtain the most economically favourable conditions, giving the best overall yield at the lowest cost. We will first consider the effect of pressure.

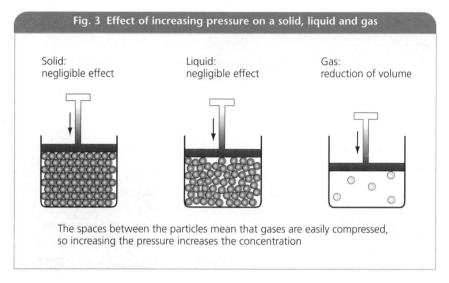

Fig. 3 Effect of increasing pressure on a solid, liquid and gas

Solid: negligible effect

Liquid: negligible effect

Gas: reduction of volume

The spaces between the particles mean that gases are easily compressed, so increasing the pressure increases the concentration

A change in overall pressure will only affect those equilibrium reactions that involve gases (Fig. 3). Since the pressure of a gas depends upon the number of gas particles in a given volume, Le Chatelier's principle indicates that if the overall pressure on a reacting gaseous system is increased, the system can minimise the imposed pressure change by forming fewer gas molecules.

Let us apply Le Chatelier's principle to the reaction:

$$PCl_3(g) + Cl_2(g) \rightleftharpoons PCl_5(g)$$

Increasing the total pressure of the system will shift the equilibrium towards $PCl_5(g)$, as shown in Fig. 4. This happens because there is just one mole on the right-hand (product) side of the equation, while there are two on the left-hand (reactant) side. The equilibrium shift to the right reduces the pressure of the reaction system.

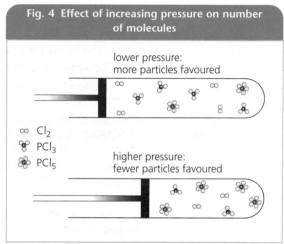

Fig. 4 Effect of increasing pressure on number of molecules

lower pressure: more particles favoured

Cl$_2$ / PCl$_3$ / PCl$_5$

higher pressure: fewer particles favoured

The change in the number of molecules (or moles) can be summarised as:

$$\Delta n = n(\text{products}) - n(\text{reactants})$$

Table 3 gives details of this effect for three reactions, and Table 4 summarises effects for any gaseous reaction.

4 How will the following equilibria change if the overall pressure is increased?

a $2CO(g) + O_2(g) \rightleftharpoons 2CO_2(g)$

b $C_2H_6(g) \rightleftharpoons C_2H_4(g) + H_2(g)$

c $H_2(g) + I_2(g) \rightleftharpoons 2HI(g)$

d $2NO_2(g) \rightleftharpoons N_2O_4(g)$

When dealing with gaseous equilibria in industry it is more convenient to use pressure measurements to derive the amounts of reactants and products, than it is to use concentration measurements.

Table 3 Effect on equilibrium position of increasing the pressure

Reaction	Change in number of gaseous molecules Δn	Effect of increasing pressure
$CO_2(g) + NO(g) \rightleftharpoons CO(g) + NO_2(g)$	$\Delta n = 0$	no effect
$C_2H_6(g) \rightleftharpoons C_2H_4(g) + H_2(g)$	$\Delta n = +1$	shift to the left
$2NO(g) + O_2(g) \rightleftharpoons 2NO_2(g)$	$\Delta n = -1$	shift to the right

Table 4 Summary of effects

Greater number of gaseous molecules	Change in pressure	Effect on equilibrium constant	Displacement in equilibrium position	Equilibrium yield of product
reactant	increase	no change	right	increased
reactant	decrease	no change	left	decreased
product	increase	no change	left	decreased
product	decrease	no change	right	increased

The SI units for pressure are **pascals** ($1\ Pa = 1\ N\ m^{-2}$). But in many textbooks, particularly when describing industrial processes, the pressure is stated in **atmospheres** (atm). To convert atmospheres to pascals, use the conversion:

$$1\ atm = 1.01 \times 10^5\ Pa \quad or \quad 101\ kPa$$

Many calculations simply use 100 kPa.

Pressure is also sometimes measured in **bars**:

$$1\ bar = 1 \times 10^5\ Pa,\ i.e.\ 100\ kPa$$

In an equilibrium mixture of gases, each gas component will contribute to the pressure, and the total pressure will be the sum of the pressures of all the gases (Fig. 5). Each gas contributes a **partial pressure** (symbol p) to the total pressure (P). However, only the total pressure of the mixture can be measured.

The mole fraction, x, of a component in a gas mixture is given by:

$$x = \frac{\text{moles of component}}{\text{total moles of all components}}$$

The partial pressure, p, of a component in a gas mixture is given by:

partial pressure = mole fraction × total pressure
i.e. $p = xP$

If the amount of substance is quoted as a mass, this must first be converted to number of moles (mass/M_r) before calculating the mole fraction.

It is easy to make and use a pressure gauge, but you cannot make a mole gauge

5 In the gas phase, 1 kg of ethene and 1 kg of steam are present at a total pressure of 5×10^6 Pa.

a Calculate the mole fraction (x) for each component.

b Calculate the partial pressure (p) of each gas.

Fig. 5 Partial pressure and total pressure

Total pressure
$P = p_A + p_B + p_C$

Partial pressure, p_A

Partial pressure, p_B

Partial pressure, p_C

gas A   gas B   gas C

Again let's take as a typical gaseous equilibrium mixture the synthesis of ammonia, at 5000 kPa (approx. 50 atm) and 750 K. As shown in Fig. 6, the mixture contains 2 molecules of ammonia for every 6 molecules of nitrogen and every 17 molecules of hydrogen, giving a total of 25.

The fraction of each gas is:

N_2 6/25, H_2 17/25 and NH_3 2/25

The partial pressure of each is therefore:

N_2 6/25 × total pressure = 6/25 × 5000 kPa = 1200 kPa
H_2 17/25 × total pressure = 17/25 × 5000 kPa = 3400 kPa
NH_3 2/25 × total pressure = 2/25 × 5000 kPa = 400 kPa

⎫ total pressure is 5000 kPa

Mole fraction and partial pressure
We refer to the fraction of particles in a given mixture as the **mole fraction**.

Fig. 6 Proportions of molecules in ammonia synthesis reaction

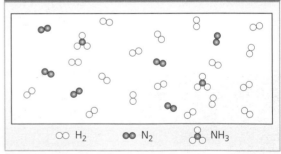

∞ H_2 ∞ N_2 NH_3

K_p
In section 2.1 we wrote a general expression for the equilibrium constant where component concentrations are known. In some homogeneous reactions, for example the reactions of question 2, all the components are in the gas phase. If the relative amounts of gases in a reaction are expressed as partial pressures, the equilibrium constant is written as K_p.

Again for the general reaction:

$$aA(g) + bB(g) \rightleftharpoons cC(g) + dD(g)$$

where a, b, c and d are numbers of moles of substances A, B, C and D, respectively, the equilibrium constant is given by:

$$K_p = \frac{p_C{}^c \, p_D{}^d}{p_A{}^a \, p_B{}^b}$$

Units of K_p

Like K_c, the units of K_p are not the same for every reaction. The units of K_p depend upon the number of terms in the expression. This depends on the number of moles of reactants and products in the balanced chemical equation. The units for K_p can be found by substituting the units of pressure for each partial pressure in the expression.

So for the ammonia synthesis described on page 16 and in Fig. 6, in which the total pressure is 5000 kPa:

$$N_2(g) + 3H_2(g) \rightleftharpoons 2NH_3(g)$$
mole fraction   6/25   17/25   2/25

$$K_p = \frac{p[NH_3]^2}{p[N_2]p[H_2]^3}$$

$$\frac{(\frac{2}{25} \times 5 \times 10^6)^2}{(\frac{6}{25} \times 5 \times 10^6) \times (\frac{17}{25} \times 5 \times 10^6)^3} = \text{units} = \frac{Pa^2}{Pa \times Pa^3}$$

$$= \frac{(2/25)^2 \times \cancel{(5 \times 10^6)^2}}{(6/25) \times (5 \times 10^6) \times (17/25)^3 \times \cancel{(5 \times 10^6)^2}}$$

$$= \frac{(2/25)^2}{(6/25) \times (17/25)^3 \times (5 \times 10^6)^2} = \frac{1}{Pa^2} = Pa^{-2}$$

$$= \frac{0.0064}{0.24 \times 0.314 \times 25 \times 10^{12}}$$

$$= 3.39 \times 10^{-15} \, Pa^{-2}$$

Whisky is made by the traditional methods of fermentation and distillation (see photos). Ethanol for the large-scale production of alcoholic drinks often comes from chemical synthesis. Drinks companies mix ethanol, water and flavourings to make the final product. They often use ethanol that has been produced by a synthetic process. The ethanol can be made by the hydrolysis of ethene in the reversible reaction:

$$C_2H_4(g) + H_2O(g) \rightleftharpoons C_2H_5OH(g) \quad \varnothing H = -25 \text{ kJ mol}^{-1}$$

About one-third of the world's ethanol supply is made from ethene, and the main source of ethene is crude oil. The naphtha fraction of crude oil contains molecules of between five and nine carbon atoms in chain length, which is cracked to give ethene.

Companies can use equilibrium data to calculate the optimum equilibrium conditions that maximise the yield of ethanol from ethene. Le Chatelier's principle shows the direction in which the equilibrium position will change when the temperature and pressure change.

Le Chatelier devised his principle simply from observing the ways in which equilibria shifted when he altered reaction conditions. The design and operation of a modern chemical plant needs to adopt a much more quantitative approach, and chemists working at the plant are expected to predict precisely the effect of changes in reaction conditions on equilibrium yield.

Referring to the equilibrium equation for ethanol production, the expression for K_p is:

$$K_p = \frac{pC_2H_5OH}{pC_2H_4 \times pH_2O}$$

Scotch whisky is still made by traditional methods, rather than using synthetically produced ethanol.

Calculating K_p

Under standard conditions (298 K and 1×10^5 kPa), 1 mole of ethene and 1 mole of steam produces 0.74 moles of ethanol at equilibrium:

$$C_2H_4(g) + H_2O(g) \rightleftharpoons C_2H_5OH(g)$$

	C_2H_4	H_2O	C_2H_5OH
Start of reaction/mol	1.00	1.00	0
Equilibrium/mol	(1.00 – 0.74)	(1.00 – 0.74)	0.74

This means that 0.74 moles of ethene has been used up so (1 – 0.74) moles of ethene remains when equilibrium is reached, as well as (1 – 0.74) moles of water, since ethene and water combine in a ratio of 1:1.

To find the value for K_p we need to convert these amounts to partial pressures. With a total pressure of 10^5 Pa (under standard conditions), the partial pressure for ethanol can be found as follows:

$$\text{partial pressure} = \frac{\text{moles of ethanol}}{\text{total moles}} \times \text{total pressure}$$

$$p_{ethanol} = \frac{\text{moles of ethanol}}{\text{moles of (ethanol + ethene + steam)}} \times \text{total pressure}$$

$$= \frac{0.74 \times 10^5}{(0.74 + 0.26 + 0.26)}$$

$$p_{ethanol} = 5.87 \times 10^4 \text{ Pa}$$

Using a similar calculation:

$$p_{ethene} = p_{steam} = \frac{0.26 \times 10^5}{(0.74 + 0.26 + 0.26)} = 2.06 \times 10^4 \text{ Pa}$$

These values are inserted into the equation to give K_p:

$$K_p = \frac{p_{ethanol}}{p_{ethene} \times p_{steam}}$$

$$= \frac{5.87 \times 10^4}{(2.06 \times 10^4) \times (2.06 \times 10^4)}$$

$$K_p = 1.38 \times 10^{-4} \text{ Pa}$$

6 The pressure for the ethene/ethanol reaction is increased to 2.02×10^5 Pa, while the temperature is maintained at 298 K, and the initial amounts are 1 mole of ethene and 1 mole of steam. This gives 0.812 moles of ethanol at equilibrium.

a Calculate the number of moles of each reactant at equilibrium.

b Calculate the equilibrium partial pressure for each gas.

c Calculate K_p.

d How does doubling the total pressure affect K_p and yield?

Remember that changing the overall pressure for an equilibrium can alter the *position* the equilibrium reaches, but it does not change the *value* of the equilibrium constant.

In the commercial production of ethanol the ethene/ethanol equilibrium is not carried out at 298 K and 100 kPa. Under these standard conditions the rate of reaction would be very slow and the overall equilibrium yield of ethanol would be low.

For *exothermic* reactions an increase in temperature will increase the rate of reaction (remember: a 10 K rise in temperature will approximately double the rate of reaction) but the equilibrium yield will be reduced. The opposing nature of these two factors needs to be balanced in commercial production. This is discussed further in the section on catalysts (page 24).

(page 24)

KEY FACTS

- Increasing the pressure of a gaseous equilibrium causes the reaction to move in the direction of fewer gas molecules.

- The total pressure of a gas mixture is equal to the sum of all the partial pressures:

$$pA + pB + pC + pD + \ldots = P$$

- Each gas in an equilibrium mixture exerts a partial pressure, p. The partial pressure of a component in a gas mixture is given by:

partial pressure = mole fraction × total pressure ($p = xP$), where x = mole fraction

- K_p is the equilibrium constant for a gaseous reaction in which the amounts of gases are expressed as partial pressures. For the general reaction:

$$aA(g) + bB(g) \rightleftharpoons cC(g) + dD(g)$$

$$K_p = \frac{p_C{}^c p_D{}^d}{p_A{}^a p_B{}^b}$$

- The units for K_p depend upon the particular equilibrium.

- At a fixed temperature, changing the overall pressure for a gaseous equilibrium changes the *position* of the equilibrium, but not the *value* of K_p.

Effect of temperature

Changing the temperature of the equilibrium changes the *position of equilibrium* and *the value* of the equilibrium constant, K_c or K_p.

Position of equilibrium

Le Chatelier's principle predicts that if the temperature is changed, the reaction will try to minimise the effect of the change. If the temperature is increased, the reaction will try to remove heat from the system, so it will move in the endothermic direction.

$$N_2(g) + 3H_2(g) \rightleftharpoons 2NH_3(g) \quad \Delta H^{\ominus} = -92 \text{ kJ mol}^{-1} \text{ (exothermic)}$$

The production of ammonia is an exothermic process, so increasing the temperature will drive the equilibrium to the left – the endothermic direction – thus decreasing the yield of ammonia. Fig. 7 shows the effect of pressure and temperature on the percentage of ammonia produced at equilibrium in the Haber process. Note that the percentage yield reduces with higher temperatures and increases with higher pressures.

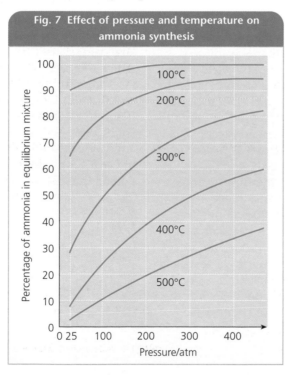

Fig. 7 Effect of pressure and temperature on ammonia synthesis

Percentage of ammonia in equilibrium mixture (y-axis: 0 to 100)
Pressure/atm (x-axis: 0 25, 100, 200, 300, 400)

Curves labelled: 100°C, 200°C, 300°C, 400°C, 500°C

The reaction now used to produce the hydrogen required for ammonia synthesis is:

$$CH_4(g) + H_2O(g) \rightleftharpoons CO(g) + 3H_2(g)$$
$$\Delta H^{\ominus} = +206 \text{ kJ mol}^{-1} \text{ (endothermic)}$$

The production of hydrogen is an endothermic process, so raising the temperature will drive the equilibrium to the right – the endothermic direction – thus increasing the equilibrium yield of hydrogen. The effects of temperature change are summarised in Table 5.

Figure 7 shows that there is a much higher percentage of ammonia in the equilibrium mixture at 100 °C than at 500 °C. Increasing the pressure has very little effect at 100 °C but has considerable effect at higher temperatures. The rate of reaction at 100 °C is very low, so ammonia is manufactured at 450 °C and 200 atm, and a catalyst is used to increase the rate.

7 In the reaction $N_2(g) + O_2(g) \rightleftharpoons 2NO(g)$, $K_c = 4 \times 10^{-31}$ at 20 °C (293 K) but at combustion temperature inside a car engine, typically 800°C (approx. 1100 K), $K_c = 8 \times 10^{-9}$.

a Is this reaction likely to happen at normal room temperature? Explain your reasoning.

b The temperature of the spark which ignites the fuel/air mixture is about 2500 K. At this temperature, how does the value of K_c change? How will this affect the yield of NO(g)?

c Is the reaction endothermic or exothermic? Explain your answer using the change in the value of the equilibrium constant.

Nitrogen oxide from petrol combustion reacts with atmospheric oxygen to give nitrogen dioxide. This gas forms the brown haze seen over polluted towns and cities. Nitrogen dioxide causes respiratory problems and acid rain.

Table 5 Effect of changing temperature on equilibrium				
Reaction	Change in temperature	Equilibrium constant	Shift in eqm. position	Eqm. yield of product
exothermic	increase	decreased	left	decreased
exothermic	decrease	increased	right	increased
endothermic	increase	increased	right	increased
endothermic	decrease	decreased	left	decreased

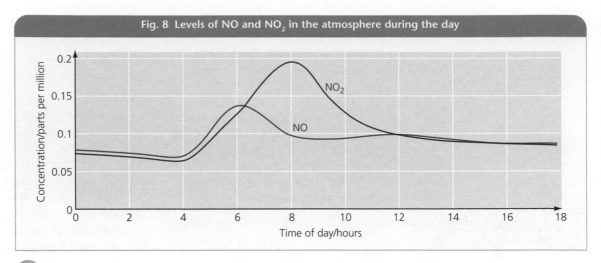

Fig. 8 Levels of NO and NO$_2$ in the atmosphere during the day

8 Explain the shape of the graph for the production of NO and NO$_2$ in Fig. 8.

Catalysis

As described above, for exothermic reactions, chemists have to balance the reduced yield produced by higher temperatures against the increased rate of reaction.

The production of ethanol from ethene provides a good example. The reaction is slightly exothermic, so the reaction mixture will naturally heat up as ethanol is formed. This has the effect of reducing the yield since, in accordance with Le Chatelier's principle, the system tries to counter the rise in temperature by absorbing heat and shifting the equilibrium to the left. Chemical engineers can cool the reaction mixture, but this slows down the *rate* at which the reaction proceeds.

The problem is helped by using a catalyst to speed up the reaction and achieve a reasonable rate of ethanol production. The catalyst speeds up both the forward and reverse reactions equally, so it does not affect the *position* of equilibrium, but it ensures that equilibrium is reached faster.

Computers are used to model the data and show the combined effects of changes in temperature and pressure on ethanol production. Computers can also calculate the likely effect on energy costs, raw material costs and plant costs, and predict the most economical, yet productive, conditions for ethanol manufacture.

The actual conditions used are a temperature of 540 K and a pressure of 6000 kPa. Ethene and steam are passed over a heated catalyst. This is phosphoric acid supported on a very porous powder to give a large surface area (the powder is the fossilised remains of microscopic sea creatures!). The resulting ratio of ethene to steam is 0.6 moles to 1.0 mole. The higher concentration of steam in the mixture shifts the position of equilibrium towards the product, ethanol, and using steam in excess ensures that it is the concentration of the more expensive ethene that is lowered.

9 a The temperature used in ethanol production is relatively high. How will this affect the ethanol yield?

b How is this factor balanced to increase ethanol yield?

Haemoglobin and gas exchange

A shortage of blood donors, and the fact that people with some infections (e.g. hepatitis, HIV) cannot donate blood, has highlighted the need for a synthetic form of blood.

Dr Kiyoshi Nagai and a team of researchers at Cambridge University looked at ways of producing synthetic blood. Dr Nagai's work involved the study of equilibria and dissociation constants.

The haemoglobin (often written as Hb) in red blood cells transports oxygen molecules from the lungs to body tissue because of the reversible reaction:

$$Hb(aq) + O_2(aq) \rightleftharpoons HbO_2(aq)$$
haemoglobin   oxyhaemoglobin

1

a Write out the expression for the equilibrium constant for this reaction.
b What are the units for K_c?
c Explain the effect on the equilibrium if you undertake strenuous exercise.

During respiration, cells of the body use oxygen and produce carbon dioxide. The oxygen is replaced in haemoglobin by carbon dioxide:

$$HbO_2(aq) + CO_2(aq) \rightleftharpoons HbCO_2(aq) + O_2(aq)$$

Carbon dioxide must be removed by the haemoglobin so that it can be exhaled.

2

a Write an expression for K_c for the exchange of oxygen and carbon dioxide in the blood cells.
b Calculate the units of K_c.

When carbon dioxide is released into the bloodstream before being exhaled in the lungs, an equilibrium is set up in the blood plasma.

$$CO_2(aq) + H_2O(l) \rightleftharpoons H^+(aq) + HCO_3^-(aq)$$

3

a Write an expression for K_c for this equilibrium.
b The expression for K_c can be simplified to:

$$K_c' = \frac{[H^+][HCO_3^-]}{[CO_2]}$$

Explain why this simplification is valid. (Hint: the amount of water is much greater in blood than the other components.)

c If $[H^+] = 2.32 \times 10^{-5}$ mol dm^{-3}, and $[CO_2] = 1.2 \times 10^{-3}$ mol dm^{-3}, calculate the value of K_c', and give its units.

At sea level atmospheric pressure is 1.01×10^5 Pa, but at an altitude of 3000 metres, atmospheric pressure is 7.0×10^4 Pa.

4

a Assuming air is 20% oxygen and 80% nitrogen, calculate the mole fractions and partial pressures of oxygen in the atmosphere at sea level and at 3000 metres.
b Some passengers on high-altitude train journeys in the Andes are provided with oxygen cylinders. Use the concept of equilibria to explain why this is so.

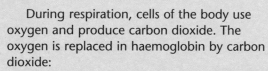

People unused to travelling in the rarefied atmosphere of the Andes mountains, at a height of over 4700 metres, are provided with oxygen if they need it.

5 If the equilibrium Hb(aq) + O$_2$(aq) $\rightleftharpoons$ HbO$_2$(aq) is exothermic, what will be the effect on oxygen uptake if your body temperature increases?

A transfusion of synthetic blood could be a life saver.

1 When 1.0 mol each of ethanol and ethanoic acid are mixed together at a fixed temperature, the reaction mixture at equilibrium is found to contain 0.66 mol of ethyl ethanoate.

a Write an equation for the reaction between ethanol and ethanoic acid. (1)
b The total volume of the reaction mixture is 0.10 dm³. Calculate the equilibrium concentrations of ethyl ethanoate and ethanol in the reaction mixture. (2)
c Write an expression for the equilibrium constant, K_c, for this reaction. (1)
d Calculate the value of the equilibrium constant for this reaction at this fixed temperature. (2)
e This reaction is affected by the addition of a few drops of concentrated sulphuric acid. Suggest a possible role for the sulphuric acid added. State the effect of sulphuric acid on the rate of reaction and on the value of the equilibrium constant. (3)
NEAB CH04 Feb 1996 Q1

2 At high temperatures, phosphorus pentachloride dissociates according to the equation

$$PCl_5(g) \rightleftharpoons PCl_3(g) + Cl_2(g)$$

At 200 °C and a final total pressure of 200 kPa, one mole of phosphorus pentachloride dissociated to produce two-thirds of a mole of chlorine at equilibrium.

a Write an expression for the equilibrium constant, K_p, for this reaction. (1)
b Calculate the total number of moles of gas in the equilibrium mixture. (2)
c Calculate the partial pressure of chlorine and of phosphorus pentachloride in the equilibrium mixture. (4)
d Calculate the value of the equilibrium constant, K_p, for this reaction, stating its units. (3)
NEAB CH04 Feb 1996 Q2

3 Dinitrogen tetroxide and nitrogen dioxide exist in the following equilibrium

$$N_2O_4(g) \rightleftharpoons 2NO_2(g)$$

When 11.04 g of dinitrogen tetroxide were placed in a vessel of volume 4.80 dm³ at fixed temperature, 5.52 g of nitrogen dioxide were produced at equilibrium under a pressure of 100 kPa. Use these data, where relevant, to answer the questions that follow.

a Calculate the equilibrium number of moles of each of the gases in the vessel (2)
b i) Write an expression for the equilibrium constant, K_c, for the above equilibrium. (1)

ii) Calculate the value of the equilibrium constant, K_c, and state its units. (3)
c i) Write an expression for the equilibrium constant, K_p, for the above equilibrium. (1)
ii) Calculate the mole fraction of NO_2 present in the equilibrium mixture. (1)
iii) Calculate the value of the equilibrium constant, K_p, and state its units. (3)
d State and explain the effect on the mole fraction of NO_2 when the pressure is increased at constant temperature. (3)
NEAB CH04 June 1997 Q3

4 Phosphorus(V) chloride dissociates at high temperatures according to the equation

$$PCl_5(g) \rightleftharpoons PCl_3(g) + Cl_2(g)$$

83.4 g of phosphorus(V) chloride are placed in a vessel of volume 9.23 dm³. At equilibrium at a certain temperature, 11.1 g of chlorine are produced at a total pressure of 250 kPa.
Use these data, where relevant, to answer the questions that follow.

a Calculate the number of moles of each of the gases in the vessel at equilibrium. (3)
b i) Write an expression for the equilibrium constant, K_c, for the above equilibrium. (1)
ii) Calculate the value of the equilibrium constant, K_c, and state its units. (3)
c i) Write an expression for the equilibrium constant, K_p, for the above equilibrium. (1)
ii) Calculate the mole fraction of chlorine present in the equilibrium mixture. (1)
iii) Calculate the partial pressure of PCl_5 present in the equilibrium mixture. (3)
iv) Calculate the value of the equilibrium constant. K_p, and state its units. (2)
NEAB CH04 June 1998 Q3

5 At a temperature of 107 °C, the reaction

$$CO(g) + 2H_2(g) \rightleftharpoons CH_3OH(g)$$

reaches equilibrium under a pressure of 1.59 MPa with 0.122 mol of carbon monoxide and 0.298 mol of hydrogen present at equilibrium in a vessel of volume 1.04 dm³.
Use these data to answer the questions that follow.

a Assuming ideal gas behaviour, determine the total number of moles of gas present. Hence calculate the number of moles of methanol in the equilibrium mixture. (3)
b Calculate the value of the equilibrium constant, K_c, for this reaction and state its units. (3)

EXAMINATION QUESTIONS

c i) Write an expression for the equilibrium constant, K_p, for this equilibrium. (1)
ii) Calculate the mole fraction of each of the three gases present in the equilibrium mixture. (3)
iii) Calculate the partial pressure of hydrogen present in the equilibrium mixture. (2)
iv) Calculate the value of the equilibrium constant, K_p, and state its units. (2)
AQA CH04 June 1999 Q5

6 Gaseous sulphur dioxide and oxygen were mixed in a 2:1 ratio and sealed in a vessel with a catalyst at 950 K. The equation for the reaction which occurred is shown below.

$$2SO_2(g) + O_2(g) \rightleftharpoons 2SO_3(g)$$

When equilibrium was reached, the total pressure in the vessel was 120 kPa and the mole fraction of $SO_3(g)$ present in the mixture was 0.9

a i) Write an expression for the equilibrium constant, K_p, for this reaction. (1)
ii) Calculate the partial pressures of each of the three gases present. (5)
iii) Calculate the value of the equilibrium constant K_p, for this reaction at 950 K and state its units. (2)
AQA CH04 March 2000 Q3

3 Acids and bases

Diabetes affects about one person in twenty in the UK. If not controlled, the disease can reduce the pH of the blood, making it more acidic, which can result in a coma, or even death.

The most common cause of diabetes in young people is a deficiency of the hormone insulin. Medical staff in hospitals have an understanding of the chemistry of the metabolic reactions involved, in order to treat the disease effectively.

Lack of insulin means that glucose, which is needed for energy, is not transferred to the muscles. Fat has to be used as an alternative energy source, and its metabolism causes the muscles to tire. When fat is broken down, metabolites can form which include keto acids. Normally, the body uses buffers to control the pH of the blood, but keto acids are much stronger than the normal acids in human blood (mainly carbonic, citric and lactic acids). Medical staff treating diabetes measure the acidity of patients' blood and compare the values with normal blood acidity.

The pH of normal blood is 7.4, but if excess keto acids form, the pH can drop to 7.0, and it is when the body cannot control this change in pH that serious damage can result. All this information is required for the correct amount of medication to be prescribed for the patient.

3.1 Acids and bases

In this chapter you will learn how to describe and calculate the strengths of acids. In addition, you will learn about titrations between acids and bases, how to select an indicator and how buffers can control pH changes.

You will already be familiar with the concept of acids and bases and of an alkali as a base that is soluble in water. In 1884, the Swedish chemist Arrhenius described an acid as a substance which contains hydrogen and releases hydrogen ions when it dissolves in water:

$HCl \rightleftharpoons H^+ + Cl^-$

$CH_3COOH \rightleftharpoons CH_3COO^- + H^+$

He also described acids such as HCl, which dissociate fully, as strong acids, and those such as ethanoic acid, which only partially dissociate, as weak acids.

For all acid dissociation reactions, an equilibrium is set up between the species involved. The HCl molecule dissociates into hydrogen ions, H^+, which are protons, and chloride ions. For HCl, the equilibrium lies well over to the right: the HCl is virtually completely ionised, and for this reason it is described as a strong acid. In contrast, only 0.1% of ethanoic acid dissociates: the equilibrium lies well over to the left, and therefore ethanoic acid is a weak acid (a value of 50% dissociation is commonly used to differentiate between strong and weak acids).

Be careful not to confuse 'strong' with 'concentrated'. Strong relates to the *proportion of molecules* that have dissociated at equilibrium, and concentrated relates to the *amount of acid* in solution. However, depending on the acid, both can lead to high concentrations of hydrogen ions.

Arrhenius's theory also states that a base is a substance which reacts with H^+ to form water. Here is an example of an insoluble base:

$MgO(s) + 2H^+(aq) \rightleftharpoons Mg^{2+}(aq) + H_2O(l)$

and a soluble base:

$Na^+(aq) + OH^-(aq) + H^+(aq) \rightleftharpoons Na^+(aq) + H_2O(l)$

A proton, H^+, has no electrons, and chemists realised that this could not exist independently in aqueous solution. (A proton, H^+, can only exist independently under special conditions in the gaseous phase.) H^+ associates with water molecules and forms the hydroxonium ion according to the equation:

$$H^+ + H_2O \rightarrow H_3O^+$$
hydrogen ion   hydroxonium ion
(proton)

H_3O^+ represents a water molecule and a proton attached to the lone pair on the water molecule. However, when writing reactions, the less precise H^+ is often used for simplicity. Instead of representing the dissociation of

Fig. 1 Dissociation of HCl and formation of a hydroxonium ion (H_3O^+)

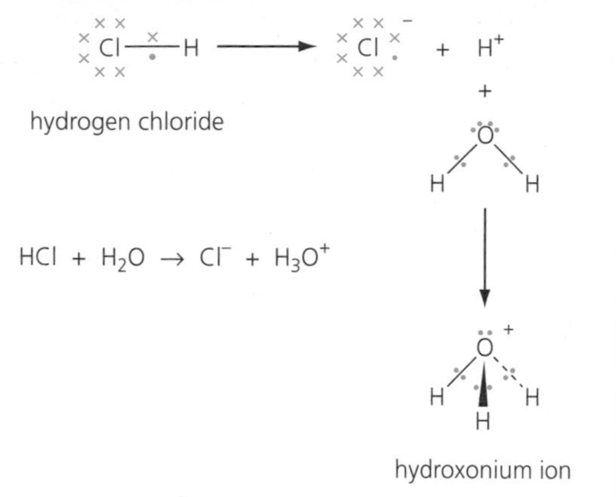

hydrogen chloride

$$HCl + H_2O \rightarrow Cl^- + H_3O^+$$

hydroxonium ion

hydrochloric and ethanoic acids as shown at the start of section 3.1, the dissociation of these acids is therefore more accurately represented as:

$$HCl(aq) + H_2O(l) \rightleftharpoons H_3O^+(aq) + Cl^-(aq)$$
$$CH_3COOH(aq) + H_2O(l) \rightleftharpoons H_3O^+(aq) + CH_3COO^-(aq)$$

These refinements recognise the important role of water as the solvent. HCl in water behaves as an acid. But if HCl is dissolved in solvents that will not accept H^+, such as methylbenzene, it will not show any acidic properties.

Brønsted–Lowry acids and bases

In 1923 the Danish scientist Brønsted and the British chemist Lowry adapted the theory of acid–base behaviour to give a more useful definition which described the behaviour of acids and bases both in aqueous and non-aqueous reactions. They defined an acid as follows:

An acid is a substance which *donates protons* in a reaction: it is a *proton donor*.

When the stoppers from hydrochloric acid and ammonia bottles are held near each other, fumes are formed which contain ammonium chloride particles

Fig. 2 The Brønsted-Lowry acid-base reaction of ammonia and hydrochloric acid

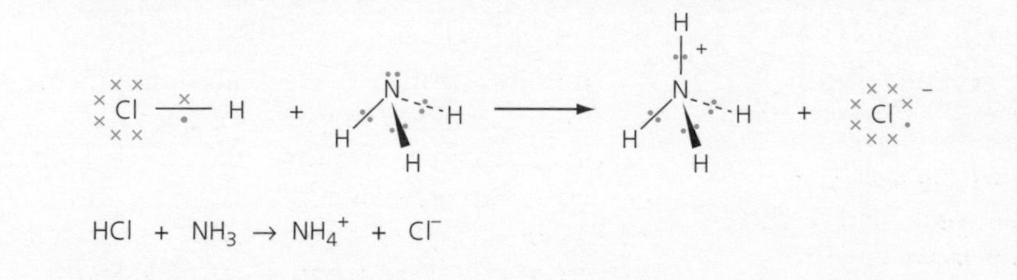

$$HCl + NH_3 \rightarrow NH_4^+ + Cl^-$$

Brønsted and Lowry also said that when an acid donates protons, a substance must accept the protons, namely a base, and that this substance behaves oppositely to an acid. They defined a base as follows:

A base is a substance which *accepts protons* in a reaction: it is a *proton acceptor*.

Brønsted–Lowry definitions therefore apply to a wide range of reactions of acids and bases, not just those that take place in aqueous solutions.

The reaction between hydrochloric acid and ammonia is an acid–base reaction which can occur in aqueous solution, in the gas phase or in non-polar solvents:

$$HCl(aq) + NH_3(aq) \rightleftharpoons NH_4^+(aq) + Cl^-(aq)$$
$$HCl(g) + NH_3(g) \rightleftharpoons NH_4^+Cl^-(s)$$

In these reactions, HCl has donated H^+ and NH_3 has accepted H^+:

$$NH_3 + H^+ \rightarrow NH_4^+$$

The aqueous and gaseous reactions are both Brønsted–Lowry acid-base reactions, but only the aqueous reaction is an Arrhenius acid-base reaction. The Brønsted–Lowry definition clearly covers a wider range of applications.

Now, consider what happens when concentrated HNO_3 and concentrated H_2SO_4 are mixed. We would normally think of both of these substances as acids, but the equation for the reaction that takes place is as follows:

$$\underset{\text{base}}{HNO_3} + \underset{\text{acid}}{H_2SO_4} \rightleftharpoons H_2NO_3^+ + HSO_4^-$$

Here, H_2SO_4 donates a proton, and therefore acts as the acid, and HNO_3 accepts this proton, and is therefore the base in this reaction.

There are other similar systems that do not include water as the solvent. For example, concentrated ethanoic acid exists as a liquid at room temperature. In the non-aqueous reaction,

$$HNO_3 + CH_3COOH \rightleftharpoons NO_3^- + CH_3COOH_2^+$$

the HNO_3 acts as the acid and the CH_3COOH as the base.

These reactions are showing us that in each system there is a proton donor *and* a proton acceptor, i.e. an acid and a base.

Also, because the reactions are reversible, the proton can be donated back again, so the donor on the right hand side of the equation must also be considered as an acid, and the acceptor as a base:

$$\underset{\text{acid}}{HCl(aq)} + \underset{\text{base}}{H_2O(l)} \rightleftharpoons \underset{\text{acid}}{H_3O^+(aq)} + \underset{\text{base}}{Cl^-(aq)}$$

$$\underset{\text{acid}}{CH_3COOH(aq)} + \underset{\text{base}}{H_2O(l)} \rightleftharpoons \underset{\text{acid}}{H_3O^+(aq)} + \underset{\text{base}}{CH_3COO^-(aq)}$$

$$\underset{\text{acid}}{H_2SO_4} + \underset{\text{base}}{HNO_3} \rightleftharpoons \underset{\text{acid}}{H_2NO_3^+} + \underset{\text{base}}{HSO_4^-}$$

An acid dissociates to form two ions, namely a proton and a negative ion, and establishes an equilibrium. The negative ion acts as a base because it recombines with the proton and re-forms the acid molecule. The negative ion is referred to as the **conjugate base** of the acid. Also, when a base such as $NH_3(g)$ reacts with H^+ to form NH_4^+, this positive ion is referred to as a **conjugate acid**.

At the foot of the this page are some examples. Note that in the last example, the right hand side shows that the same species can act both as an acid and a base. Pure water will always contain some hydrogen ions and hydroxide ions: it is amphoteric, and can be described as **self-protonating**:

$$2H_2O \rightleftharpoons H_3O^+ + OH^-$$

In water, any species providing H_3O^+ is an acid, and any species providing OH^- is a base.

Other solvents such as liquid ammonia also exhibit this behaviour:

$$2NH_3 \rightleftharpoons NH_4^+ + NH_2^-$$

In liquid ammonia, any species providing NH_4^+ is an acid, and any species providing NH_2^- is a base.

$CH_3COOH(l)$ ethanoic acid	+	$OH^-(aq)$ base	$\rightleftharpoons$	$CH_3COO^-(aq)$ ethanoate ion: conjugate base	+	$H_2O(l)$ conjugate acid
donates proton		accepts proton		accepts proton		donates proton
$HCl(g)$ acid	+	$NH_3(g)$ base	$\rightleftharpoons$	NH_4^+ conjugate acid	+	$Cl^-(s)$ conjugate base
donates proton		accepts proton		donates proton		accepts proton
$HCl(g)$ acid	+	$H_2O(l)$ base	$\rightleftharpoons$	$H_3O^+(aq)$ conjugate acid	+	$Cl^-(aq)$ conjugate base
donates proton		accepts proton		donates proton		accepts proton
$H_3O^+(aq)$ acid	+	$CO_3^{2-}(aq)$ base	$\rightleftharpoons$	$HCO_3^-(aq)$ conjugate acid	+	$H_2O(l)$ conjugate base
donates proton		accepts proton		donates proton		accepts proton
$H_3O^+(aq)$ acid	+	$NH_3(aq)$ base	$\rightleftharpoons$	$NH_4^+(aq)$ conjugate acid	+	$H_2O(l)$ conjugate base
donates proton		accepts proton		donates proton		accepts proton
$H_3O^+(aq)$ acid	+	$OH^-(aq)$ base	$\rightleftharpoons$	$H_2O(l)$ conjugate acid	+	$H_2O(l)$ conjugate base
donates proton		accepts proton		donates proton		accepts proton

3.2 Ionic equations

In aqueous solution, an acid and a base react to form a salt and water only. This is a **neutralisation reaction**. The base accepts one or more protons from the acid:

$$HCl(aq) + NaOH(aq) \rightarrow NaCl(aq) + H_2O(l)$$

This equation does not show what happens to the ions in the solution. When we rewrite it showing all the ions present, some ions appear on both sides of the equation. The above is a reaction of a strong acid with a strong base, so all the reactant species are fully dissociated.

$$H^+(aq) + Cl^-(aq) + Na^+(aq) + OH^-(aq) \rightleftharpoons Na^+(aq) + Cl^-(aq) + H_2O(l)$$

Because sodium ions and chloride ions appear on both sides of the equation, we can say that the sodium ions and chloride ions do not take part in the reaction. Omitting them allows us to write a simplified equation, the **ionic equation**, for the neutralisation reaction:

$$H^+(aq) + OH^-(aq) \rightleftharpoons H_2O(l)$$

Other strong acids and strong bases react similarly and they can all be represented by the ionic equation for water.

Table 1 Some common strong acids and strong bases

Strong acids	Strong bases
hydrochloric acid HCl	sodium hydroxide NaOH
sulphuric acid H_2SO_4	potassium hydroxide KOH
nitric acid HNO_3	
phosphoric acid H_3PO_4	

The solvent liquid ammonia can be considered in the same way:

$$NH_4^+ + NH_2^- \rightleftharpoons 2NH_3$$

Any species containing NH_4^+ is an acid, any species containing NH_2^- is a base, and neutralisations will produce $NH_3(l)$.

1 Write full equations and ionic equations for the reactions between:

a hydrochloric acid and potassium hydroxide solution,

b sulphuric acid and sodium hydroxide solution.

2

a Explain why methane does not act as a proton donor, but methanoic acid (HCOOH) does.

b Write an equation to show how methanoic acid ionises in water.

3 Identify the acids and bases in the following reactions:

a $CuO(s) + H_2SO_4(aq) \rightleftharpoons CuSO_4(aq) + H_2O(l)$

b $NH_4^+(aq) + OH^-(aq) \rightleftharpoons NH_3(aq) + H_2O(l)$

c $CH_3COO^-(aq) + H_3O^+(aq) \rightleftharpoons CH_3COOH(aq) + H_2O(l)$

KEY FACTS

- Acids are proton donors.
- Bases are proton acceptors.
- Acid–base reactions involve the transfer of protons.
- Water is self-protonating.

3.3 Acid dissociation constants

Most acids are weak acids. When these dissociate and an equilibrium is set up, significant amounts of all the molecules and ions are present:

$$CH_3COOH(aq) + H_2O(l) \rightleftharpoons CH_3COO^-(aq) + H_3O^+(aq)$$
ethanoic acid   ethanoate ion

The strength of the acid is determined by the position of equilibrium. If the equilibrium lies to the left the acid is weak (less dissociated); the further to the right the equilibrium lies, the stronger the acid (more dissociated). Strong acids such as hydrochloric and sulphuric have an equilibrium that lies well over to the right. They are more or less fully dissociated.

The same principle applies to bases:

$$NH_3(aq) + H_2O(l) \rightleftharpoons NH_4^+(aq) + OH^-(aq)$$

When ammonia is dissolved in water, only a small proportion of ammonia molecules dissociate. Ammonia is therefore a weak base, with the equilibrium well over to the left.

When sodium hydroxide dissolves in water it dissociates completely, releasing sodium ions and hydroxide ions, so it is a strong base with the equilibrium very much over to the right:

$$Na^+OH^-(s) \rightleftharpoons Na^+(aq) + OH^-(aq)$$

The vinegar commonly put on chips is a dilute aqueous solution of ethanoic acid with colouring added

Acid dissociation constants

As with other equilibria (including reaction equilibria, see Chapter 2) acid-base equilibria can be described using equilibrium constants. These are called acid dissociation constants and can be used to define the strength of the acid. For acids, they are a measure of how readily hydrogen ions are released. Dissociation constants are calculated in a similar way to equilibrium constants for reversible reactions with products (right hand side of equation) divided by reactants (left hand side).

Consider this equilibrium between the undissociated acid and the ethanoate ions in aqueous solution:

$$CH_3COOH(aq) + H_2O(l) \rightleftharpoons CH_3COO^-(aq) + H_3O^+(aq)$$
ethanoic acid   ethanoate ion

The extent of dissociation is given by the **equilibrium constant, K_c,** for the process:

$$K_c = \frac{[CH_3COO^-(aq)]\ [H_3O^+(aq)]}{[CH_3COOH(aq)]\ [H_2O(l)]}$$

The units for concentrations cancel out, so K_c has no units.

Because it is assumed that the water molecules are in vast excess and because they ionise only very slightly, the concentration of the water molecules hardly changes. We can simplify the expression and rewrite the equilibrium using K_a as the acid dissociation constant:

$$K_a = \frac{[CH_3COO^-(aq)]\ [H_3O^+(aq)]}{[CH_3COOH(aq)]}$$

For ethanoic acid, a weak acid, K_a is 1.7×10^{-5} mol dm⁻³.

For the units for K_a, we insert the units for the terms in the expression for the dissociation constant. Where the acid releases one hydrogen ion only, the units are:

$$\frac{\text{mol dm}^{-3} \times \text{mol dm}^{-3}}{\text{mol dm}^{-3}} \quad \text{i.e. mol dm}^{-3}$$

We can now give a general expression for the dissociation constant K_a of any weak acid, of general formula HA, as:

$$K_a = \frac{[H^+]\ [A^-]}{[HA]}$$

where all concentrations are those at equilibrium and the units are mol dm⁻³.

The less an acid dissociates, the smaller the value, while the stronger the acid, the higher is the value of its dissociation constant.

Table 2 Dissociation constants of some common acids	
Acid	**K_a/mol dm⁻³**
HCl	1.0×10^7
HNO₃	4.0×10^1
HF	5.6×10^{-4}
HNO₂	4.7×10^{-4}
HCOOH	1.6×10^{-4}
C₆H₅COOH	6.5×10^{-5}
CH₃COOH	1.7×10^{-5}
C₂H₅COOH	1.3×10^{-5}
HClO	3.7×10^{-8}
HCN	4.9×10^{-10}

4   Write an expression for the acid dissociation constant of butanoic acid, $CH_3(CH_2)_2COOH$.

Writing expressions for K_a

Ethanoic acid dissociates only slightly (approximately 0.1% of the molecules dissociate, roughly one in a thousand) so the equilibrium lies well over to the left. We can use this fact to simplify the calculation of K_a. The weak dissociation of ethanoic acid means that the concentration [CH_3COOH] hardly changes because so few molecules dissociate to form CH_3COO^- and H^+ ions. The equilibrium concentration of CH_3COOH can therefore be taken as equal to its *initial* concentration (this is only valid when a weak acid alone is in aqueous solution). This concentration term can be written as:
[CH_3COOH]$_{initial}$

Also for every H^+ released a CH_3COO^- is released, so their concentrations are equal:

[CH_3COO^-] = [H^+]

We can then rewrite the expression for K_a as:

$$K_a = \frac{[H^+]^2}{[CH_3COOH]_{initial}} \text{ (neglecting } H^+ \text{ from water)}$$

The keto acid 3-oxobutanoic acid can form in the liver of diabetics (see Opener). It dissociates weakly according to the equation:

$$CH_3COCH_2COOH(aq) \rightleftharpoons CH_3COCH_2COO^-(aq) + H^+(aq)$$

The equation for the dissociation constant of this equilibrium is:

$$K_a = \frac{[CH_3COCH_2COO^-][H^+]}{[CH_3COCH_2COOH]_{initial}}$$

which according to the above approximation is:

$$= \frac{[H^+]^2}{[CH_3COCH_2COOH]_{initial}}$$

If 0.010 mol dm^{-3} solution of the acid contains 1.62×10^{-3} mol dm^{-3} of hydrogen ions, then the value for K_a can be calculated from the expression above:

$$K_a = \frac{[H^+]^2}{[CH_3COCH_2COOH]_{initial}}$$

$$= \frac{(1.62 \times 10^{-3})^2}{0.010}$$

$$= 2.62 \times 10^{-4} \text{ mol dm}^{-3}$$

In these examples we have only considered the release of a single H^+ ion. Acids such as CH_3COOH that have only one H^+ to donate are called monoprotic acids. Many acids can release more than one H^+ ion. For example, sulphuric(IV) acid, H_2SO_3, is a weak acid. Two equilibria are involved and there is an equilibrium constant for each of these:

acid → hydrogensulphite ion:

$$H_2SO_3(aq) + H_2O(l) \rightleftharpoons HSO_3^-(aq) + H_3O^+(aq)$$
$$K_a = 1.5 \times 10^{-2} \text{ mol dm}^{-3}$$

hydrogensulphite ion → sulphite ion:
$$HSO_3^-(aq) + H_2O(l) \rightleftharpoons SO_3^{2-}(aq) + H_3O^+(aq)$$
$$K_a = 6.2 \times 10^{-8} \text{ mol dm}^{-3}$$

Fig. 3 General expression for the dissociation constant of a weak acid

A weak acid can be represented by the formula HA and dissociates according to the equilibrium:

$HA(aq) \rightleftharpoons H^+(aq) + A^-(aq)$

The expression for its dissociation constant can be written as follows:

$$K_a \approx \frac{[H^+][A^-]}{[HA]}$$

If HA is a weak acid the concentration of HA will not change significantly at equilibrium *relative to* [A^-] and [H^+], so for [HA] we can use the *total* concentration of the acid. Also [A^-] will be equal to [H^+], so that we can substitute [H^+] for [A^-]:

$$K_a \approx \frac{[H^+]^2}{[HA]}$$

5 For the two equilibria for sulphuric(IV) acid in water, which reaction indicates the stronger acid? Explain how you used the values of K_a to decide upon your choice.

- The strength of an acid is defined in terms of its acid dissociation constant.

- A weak acid dissociates only slightly in water, and has a small dissociation constant.

- A strong acid dissociates fully in water, and has a large dissociation constant.

- The dissociation constant, K_a, for a weak acid, HA, can be calculated using:

$$K_a = \frac{[H^+][A^-]}{[HA]_{eqm}} \text{ and this can be approximated as}$$

$$K_a = \frac{[H^+]^2}{[HA]_{initial}}$$

3.4 pH

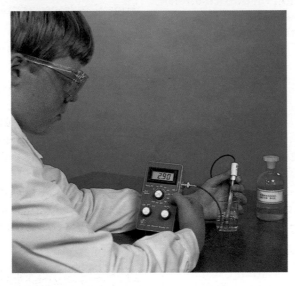

The changing colour of chemical indicators gives a rough measure of the pH of a solution. A more accurate measure is given by a pH meter.

We could use hydrogen ion concentration as a measure of acidity, but where the values of $[H^+]$ are very small (see Table 2 for examples), the numbers can be cumbersome to work with, and using a log scale makes them easier to handle.

The pH of a solution is $-\log_{10}$ of the molar hydrogen ion concentration in that solution.

The H refers to the hydrogen ion concentration, $[H^+]$, and the symbol p is related to the power to which 10 is raised in the expression $-\log_{10}$ to give that concentration.

$$pH = -\log_{10} [H^+]$$

Because pH values are logarithms, they have no units.

Calculating pH of strong acids from concentrations

We can apply the definition above to calculate the pH of a solution of hydrochloric acid containing 0.100 mol dm⁻³.

Hydrochloric acid is a strong acid. All the

hydrogen chloride molecules become fully dissociated into hydrogen ions and chloride ions. So in 1 dm³ of 0.100 mol dm⁻³ hydrochloric acid there are 0.100 moles of H^+ ions and 0.100 moles of Cl^- ions. 0.100 is the same as writing 1×10^{-1} or 10^{-1}; in other words, $[H^+] = 10^{-1}$. Therefore, since $pH = -\log_{10} [H^+]$:

$$pH = -\log_{10} 10^{-1}$$
$$= -(-1)$$

Therefore, the pH of hydrochloric acid of concentration 0.100 mol dm⁻³ = 1.

It is a lot easier to talk about 'a pH of 1' than to say 'a hydrogen ion concentration of 0.100 mol dm⁻³' or '10⁻¹ mol dm⁻³'. The pH of 0.010 (i.e. 10⁻²) mol dm⁻³ hydrochloric acid would be $-\log_{10} 10^{-2} = -(-2) = 2$, a lower $[H^+]$, so a higher pH value.

Moving up or down the pH scale by 1 unit means that the hydrogen ion concentration changes by a factor of 10 (= 10¹).

The calculation above also indicates that it is possible to have pH values outside the often quoted range of 1–14. A strong acid of concentration 1.000 mol dm⁻³ would have a pH of $-\log_{10} 1 = 0$, so the pH would be 0. Any strong acid with a concentration of greater than 1 mol dm⁻³ will have a pH of less than 0, and therefore a minus value. At the other end of the pH range, solutions with very low concentrations of hydrogen ions, for example, $[H^+] = 1 \times 10^{-15}$ mol dm⁻³, the pH would be 15.

Such calculations can be performed quickly using the expression:

$$[H^+] = 10^{-pH}$$

Remember: pH is the power to which 10 is raised to give the $[H^+]$, and the sign is changed.

Table 3 Comparison of $[H^+]$ values with corresponding pH values	
$[H^+]$/mol dm⁻³	pH
10⁻⁴	4
10⁻⁵	5
10⁻⁶	6
10⁻⁷	7
10⁻⁸	8
10⁻⁹	9

6 Calculate the pH of:

a 0.2 mol dm⁻³ nitric acid (a strong acid, i.e. fully ionised)

b $[H^+] = 1.2 \times 10^{-2}$ mol dm⁻³

c $[H^+] = 6.7 \times 10^{-5}$ mol dm⁻³

7 What is the pH of a 2.00 mol dm⁻³ solution of hydrochloric acid?

Calculating concentration from pH

Given a pH, we can work the other way and calculate the hydrogen ion concentration in the solution. What is the hydrogen ion concentration in a solution of pH 2.61?

Since $pH = -\log_{10} [H^+]$
$$[H^+] = 10^{-pH}$$
$$[H^+] = 10^{-2.61}$$

Therefore, $[H^+] = 0.00245$
$$= 2.45 \times 10^{-3} \text{ mol dm}^{-3}$$

8

a A solution has a pH of 5.49. What is the value of $[H^+]$?

b What is the hydrogen ion concentration in a solution of pH 2.75?

c What is the hydrogen ion concentration in a solution of pH −0.50?

Calculating pH from K_a values for weak acids

We can calculate the pH of a solution of a weak acid if we are given the following information:

the concentration of the weak acid, the value for its dissociation constant, and the expression: $K_a = \dfrac{[H^+][A^-]}{[HA]}$

This is illustrated in the following example.

Lactic acid forms in muscle cells if insufficient oxygen reaches the muscles during exercise:

$$CH_3CH(OH)COOH(aq) \rightleftharpoons H^+(aq) + CH_3CH(OH)COO^-(aq)$$
lactic acid

The lactic acid concentration is typically 0.01 mol dm⁻³, and its dissociation constant is $K_a = 1.287 \times 10^{-4}$ mol dm⁻³.

$$K_a = \frac{[H^+(aq)][CH_3CH(OH)COO^-(aq)]}{[CH_3CH(OH)COOH(aq)]}$$

$$K_a = \frac{[H^+]^2}{[CH_3CH(OH)COOH(aq)]}$$

Rearranging this equation gives:

$$[H^+]^2 = [CH_3CH(OH)COOH(aq)] \times K_a$$

Substituting the values given above:

$$[H^+]^2 = 0.01 \times 1.287 \times 10^{-4}$$
$$= 1.287 \times 10^{-6} \text{ mol}^2 \text{ dm}^{-6}$$

$$[H^+] = \sqrt{1.287 \times 10^{-6}}$$
$$= 1.135 \times 10^{-3} \text{ mol dm}^{-3}$$

The pH of the solution can be found using:

$$pH = -\log_{10} [H^+]$$
$$= -\log_{10} 1.135 \times 10^{-3}$$
$$= 2.95$$

Increasing the concentration of a weak acid increases the hydrogen ion concentration, which decreases the pH of the solution. A higher value of K_a indicates a stronger acid and therefore a lower pH.

9 Calculate the pH of a 0.01 mol dm⁻³ solution of propanoic acid which has a dissociation constant of 1.34×10^{-5} mol dm⁻³.

Strenuous exercise can increase the concentration of lactic acid in muscles and this can give rise to painful cramp.

3.5 pH of water

We have seen that water can act as an acid and a base. It acts as a proton acceptor (base) when hydrogen chloride dissolves in it:

$$HCl(g) + H_2O(l) \rightarrow H_3O^+(aq) + Cl^-(aq)$$

It acts as a proton donor (acid) when ammonia dissolves in it:

$$NH_3(g) + H_2O(l) \rightleftharpoons NH_4^+(aq) + OH^-(aq)$$

Pure water itself is very weakly ionised:

$$H_2O(l) \rightleftharpoons H^+(aq) + OH^-(aq)$$

The equilibrium constant for this process is:

$$K_c = \frac{[H^+][OH^-]}{[H_2O]}$$

The concentration of water molecules in the equilibrium is very high. Because of the low degree of ionisation of water we can consider $[H_2O]$ constant. This gives us a useful expression for the value of the **ionic product of water**, K_w:

$$K_w = [H^+][OH^-]$$

At 298 K, K_w is 10^{-14} mol^2 dm^{-6}.

We can calculate the pH of pure water, using the fact that it contains equal numbers of hydrogen ions and hydroxide ions:

$$[H^+] = [OH^-]$$

At 298 K, from K_w:

$$K_w = [H^+][OH^-] = [H^+]^2 = 10^{-14} \text{ mol}^2 \text{ dm}^{-6}$$
$$[H^+] = 10^{-7} \text{ mol dm}^{-6}$$

(To get the square root of a power, divide by 2.)

$$pH = -\log_{10}[H^+]$$
$$= -\log_{10} 10^{-7}$$
$$= 7$$

The pH of pure water is 7 at 298 K.

10 What is the concentration of hydroxide ions in pure water at 298 K?

Experiments show that the ionic product of water increases slightly with increasing temperature, which means that the equilibrium:

$$H_2O(l) \rightleftharpoons H^+(aq) + OH^-(aq)$$

moves to the right. Le Chatelier's principle tells us that imposing a stress on a system causes the system to change to remove the stress. The fact that a temperature increase moves the equilibrium to the right tells us that the dissociation of water is an endothermic process.

When the temperature is increased, the concentration of hydrogen ions increases and so the value of the pH will decrease (Table 4). However, the increase in $[H^+]$ is matched by an identical increase in $[OH^-]$, so the water can still be described as neutral – there are equal concentrations of H^+ and OH^-.

Bases and pH

When bases dissolve in water they produce hydroxide ions, $[H^+] < [OH^-]$, and the hydroxide ions shift the dissociation of water equilibrium to the left (Fig. 4). But the ionic product of water remains constant at 1×10^{-14}.

$$H_2O(l) \rightleftharpoons H^+(aq) + OH^-(aq)$$
$$K_w = 1 \times 10^{-14} \text{ at 298 K}$$

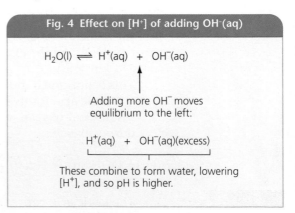

Fig. 4 Effect on [H⁺] of adding OH⁻(aq)

$$H_2O(l) \rightleftharpoons H^+(aq) + OH^-(aq)$$

Adding more OH⁻ moves equilibrium to the left:

$$H^+(aq) + OH^-(aq)(excess)$$

These combine to form water, lowering [H⁺], and so pH is higher.

We can calculate the pH of a strong alkali. It is a little more complex than calculating the pH of a strong acid.

Sodium hydroxide ionises fully in solution:

$$NaOH(aq) \rightarrow Na^+(aq) + OH^-(aq)$$

Table 4 The pH of water at different temperatures				
Temperature/°C	$-\log_{10} K_w$	K_w/mol dm^{-3}	[H⁺]	pH
0	14.94	1.15×10^{-15}	3.39×10^{-8}	7.47
20	14.17	6.76×10^{-15}	8.22×10^{-8}	7.09
25	14.00	1.00×10^{-14}	1.00×10^{-7}	7.00
30	13.83	1.48×10^{-14}	1.22×10^{-7}	6.91
40	13.54	2.88×10^{-14}	1.70×10^{-7}	6.77
50	13.26	5.50×10^{-14}	2.35×10^{-7}	6.63
60	13.02	9.55×10^{-14}	3.09×10^{-7}	6.51

So in 1 dm³ of 0.100 mol dm⁻³ sodium hydroxide solution, there are 0.100 moles of hydroxide ions. We use the ionic product of water to calculate the hydrogen ion concentration.

Stage 1

Calculate $[H^+]$ from $K_w = [H^+][OH^-]$:

$[OH^-] = 0.100$ and, at 298 K, K_w is 10^{-14} mol² dm⁻⁶

$$[H^+] = \frac{K_w}{[OH^-]} = \frac{10^{-14}}{0.100}$$

$$= 10^{-13} \text{ mol dm}^{-3}$$

Stage 2

Calculate pH from $[H^+]$:

$$pH = -\log_{10}[H^+] = -\log_{10}10^{-13} = -(-13)$$
$$= 13$$

The pH of a 0.100 mol dm⁻³ solution of sodium hydroxide at 298 K is 13. Try repeating this calculation for a higher concentration of hydroxide ions. You will see that the pH increases.

11 Calculate the pH of:

a 0.01 mol dm⁻³ sodium hydroxide solution

b 0.3 mol dm⁻³ potassium hydroxide solution

3.6 pK_a values

Values of K_a, the acid dissociation constant, can be very small, so again it would be more convenient to deal with them in the same way that we deal with hydrogen ion concentrations. To make $[H^+]$ values more manageable, the numbers are converted to pH, and in a similar manner K_a values can be converted to pK_a values:

$pK_a = -\log_{10} K_a$

K_a for ethanoic acid is 1.754×10^{-5} mol dm⁻³ at 298 K, so

$pK_a = -\log_{10} K_a$
$\quad\quad = -\log_{10}(1.754 \times 10^{-5})$
$\quad\quad = 4.76$ (again, there are no units for pK_a values, because they are logarithms)

Calculating pH from pK_a

We can calculate the pH of a solution of known concentration using pK_a values.

Let us calculate the pH of a 0.125 mol dm⁻³ solution of ethanoic acid. The pK_a value for ethanoic acid is 4.76. As before:

$$K_a = \frac{[CH_3COO^-][H^+]}{[CH_3COOH]}$$

$$= \frac{[H^+]^2}{[CH_3COOH]}$$

$$[H^+]^2 = [CH_3COOH] \times K_a$$

$K_a = 10^{-pK_a}$, therefore:

$$[H^+]^2 = [CH_3COOH] \times 10^{-pK_a}$$
$$= 0.125 \times 10^{-4.76}$$
$$= 2.17 \times 10^{-6}$$

$[H^+] = 1.47 \times 10^{-3}$ mol dm⁻³
pH = 2.83

12 Calculate the pH of a 0.45 mol dm⁻³ solution of propanoic acid. The pK_a value of propanoic acid is 4.87.

Table 5 pK_a values for some common acids					
Acid	**K_a/mol dm⁻³**	**pK_a**	**Acid**	**K_a/mol dm⁻³**	**pK_a**
HCl	1.0×10^7	−7.00	C_6H_5COOH	6.5×10^{-5}	4.19
HNO_3	4.0×10^1	−1.60	CH_3COOH	1.7×10^{-5}	4.77
HF	5.6×10^{-4}	3.25	C_2H_5COOH	1.3×10^{-5}	4.89
HNO_2	4.7×10^{-4}	3.33	HClO	3.7×10^{-8}	7.43
HCOOH	1.6×10^{-4}	3.80	HCN	4.9×10^{-10}	9.31

Section 3.1 showed that a protonated weak base behaves as a weak acid (it is the conjugate acid of the weak base). For example, NH_4^+ is the conjugate acid of the weak base NH_3.

$$NH_4^+ \rightleftharpoons H^+ + NH_3$$
$$NH_4^+ + H_2O \rightleftharpoons H_3O^+ + NH_3$$

Calculate the pH of a 0.20 mol dm^{-3} solution of ammonium chloride. The pK_a value for the ammonium ion is 9.25.

$$K_a \approx \frac{[H^+]^2}{[NH_4Cl]_{total}}$$
$$[H^+] = \sqrt{K_a \times [NH_4Cl]_{total}}$$
$$= \sqrt{10^{-9.25} \times 0.20}$$
$$= 1.061 \times 10^{-5}$$
$$pH = 4.97$$

This behaviour is typical of ions formed when a weak base reacts with a strong acid. The products are referred to as strong acid salts of weak bases and they will be discussed in Section 3.7. The pH of solutions of these salts is less than 7.

Salts formed when strong bases react with weak acids (strong base salts of weak acids) will behave as weak bases and the pH of such solutions will be greater than 7.

Learn the expressions for pH, K_a and pK_a (Fig. 5), because, if you know the concentration of an acid, you can calculate the value of either pH, K_a or pK_a (Fig. 6).

Fig. 5 Expressions for K_a, pK_a and pH

$$K_a = \frac{[H^+][A^-]}{[HA]}$$
$$pK_a = -\log_{10} K_a$$
$$pH = -\log_{10} [H^+]$$

Fig. 6 Calculating [H⁺] and pH from K_a

Lactic acid forms in muscle cells if insufficient oxygen reaches the muscles during exercise.

$$CH_3CH(OH)COOH(aq) \rightleftharpoons$$
$$CH_3CH(OH)COO^-(aq) + H^+(aq)$$
lactic acid

The lactic acid typically has a concentration of 0.01 M, and its dissociation constant is $K_a = 1.287 \times 10^{-4}$ mol dm^{-3}
So:

$$K_a = \frac{[CH_3CH(OH)COO^-][H^+]}{[CH_3CH(OH)COOH]}$$
$$= \frac{[H^+]^2}{[CH_3CH(OH)COOH]}$$

Rearranging this equation gives:

$$[H^+]^2 = [CH_3CH(OH)COOH] \times K_a$$

Substituting for values given above:

$$[H^+]^2 = 0.01 \times 1.287 \times 10^{-4}$$
$$= 1.287 \times 10^{-6} \text{ mol}^2 \text{ dm}^{-6}$$
$$[H^+] = \sqrt{1.287 \times 10^{-6}}$$
$$= 1.135 \times 10^{-3} \text{ mol dm}^{-3}$$

The pH of the solution can be found using:

$$pH = -\log_{10} [H^+]$$
$$= -\log_{10} 1.135 \times 10^{-3}$$
$$= 2.95$$

13

a Calculate K_a for a 0.01 mol dm^{-3} solution of benzoic acid if $[H^+] = 7.94 \times 10^{-4}$ mol dm^{-3}.
$$C_6H_5COOH(aq) \rightleftharpoons C_6H_5COO^-(aq) + H^+(aq)$$

b Use your answer to (a) to decide whether benzoic acid is a strong or a weak acid.

c Calculate the pK_a value for benzoic acid.

14 Citric acid has a pK_a value of 3.075 (for the first ionisation). For a 0.01 mol dm^{-3} solution of the acid and for this ionisation, calculate:

a K_a

b $[H^+]$

c pH

15 Which will have the higher pK_a value: a weak acid or a strong acid?

KEY FACTS

■ pK_a values are a convenient way of dealing with strengths of weak acids.

■ The expression for the pK_a of a weak acid is:
$$pK_a = -\log_{10} K_a$$

■ pK_a values do not have units.

■ The acid dissociation constant of a weak acid HA is given by the expression: $K_a = \frac{[A^-(aq)][H^+(aq)]}{[HA(aq)]}$

■ The pH of a weak acid can be calculated from its dissociation constant and its concentration in mol dm^{-3}, using the approximation:
$$K_a = \frac{[H^+]^2}{[HA]}; \text{ and } pH = -\log_{10} [H^+]$$

3.7 pH curves

In a typical titration, an alkaline solution is gradually added to an acid solution (or an acid to an alkali) of known volume and concentration and containing an indicator. Its change in colour shows when the alkali and acid reach stoichiometric equivalence.

During a laboratory titration, the pH changes as one solution is added to the other. A change in indicator colour signals the **equivalence point** of the titration (also called end-point, but see page 46): this corresponds to the mixing of stoichiometrically equivalent amounts of acid and base. Changes in pH at points before and after equivalence can be measured using a pH meter.

If a base is added to an acid, the pH rises as the acid is neutralised. However, the pH change is not directly proportional to the amount of alkali added. This is seen clearly when a graph of measured pH against quantity of solution added is plotted. It produces a plot known as a **pH curve**.

Shapes of pH curves

Strong acid–strong base
Strong acid-strong base curves looks like the ones in Figs. 7 and 8. They are simply mirror images of one another, depending upon whether the solution in the flask is an acid or a base. Strong acid-strong base curves are typical of reactions such as:

$$HCl(aq) + NaOH(aq) \rightarrow NaCl(aq) + H_2O(l)$$

Both the acid and the base are fully dissociated, existing as individual aqueous

ions. When a small volume of base is added to the acid at the start of the titration, the *proportion* of hydrogen ions removed by that base is quite small compared to the total amount present. This produces a very small increase in pH. As more base is added, the proportion of hydrogen ions being removed each time (relative to the total number) increases, so there is a larger change in pH. This change is greatest near the equivalence point, where the curve is steepest. The equivalence point for this curve is pH 7 – the pH of pure water. This is because the salt ions $Na^+(aq)$ and $Cl^-(aq)$ have no acid/base properties and so do not affect the $H^+(aq) + OH^-(aq) \rightleftharpoons H_2O(l)$ equilibrium.

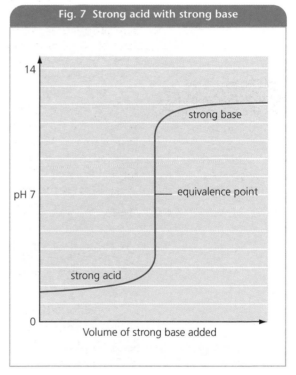

Fig. 7 Strong acid with strong base

To find the equivalence point on a pH curve, extrapolate the lines at the beginning and end of the curves where the pH is not changing as steeply, then find the midpoint on this vertical section of the graph. The equivalence point for a strong acid with a strong base lies midway between 2 and 12, namely at 7. For a weak acid with a strong base it lies midway between pH 5.8 and 12, namely at 8.9.

All the diagrams for Figs. 7 to 11 apply to titrations involving a monobasic acid with a monoacidic base in which one proton only is

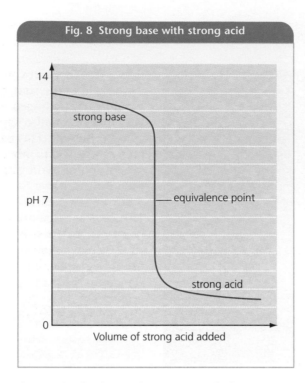

Fig. 8 Strong base with strong acid

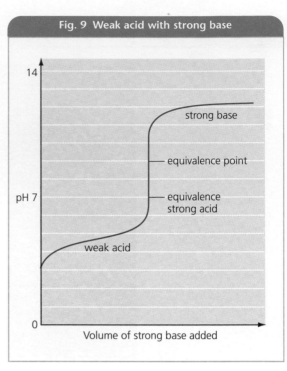

Fig. 9 Weak acid with strong base

donated. The first solution in each figure caption is of known concentration and is in the flask, and the second, of unknown concentration, is added from the burette.

16

a Sketch a pH curve for the titration of a strong base with a weak acid.

b What is the pH at the equivalence point in your diagram?

Weak acid–strong base

The pH curve for the titration of a weak acid with a strong base (Fig. 9) differs from the curve in Fig. 7.

Consider ethanoic acid reacting with sodium hydroxide:

$$CH_3COOH(aq) + NaOH(aq) \rightarrow Na^+(aq) + CH_3COO^-(aq) + H_2O(l)$$

The pH curve starts at a higher pH value, since the weak acid is only slightly dissociated, producing a low $[H^+]$ at equilibrium. The ethanoate ion acts as a weak conjugate base and recombines with $H^+(aq)$ to form $CH_3COOH(aq)$, and the greater the $[CH_3COO^-(aq)]$, the greater is the tendency to do this. This resistance to change in $[H^+]$, despite increasing acid concentration, is called **buffering action** and will be discussed later.

In the titration of strong or weak monobasic acids, where the number of moles of acid is the same, the total number of hydrogen ions that react will be the same

because, despite partial dissociation, all the available hydrogens are eventually released to react with a stoichiometric equivalent amount of base. Therefore, strong and weak acids give the same quantities in a calculation, but the pH of the *equivalence point* for weak acid-strong base titrations will have a higher value than for strong acid-strong base titrations: it is greater than 7. After the equivalence point, the curve follows the pattern for a strong acid-strong base titration.

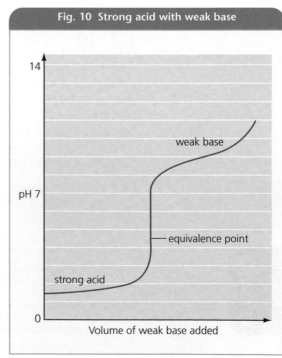

Fig. 10 Strong acid with weak base

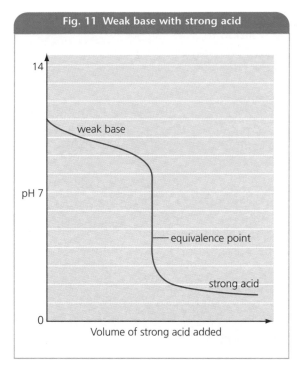

Fig. 11 Weak base with strong acid

pH 7

weak base

equivalence point

strong acid

Volume of strong acid added

The pH curve for the titration of a strong base with a weak acid is the mirror image and the pH of equivalence is still greater than 7.

Strong acid–weak base

For a strong acid-weak base titration, the curve (Fig. 10) follows the pattern for strong acid initially, but the equivalence point is at a pH of less than 7. The reaction between $HCl(aq)$ and $NH_3(aq)$ shows this:

$$HCl(aq) + NH_3(aq) \rightarrow NH_4^+(aq) + Cl^-(aq) + H_2O(l)$$

An equilibrium will be set up between the weak base $NH_3(aq)$ and its conjugate acid $NH_4^+(aq)$:

$$H^+(aq) + NH_3(aq) \rightleftharpoons NH_4^+(aq)$$

This increases the concentration of $H^+(aq)$ and lowers the pH compared to that for a strong base solution.

pH curves can be plotted for all the combinations of strong and weak acids with strong and weak bases. For the titration of a weak base with a strong acid (Fig. 11) the curve will start at a high pH but will have an equivalence point with a pH less than that for the titration of a strong base with a strong acid, i.e. less than 7.

Weak base–weak acid

When a weak base is titrated with a weak acid, the variation in pH with volume of titre is more gradual, and it is not possible to detect an accurate equivalence point (Fig. 12).

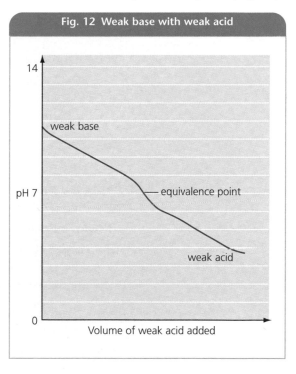

Fig. 12 Weak base with weak acid

pH 7

weak base

equivalence point

weak acid

Volume of weak acid added

Calculating amounts

When the volume of added solution to produce the equivalence point is determined, its concentration can be calculated.

Example: Calculation for strong acid-strong base

Calculate the concentration of sodium hydroxide if 25.0 cm³ of the solution just neutralises 20.0cm³ of 0.100 mol dm⁻³ hydrochloric acid solution.

1   Calculate the number of moles of the solution of known concentration, HCl(aq): The number of moles of a reacting substance can be calculated using the expression:

$$\text{number of moles in solution (mol)} = \text{concentration of solution (mol dm}^{-3}) \times \text{volume of solution (dm}^3)$$

(or $m = M \times V/1000$ for volumes in cm³)

$$\text{number of moles of HCl(aq) (mol)} = \text{concentration of HCl(aq) (mol dm}^{-3}) \times \frac{\text{vol. in cm}^3 \text{ of HCl(aq)}}{1000} \text{ (dm}^3)$$

$$= 0.100 \times 20.0/1000$$

$$= 0.002 \text{ mol}$$

2   Write the equation to find the stoichiometry:

$$HCl(aq) + NaOH(aq) \rightarrow NaCl(aq) + H_2O(l)$$

Therefore, 1 mole of HCl(aq) needs 1 mole of NaOH(aq) for neutralisation (stoichiometry is 1:1)

$$\text{number of moles of NaOH(aq)} = \text{number of moles of HCl(aq)}$$
$$= 0.002 \text{ mol}$$

3   Calculate the concentration of NaOH(aq):

$$\text{concentration of NaOH(aq)} = \frac{\text{number of moles of NaOH(aq)}}{\text{vol. in cm}^3 \text{ NaOH(aq)}/1000}$$

$$= \frac{0.002}{0.025}$$

$$= 0.080 \text{ mol dm}^{-3}$$

If the stoichiometry of a reaction is 1:2, this factor is introduced into the calculation.

Example: Calculation for a reaction of 1:2 stoichiometry

If 34.0 cm³ of 0.125 mol dm⁻³ sodium hydroxide just neutralises 25.0 cm³ of sulphuric acid, calculate the concentration of the acid.

1   Calculate the number of moles of alkali:

$$\text{number of moles} = \text{concentration} \times \text{volume in cm}^3/1000$$
$$= 0.125 \times 34.0/1000$$
$$= 4.25 \times 10^{-3} \text{ mol}$$

2   Use the balanced equation to find the stoichiometry of the reaction:

$$2NaOH(aq) + H_2SO_4(aq) \rightarrow Na_2SO_4(aq) + H_2O(l)$$
ratio 2NaOH(aq):1H₂SO₄(aq)

3   Calculate the number of moles of acid required to neutralise 1 mole of base:

$$\text{number of moles of acid} = 0.5 \text{ the number of moles of alkali}$$
$$= 0.5 \times 4.25 \times 10^{-3} \text{ mol}$$
$$= 2.125 \times 10^{-3} \text{ mol}$$

4   Calculate the concentration of the acid:

$$\text{number of moles} = \text{concentration} \times \text{volume in cm}^3/1000$$

$$\text{concentration} = \frac{\text{number of moles}}{\text{vol. in cm}^3/1000}$$

$$= 2.125 \times 10^{-3}/0.025$$
$$\text{concentration of the sulphuric acid} = 0.085 \text{ mol dm}^{-3}$$

Similar calculations can be done for diprotic bases.

Example: Calculation for diprotic bases

If 27.3 cm³ of a 0.31 mol dm⁻³ solution of Ba(OH)₂(aq) are required to neutralise 25.0 cm³ of a solution of HCl(aq), calculate the concentration of the HCl(aq) solution.

1   Calculate the number of moles of alkali:

$$\text{number of moles} = \text{concentration} \times \text{volume in cm}^3/1000$$
$$= 0.31 \times 27.3/1000$$
$$= 8.463 \times 10^{-3} \text{ mol}$$

2   Use the balanced equation to find the stoichiometry of the reaction:

$$Ba(OH)_2(aq) + 2HCl(aq) \rightarrow BaCl_2(aq) + 2H_2O(l)$$
ratio = 1Ba(OH)₂(aq) : 2HCl(aq)

3   Calculate the number of moles of acid required:

$$\text{number of moles of acid} = 2 \times \text{the number of moles of alkali}$$
$$= 2 \times 8.463 \times 10^{-3} \text{ mol}$$
$$= 1.693 \times 10^{-2} \text{ mol}$$

4 Calculate the concentration of the acid:

number of moles = concentration × volume in cm³/1000

$$\text{concentration} = \frac{\text{number of moles}}{\text{volume}/1000}$$

$$= 1.693 \times 10^{-2}/0.025$$

concentration of HCl = 0.677 mol dm⁻³

Note: It is easy to make a mistake when using the factor of 2 in reactions with 2:1 stoichiometry, so it is worth doing an approximate check. If the volumes of each solution are similar, then the concentration of the substance needing 2 moles in the equation should be approximately double. In the last example, $1Ba(OH)_2(aq):2HCl(aq)$, the concentrations were 0.31 mol dm⁻³: 0.667 mol dm⁻³.

pH calculations

Strong acid–strong base titrations
When calculating the pH of a solution at a point during a titration *which is not the equivalence point*, use the equations in the worked examples above. The procedure is as follows:

- Find the number of moles H⁺ and OH⁻ present from both solutions in the mixture.
- Decide which one is in excess.
- Calculate the total volume of solution present.
- Calculate the concentration of the component in excess.
- Calculate the [H⁺] (if OH⁻ is in excess use $K_w = [H^+][OH^-] = 10^{-14}$).
- Calculate the pH.

Example: Calculating pH for strong acid–strong base
Calculate the pH in the titration of 20.0 cm³ of 0.15 mol dm⁻³ HCl(aq) at the point when 10.0cm³ of 0.20 mol dm⁻³ of NaOH(aq) have been added.

Calculate the number of moles of H⁺ present in the original solution:

number of moles = 0.15 × 20.0/1000
= 0.003 mol

Calculate the number of moles of OH⁻ added to the original solution:

number of moles = 0.20 × 10.0/1000
= 0.002 mol

Calculate the number of moles in excess:

number of moles = 0.003 − 0.002
= 0.001 mol (H⁺ is in excess here)

The total volume of solution is 20.0 + 10.0 = 30.0 cm³.

Calculate the concentration of the H⁺ in excess:

$[H^+] = 0.001/0.030 = 0.033$ mol dm⁻³
$pH = -\log_{10}[H^+] = -\log_{10}0.033$
$= 1.48$

Example: Calculating pH for strong diprotic acid–strong base
Calculate the pH in the titration of 25.0 cm³ of 0.150 mol dm⁻³ of NaOH(aq) at the point when 10.5 cm³ of 0.12 mol dm⁻³ ethanedioic acid $H_2C_2O_4(aq)$ (a diprotic acid) have been added.

Calculate the number of moles of OH⁻ present in the original solution:

number of moles = 0.15 × 25.0/1000
= 0.00375 mol

Calculate the number of moles of H⁺ added to the original solution (remember: there are two H⁺ per molecule of acid):

number of moles = 2 × 0.120 × 10.5/1000
= 0.00252 mol

Calculate the number of moles in excess:

number of moles = 0.00375 − 0.00252
= 0.00123 mol (OH⁻ is in excess here)

The total volume of solution is 25.0 + 10.5 = 35.5 cm³.

Calculate the concentration of the OH⁻ in excess:

$[OH^-] = 0.00123/0.0355$
$= 0.0347$ mol dm⁻³

$[H^+] = 10^{-14}/[OH^-] = 10^{-14}/0.0347$
$= 2.882 \times 10^{-13}$

$pH = -\log_{10}[H^+] = -\log_{10}2.882 \times 10^{-13}$
$= 12.54$

Weak acid–strong base titrations
The calculations for these examples are similar but will depend on how far the titration has progressed.

- If the weak acid is in excess, the relative proportions of HA and A⁻ have to be determined using K_a.

- If the strong base is in excess, the calculation is similar to that in strong acid–strong base titrations (see above).

Example 1: Calculating pH for weak acid–strong base

Calculate the pH in the titration when 10.0 cm^3 of 0.20 mol dm^{-3} NaOH(aq) have been added to 30.0 cm^3 of 0.15 mol dm^{-3} CH_3COOH(aq). $K_a = 1.76 \times 10^{-5}$ mol dm^{-3}.

Calculate the number of moles of CH_3COOH present in the original solution:

number of moles = 0.15 × 30.0/1000
= 0.0045 mol

Calculate the number of moles of OH^- added to the original solution:

number of moles = 0.20 × 10.0/1000
= 0.002 mol
(this is also the number of moles of CH_3COO^- formed)

Calculate the number of moles CH_3COOH in excess:

number of moles = 0.0045 − 0.002
= 0.0025 mol

Since both CH_3COOH remaining and the CH_3COO^- formed are in the same total volume of solution, then:

concentration ratio = mole ratio (there is no need to calculate the total volume)

$$[H^+] = \frac{K_a \times [CH_3COOH]}{[CH_3COO^-]}$$

$$= \frac{1.76 \times 10^{-5} \times 0.0025}{0.002}$$

$$= 2.200 \times 10^{-5} \text{ mol dm}^{-3}$$

pH = $-\log_{10}[H^+] = -\log_{10} 2.200 \times 10^{-5}$
= 4.66

Example 2: Calculating pH for weak acid–strong base

Calculate the pH in the titration of 25.0 cm^3 of 0.150 mol dm^{-3} NaOH(aq) at the point when 10.5 cm^3 of 0.12 mol dm^{-3} CH_3COOH have been added.

Calculate the number of moles of OH^- present in the original solution:

number of moles = 0.15 × 25.0/1000
= 0.00375 mol

Calculate the number of moles of CH_3COOH added to the original solution:

number of moles = 0.120 × 10.5/1000
= 0.00126 mol

Calculate the number of moles in excess:

number of moles = 0.00375 − 0.00126
= 0.00249 mol ([OH^-] in excess)

Since the strong base is in excess, the calculation is just the same as for strong acid–strong base. The total volume of solution is 25.0 + 10.5 = 35.5 cm^3.

Calculate the concentration of the [OH^-] in excess:

[OH^-] = 0.00249/0.0355
= 0.0701 mol dm^{-3}

[H^+] = 10^{-14}/[OH^-] = 10^{-14}/0.0701
= 1.427×10^{-13} mol dm^{-3}

pH = $-\log_{10}[H^+] = -\log_{10} 1.427 \times 10^{-13}$
= 12.85

pH of a weak acid at half equivalence

Half equivalence is the point at which half the amount of a strong base needed to neutralise a weak acid has been added. At this point, [HA] = [A^-] for the weak acid, so the ratio of [A^-] : [HA] = 1:1 in the expression:

$$K_a = \frac{[H^+][A^-]}{[HA]}$$

so K_a = [H^+] and pK_a = pH

Standardising solutions

Sodium carbonate can be used to standardise solutions of acids. The equation for the reaction with hydrochloric acid is:

Na_2CO_3(aq) + 2HCl(aq) →
2NaCl(aq) + H_2O(l) + CO_2(g)

The carbonate ion acts successively as two Brønsted–Lowry bases, successively accepting two protons and giving two equilibria:

CO_3^{2-}(aq) + H^+(aq) ⇌ HCO_3^-(aq)
HCO_3^-(aq) + H^+(aq) ⇌ H_2CO_3(aq) →
H_2O(l) + CO_2(g)

The CO_3^{2-}(aq) and the HCO_3^-(aq) on the left hand side are reacting with H^+(aq), so are behaving as bases, the two reverse reactions will have values for acid dissociation constants.

When sodium carbonate is titrated with a monoprotic acid, two equivalence points can be seen on the pH curve. The curve shows two regions where the changes in pH are slightly steeper (Fig. 13). This is because the carbonate ions (CO_3^{2-}) react first with one hydrogen ion, and then another. The first one occurs at pH 8.31 and the second at pH 3.69.

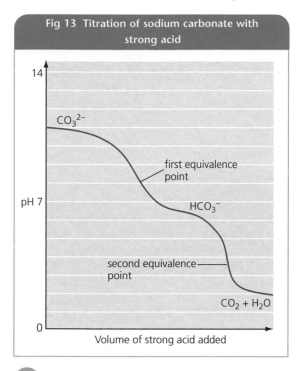

Fig 13 Titration of sodium carbonate with strong acid

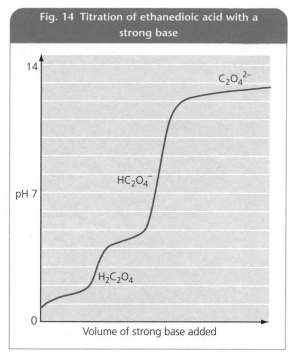

Fig. 14 Titration of ethanedioic acid with a strong base

17

a Calculate the volume of sulphuric acid (0.125 M) needed to neutralise 25.0 cm³ of 0.1 mol dm⁻³ sodium carbonate. (Write a balanced equation first.)

b Calculate the concentration of hydrochloric acid if 35.0 cm³ of acid neutralises 25.0 cm³ of 0.1 mol dm⁻³ sodium carbonate.

Diprotic acids

The pH curves for the titration of diprotic acids with bases also have two equivalence points. Diprotic acids, e.g. ethanedioic acid ($H_2C_2O_4$), contain two proton-donating groups, so that each group can consecutively act as a Brønsted-Lowry acid.

If a diprotic acid, such as ethanedioic acid, is titrated with a strong base that accepts only one hydrogen ion (e.g. sodium hydroxide), first one carboxylic group donates a hydrogen ion to the base, and then the second carboxylic group donates a hydrogen ion:

$$H_2C_2O_4(aq) \rightleftharpoons HC_2O_4^-(aq) + H^+(aq)$$
$$HC_2O_4^-(aq) \rightleftharpoons C_2O_4^{2-}(aq) + H^+(aq)$$

This produces a pH curve (Fig. 14) similar in shape (almost a mirror image) to the titration of sodium carbonate with hydrochloric acid (Fig. 13) and there will be two equivalence points. The pK_a values are 1.23 for the first proton and 4.19 for the second proton, so ethanedioic acid is behaving as a weak acid for both dissociations. The equivalence points are at pH 2.7 for the first proton and 8.4 for the second.

Indicators

Acid-base indicators are solutions which show a colour change at a particular range of $[H^+]$ changes.

Indicators are weak acids (usually written HIn) where the dissociated form (In⁻) and undissociated forms (HIn) have different colours. There is an equilibrium for this reaction:

$$HIn \rightleftharpoons H^+ + In^-$$

If acid is added or neutralised during a titration, then $[H^+]$ changes and this will affect the indicator equilibrium. The indicator should change colour over a narrow pH range. The most significant change in $[H^+]$ in a titration is at the equivalence point, so you need the indicator to change from almost completely one colour to the other when there is this large change in pH. The colour of at least one form of the indicator needs to be quite intense, so that only a small amount

Table 6 Colour changes for some indicators			
Indicator	Colour in acid (HIn)	Colour in alkali (In⁻)	pH range for colour change
methyl orange	red	yellow	2.9–4.0
methyl red	red	yellow	4.2–6.3
litmus	red	blue	5.0–8.0
bromothymol blue	yellow	blue	6.0–7.6
phenol red	yellow	red	6.6–8.0
thymol blue	yellow	blue	9.1–9.6
phenolphthalein	colourless	purple	8.2–10.0
alizarin yellow	yellow	red	10.1–12.0

(1 or 2 drops) needs to be used to give a clearly visible change without significantly affecting the [H⁺] itself.

When the concentrations of the acid and base forms of the indicator are equal, then this is referred to as the **end-point** of the titration. The **equivalence point** is not necessarily the same, but the indicator will give the most precise result for the titration when the two (end-point and equivalence point) coincide.

Choosing indicators for titrations

Different indicators change colour at different pH values (Table 6). This property can be used to select a suitable indicator for a particular titration. A suitable indicator must change colour in the pH range that corresponds to the steep part of the pH curve. Here there is a marked change in the pH, so that the indicator will change colour completely. The best way to do this is to look at the pH curve for the particular titration and choose one where the pH range of activity for the indicator matches the equivalence point for the titration (see Figs 15 and 16).

For the titration of a strong acid with a strong base any of the indicators in Table 6 would be suitable, because the steep part of the pH curve runs from pH 2 to pH 12 and all the indicators fall in this range.

For a weak acid being titrated with a strong base the pH at equivalence is 8.8. This applies for solutions of 0.1 mol dm⁻³ ethanoic acid with 0.1 mol dm⁻³ sodium hydroxide. An indicator such as methyl orange changes colour in the pH range 2.9–4.0 and would not be suitable since it would change colour before equivalence is reached. The most suitable indicator in the list would be phenolphthalein, which changes colour between pH 8.2 and 10.0.

The pH equivalence for a strong base with a strong acid is 7, so again any indicator would be suitable. For a weak base with a strong acid, for example 0.1 mol dm⁻³ aqueous ammonia acid with 0.1 mol dm⁻³ hydrochloric acid, the pH at equivalence is 5.2, so methyl red (pH range 4.2–6.3) would be the most suitable.

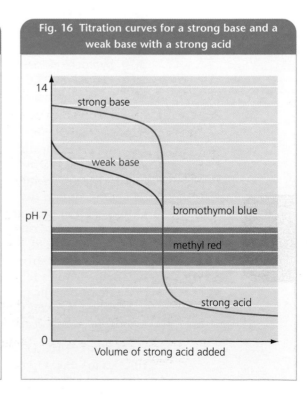

Fig. 15 Strong acid and weak acid with strong base

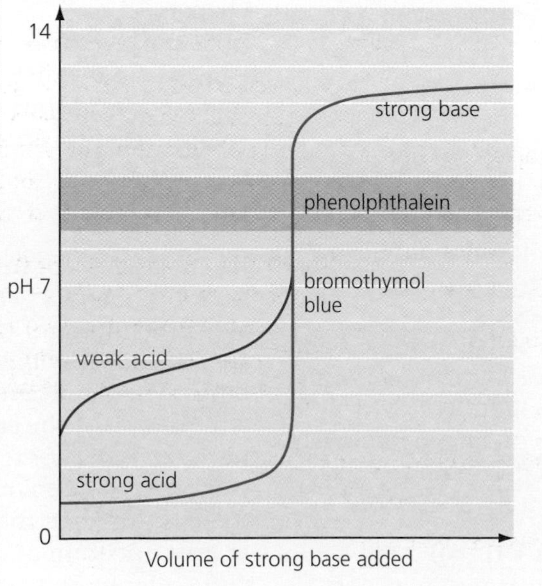

Fig. 16 Titration curves for a strong base and a weak base with a strong acid

Table 7 shows the pH at equivalence for a range of titrations using 1 M solutions.

Table 7 pH range at equivalence			
Acid	Base	pH range at equivalence	Suitable indicators
HCl (strong)	NaOH (strong)	3–11	any: the pH change is over a very wide range
CH$_3$COOH (weak)	NaOH (strong)	7–11	any from phenol red downwards
HCl (strong)	NH$_3$ (weak)	3–7	methyl orange and methyl red
CH$_3$COOH (weak)	NH$_3$ (weak)	no sharp change	no suitable indicator, as there is no sharp change even a pH meter will be no use
HCl (strong)	Na$_2$CO$_3$	2.5–5.5	methyl orange or methyl red
		6.5–9.5	phenol red or phenolphthalein
H$_2$C$_2$O$_4$	NaOH (strong)	1.5–3.5	methyl orange or thymol blue but both are only just in range; better is a pH meter
		5–11	any from bromothymol blue downwards

18 Choose a suitable indicator for the following titrations. Use Tables 6 and 7 to help you.

a Adding ammonia (weak base) to nitric acid (strong acid).

b Adding potassium hydroxide (strong base) to ethanoic acid (weak acid).

c Adding ammonia to ethanoic acid.

d Sketch the pH curve for each titration.

The curve in Fig. 12 shows that during the titration of a weak acid with a weak base the change in pH is so gradual that this reaction cannot be used in the quantitative estimation of concentration.

KEY FACTS

- An indicator can be used to determine the equivalence point in a titration.
- The end point of a titration is where [HIn] = [In⁻].
- Different indicators change colour over different pH ranges.
- Indicators change colour over a narrow pH range.
- pH curves must be used to select a suitable indicator for a particular titration.
- The equation:

$$\text{number of moles in a solution (mol)} = \text{conc. of solution (mol dm}^{-3}） \times \text{volume of solution (dm}^3)$$

can be used to calculate concentrations and volumes in titrations.

- Some bases can accept more than one hydrogen ion, to act consecutively as multiple Brønsted-Lowry bases, e.g. sodium carbonate; the titration of such a base with a monoprotic acid produces a pH curve with more than one equivalence point.
- Diprotic acids have two hydrogen ion donating groups in each molecule, so they act consecutively as two Brønsted-Lowry acids; the titration of such an acid with a monoprotic base produces a pH curve with two equivalence points.

3.8 Buffer solutions

A solution that can resist a change in pH when small amounts of acid or base are added is called a **buffer**.

Even though there are many different buffers, they all work according to the same principle. Generally, buffers contain either:

- weak acid and a large amount of a strong base salt of the acid (an acid buffer), e.g. ethanoic acid with sodium ethanoate; or

- weak base and a large amount of a strong acid salt of the base (an alkaline buffer), e.g. ammonia with ammonium chloride.

The buffering action works because the weak acid or base maintains a store of associated ions (e.g. ethanoate or ammonium) which are released to stabilise pH.

A mixture of ethanoic acid (CH_3COOH) and ethanoate CH_3COO^- has a buffering action. Ethanoic acid, written here as HAc, is a weak acid ('Ac' = 'acetate' from the old name for the acid), while sodium ethanoate dissociates completely in aqueous solution:

$$HAc(aq) + H_2O(l) \rightleftharpoons Ac^-(aq) + H_3O^+(aq)$$
$$NaAc\ (aq) \rightarrow Na^+\ (aq) + Ac^-(aq)$$

An acid buffer normally contains relatively high concentrations of both undissociated acid and acid anions (conjugate base).

We can write the acid dissociation constant:

$$K_a = \frac{[Ac^-(aq)]\ [H_3O^+(aq)]}{[HAc(aq)]}$$

The presence of ethanoate ions pushes the ethanoic acid equilibrium towards undissociated acid:

$$HAc(aq) + H_2O(l) \rightleftharpoons Ac^-(aq) + H_3O^+(aq)$$

$Ac^-(aq)$ present in large amounts from the salt drives the equilibrium to the left.

If acid is added to the buffer, the equilibrium adjusts as H_3O^+ combines with some of the large amount of Ac^- to give more undissociated acid:

$$HAc(aq) + H_2O(l) \rightleftharpoons Ac^-(aq) + H_3O^+(aq)$$
$$\leftarrow \text{equilibrium moves to the left}$$

If alkali is added to the buffer, the equilibrium adjusts:

$$HAc(aq) + H_2O(l) \rightleftharpoons Ac^-(aq) + H_3O^+(aq)$$
$$\text{equilibrium moves to the right} \rightarrow$$

$OH^-(aq)$ reacts with $H_3O^+(aq)$, neutralising it, while some of the large amount of HAc present dissociates to give more H_3O^+. The buffer maintains the equilibrium reaction and the pH remains steady, except in the presence of large quantities of acid or alkali. K_a remains constant throughout.

K_a can be used to show that $[H^+]$ remains fairly constant.

$$K_a = \frac{[Ac^-(aq)]\ [H_3O^+(aq)]}{[HAc(aq)]}$$

$$[H_3O^+(aq)] = \frac{K_a \times [HAc(aq)]}{[Ac^-(aq)]}$$

If both $[Ac^-(aq)]$ and $[HAc(aq)]$ are large, then small changes in their concentrations will not affect the overall ratio significantly, so $[H_3O^+(aq)]$ remains fairly constant.

If equal amounts of salt and acid are present (half-neutralisation) in the buffer, then $[Ac^-(aq)] = [HAc(aq)]$

$$Ka = \frac{\cancel{[Ac^-(aq)]}\ [H_3O^+(aq)]}{\cancel{[HAc(aq)]}} = [H_3O^+(aq)]$$

so pK_a (of the acid) = pH (of the buffer)

Buffer region and buffer range

If the half equivalence point is drawn on a titration curve for a weak acid with a strong base (see Fig. 9), the pH in that region stays fairly constant (within ±0.1 pH units) even if strong base is added. This is called the **buffer region** for the weak acid. The **buffer range** is the range of pH values in which a buffer for a given weak acid can be prepared. The range is usually within pK_a ±1. Ethanoic acid has a pK_a value of 4.7, so the buffer range would normally be between pH 3.7 and 5.7 (Fig. 17).

Using $K_a = \dfrac{[Ac^-(aq)]\ [H_3O^+(aq)]}{[HAc(aq)]}$

allows us to calculate the pH of any buffer mixture we make up.

The pH of a buffer mixture containing 0.2 mol dm^{-3} of ethanoate ions and 0.25 mol dm^{-3} ethanoic acid, $K_a = 1.7 \times 10^{-5}$, can be calculated as follows:

$$K_a = \frac{[Ac^-(aq)]\ [H_3O^+(aq)]}{[HAc(aq)]}$$

$$[H_3O^+(aq)] = \frac{1.76 \times 10^{-5} \times 0.25}{0.2}$$

$$= 2.20 \times 10^{-5}$$
$$pH = 4.66$$

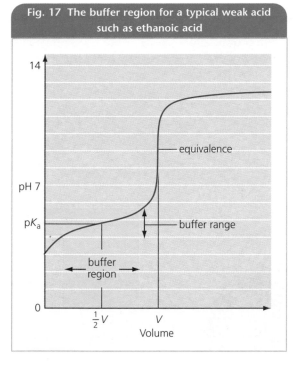

Fig. 17 The buffer region for a typical weak acid such as ethanoic acid

Example: Calculating buffer pH

Calculate the pH of a buffer made by mixing 35.0 cm³ of 2.0 mol dm⁻³ ethanoic acid, $K_a = 1.76 \times 10^{-5}$ with 30.0 cm³ of 1.5 mol dm⁻³ sodium ethanoate.

number of moles of HAc = 2.0 × 35.0/1000
$$= 0.070 \text{ mol}$$

number of moles of Ac⁻ = 1.5 × 30.0/1000
$$= 0.045 \text{ mol}$$

Using the expression above:

$$[H_3O^+(aq)] = \frac{K_a \times [HAc(aq)]}{[Ac^-(aq)]}$$

$$= \frac{1.76 \times 10^{-5} \times 0.070}{0.045}$$

$$= 2.738 \times 10^{-5}$$

$$pH = -\log_{10}[H^+]$$
$$= -\log_{10} \times 2.738 \times 10^{-5}$$
$$= 4.56$$

19
a Explain how an NH_3/NH_4Cl buffer works.

b Calculate the pH of a propanoic acid/propanoate buffer containing 0.2 mol dm⁻³ salt and 0.05 mol dm⁻³ acid, $K_a = 1.3 \times 10^{-5}$.

Example: Calculating the change in pH in buffer solutions

Calculate the change in pH of the buffer solution in the last example if 10.0 cm³ of 0.10 mol dm⁻³ HCl is added.

When the acid is added the equilibrium shifts to the left, removing H⁺ and with it Ac⁻ and an equal amount of HAc is formed.

$$HAc(aq) + H_2O(l) \rightleftharpoons Ac^-(aq) + H_3O^+(aq)$$

number of moles of:

H⁺ added = 0.1 × 10.0/1000 = 0.001 mol
Ac⁻ present in original buffer = 0.045 mol
Ac⁻ remaining = 0.045 – 0.001 = 0.044 mol
HAc present in original buffer = 0.070 mol
HAc present now = 0.070 + 0.001 = 0.071 mol

All the solutions are in the same total volume so the concentration ratio = the mole ratio.

$$[H_3O^+(aq)] = \frac{K_a \times [HAc(aq)]}{[Ac^-(aq)]}$$

$$= \frac{1.76 \times 10^{-5} \times 0.071}{0.044}$$

$$= 2.84 \times 10^{-5}$$

$$pH = -\log_{10}[H^+] = -\log_{10} 2.84 \times 10^{-5}$$
$$= 4.55$$

The new pH is 4.55, and before addition of acid it was 4.56, so the pH has changed by –0.01 pH units: a very small change.

20
a Calculate the pH of a buffer made by mixing 25.0 cm³ of 2.0 mol dm⁻³ propanoic acid, $K_a = 1.34 \times 10^{-5}$ with 20.0 cm³ of 1.5 mol dm⁻³ sodium propanoate.

b Calculate the change in pH of the buffer solution in (a) above if 10.0 cm³ of 0.05 mol dm⁻³ HNO_3 are added.

APPLICATION

The cleaning chemicals in hair-care products

Cosmetic scientists who design products for hair care take account of the acidity of their preparations. The wrong formulation could quite easily cause damage to the hair or skin, while a good shampoo will:

- clean hair well
- leave hair in good condition
- produce a pleasant lather that rinses out easily
- be safe to use
- leave hair smelling fragrant

Everyone's skin produces sweat, grease and oil, and this collects on hair, together with dust and dirt from the atmosphere. Cleaning off these deposits is what we expect a shampoo to do. One of the most effective chemicals for removing grease is sodium hydroxide, which is a strong base and an alkali.

1 What do these terms mean in that sentence 'it is a *strong base* and an *alkali*'.

Stearic acid $CH_3(CH_2)_{16}COOH$ is often used to make soap, which contains sodium stearate.

2 Write an equation showing stearic acid acting as a weak acid, and write a possible equation for the formation of sodium stearate.

3 Write an expression for the acid dissociation constant for stearic acid, and calculate the pH of a 0.5 mol dm^{-3} solution of stearic acid if its $K_a = 3.98 \times 10^{-6}$ mol dm^{-3}?

Soap is too alkaline for cleaning hair and it can damage it, so manufacturers need to check the pH of their shampoos.

4 Use a pH titration curve to explain why soap will have an alkaline pH.

Shampoos are made by neutralising lauryl sulphonic acid $C_{12}H_{25}OSO_2OH$, a much stronger acid than stearic acid.

5 Will the dissociation constant of lauryl sulphonic acid be higher or lower than that for stearic acid?

6

a 25.0 cm^3 of a 0.270 mol dm^{-3} solution of lauryl sulphonic acid is titrated with 0.15 mol dm^{-3} sodium hydroxide. Calculate the volume of sodium hydroxide needed to just neutralise the lauryl sulphonic acid.

b If lauryl sulphonic acid behaves as a strong acid, what will be the pH of the mixture at equivalence and what would be a suitable indicator for the titration?

Human skin is weakly acidic, because its outermost layer contains a mixture of water-soluble acids and salts, which are secreted in sweat. The skin secretions form a natural buffer. The buffering action of skin is mainly due to a mixture of lactic acid (2-hydroxypropanoic acid, $CH_3CH(OH)COOH$) and a lactate. Lactic acid, a weak acid, can be written here as LaH and the lactate ion as La$^-$.

7 Explain what is meant by the term *buffer*.

8 Explain how the secretions can buffer the skin when different shampoos are applied that are:

a acidic b basic

9 The acid dissociation constant for LaH is 8.4×10^{-4} mol dm^{-3}. Calculate the pH of skin if the surface moisture contains 0.012 millimoles dm^{-3} of LaH and 3.20 millimoles dm^{-3} of La$^-$ (millimole = 10^{-3} mole).

10 Calculate the change in skin pH (from question 9) if 10 cm^3 of some shampoo containing 0.2 mol dm^{-3} of H$^+$ were used to wash your hair.

1

a Write an expression for the ionic product of water, K_w, and state the value of the quantity pK_w ($-\log_{10} K_w$) at 298 K. (2)

b What is meant by the term *strong* when applied to an acid or a base? (1)

c Answer the questions below and hence deduce the effect of increasing temperature on the value of pK_w.

i) Write an ionic equation for the reaction between a strong acid and a strong base. (1)

ii) State whether this reaction is exothermic or endothermic. (1)

iii) Deduce whether the ionic equilibrium to which K_w refers is endothermic or exothermic. (1)

iv) Deduce, giving a reason, the effect on the value of K_w if the temperature rises above 298 K. (2)

v) Deduce the effect on the value of pK_w if the temperature rises above 298 K. (1)

NEAB CH04 June 1997 Q4

2

a Define the terms *pH* and *pK_a*. (2)

b What property of methyl orange enables it to act as an indicator in an acid-base titration? (2)

c The pK_a of propanoic acid is 4.85. Calculate the pH of a 0.45 mol dm^{-3} solution of propanoic acid. (5)

NEAB CH04 Feb 1996 Q3

3 In aqueous solution, hydrofluoric acid, HF, is a weak acid.

a Given that ionisation of HF in aqueous solution is an endothermic reaction, state whether K_a increases or decreases as the temperature is raised and explain your answer. (2)

b 10 cm^3 of 0.50 mol dm^{-3} aqueous sodium hydroxide are added to 100 cm of water at 298 K. Calculate the number of moles of sodium hydroxide added and hence the pH of the resulting solution. At 298 K, $K_w = 1.0 \times 10^{-14}$ mol^2 dm^{-6}. (4)

c 10 cm^3 of 0.50 mol dm^{-3} aqueous sodium hydroxide are added to solution A which comprises 100 cm^3 of an aqueous solution containing hydrofluoric acid (0.50 mol dm^{-3}) and sodium fluoride (0.5 mol dm^{-3}). In aqueous solution, NaF is fully ionised. At 298 K, hydrofluoric acid has $pK_a = 3.45$.

i) Write an expression for K_a for aqueous HF. Hence calculate the pH of solution **A** before the sodium hydroxide has been added. (3)

ii) Calculate the concentration of HF present in the solution formed after adding the sodium hydroxide. (2)

iii) Calculate the concentration of fluoride ions present in the solution formed after adding the sodium hydroxide. (2)

iv) Use your answers above to determine the pH of the solution formed after adding the sodium hydroxide. (3)

NEAB CH04 March 1998 Q2

4 The pH of solution A, a 0.15 mol dm^{-3} solution of a weak monoprotic acid HX, is 2.69.

a Calculate [H$^+$] in solution A and hence determine the value of the acid dissociation constant, K_a, of HX. (3)

b i) A 25 cm^3 sample of A is titrated with 0.25 mol dm^{-3} sodium hydroxide. Calculate the volume of sodium hydroxide needed to reach equivalence in the titration. Give the best estimate you can of the pH of the neutralised solution, stating a reason. (3)

ii) Calculate the pH of the titration solution when HX is exactly half-neutralised and [HX] = [X$^-$]. (2)

iii) Calculate the pH of the titration solution when a total of 25 cm^3 of 0.25 mol dm^{-3} sodium hydroxide has been added. (3)

NEAB CH04 June 1998 Q5

5

a Giving appropriate examples, explain what is meant by the terms *acidic buffer* and *basic buffer*. Explain how acidic buffers and basic buffers perform their function. (12)

b How would you make a basic buffer if you were given 250 cm^3 each of the following 0.5 mol dm^{-3} solutions and no other chemicals?

hydrochloric acid
ethanoic acid
ammonia
sodium hydroxide

Give approximate volumes of the solutions which you would use to make as much as possible of this buffer. You may not need to use all of the solutions listed above. (4)

NEAB CH04 June 1997 Q8

6

a State two identical features of and two differences between the titration curves for the separate addition of 0.75 mol dm^{-3} sodium hydroxide to 25 cm^3 samples of

i) 1.5 mol dm^{-3} hydrochloric acid

ii) 1.5 mol dm^{-3} ethanoic acid (4)

b What characteristic features must all acid-base indicators possess? Name an indicator that could be used for both of the titrations in **(a)** above and explain the reasons for your choice. (4)

c An aqueous solution of the weak acid HA, which has an initial pH of 2.50, is titrated against sodium hydroxide. When the pH reaches 4.30, [HA] = [A$^-$]. Calculate the value of the acid dissociation constant of HA and hence the original concentration of the acid in solution. (7)

NEAB CH04 June 1997 Q7

7

a The pH of a 0.15 mol dm^{-3} solution of a weak acid, HA, is 2.82 at 300 K.

i) Write an expression for the acid dissociation constant, K_a, of HA, and determine the value of K_a for this acid at 300 K, stating its units. (5)

ii) The dissociation of HA into its ions in aqueous solution is an endothermic process. How would its pH change if the temperature were increased? Explain your answer. (3)

b Solution A contains n moles of a different weak acid, HX. The addition of some sodium hydroxide to A neutralises one third of the HX present to produce Solution B.

i) In terms of the amount, n, how many moles of HX are present in Solution B? (1)

ii) Determine the ratio $\dfrac{[HX]}{[X^-]}$ in Solution B. (2)

iii) Solution B has a hydrogen ion concentration of 4.2×10^{-4} mol dm^{-3}. Use this information to determine the value of the acid dissociation constant of HX. (2)

	Start	Half equivalence	Equivalence	Double equivalence
Volume/cm³ Ba(OH)₂ solution added	0.0			
pH for titration of **S**	0.80		7	
pH for titration of **W**	2.74		8.5	

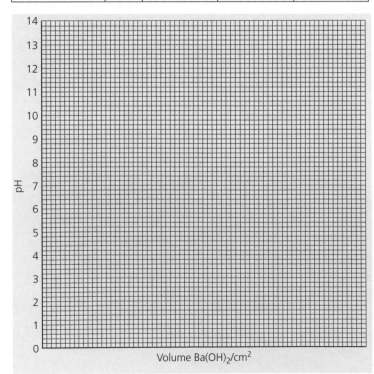

pH / Volume Ba(OH)₂/cm² graph

c Why is methyl orange not suitable as an indicator for the titration of HX with sodium hydroxide? (2)

d Solution B can act as a buffer. Explain what this means and write an equation that shows how Solution B acts as a buffer if a little hydrochloric acid is added. (3)

AQA CH04 March 1999 Q1

8 Solution S is 0.16 mol dm^{-3} hydrochloric acid, HI, a strong acid.
Solution W is 0.16 mol dm^{-3} HX, a weak monoprotic acid. It has a pH of 2.74.
Solution Z is 0.12 mol dm^{-3} barium hydroxide, Ba(OH)₂, a strong base.

a i) Explain the terms *weak* and *strong* as applied to acids or bases.

ii) Determine a value of the acid dissociation constant of the weak acid HX using the expression $K_a \approx \dfrac{[H^+]^2}{c}$ where c is the original concentration of HX. Explain why it is reasonable to use this approximation. (6)

b Show details of all calculations in answering this part of the question.
At a temperature of 25 °C, 18 cm^3 of solution S are titrated with solution Z. The titration is repeated using 18 cm^3 of solution W.

i) Determine the equivalence volume (end-point) for these titrations and enter your result into the appropriate space in the incomplete table on the left. Enter also the half-equivalence volume and the double-equivalence volume in the appropriate space.

ii) Calculate the missing pH values for each of the two titration solutions and enter these into the table.
(You should assume that, at half-equivalence, [HX] = [X$^-$] for the weak acid HX, and that, for both acids at double-equivalence, only the alkali that is in excess contributes to the pH of the resulting solution.)

iii) Plot these results on a copy of the graph opposite and use the points you have plotted to sketch the complete titration curves for solution **S** and solution **W** titrated with solution **Z**. (18)

c i) Explain what is meant by the term *buffer solution*. Suggest how solution **W**, when half-neutralised, can behave as a buffer solution.

ii) State the difference between *acidic* and *basic* buffers. To which of these two types of buffer does a half-neutralised solution of **W** belong? What might you use to make a buffer solution of the other type? (6)

AQA CH04 June 1999 Q6

9

a The graph below shows how the pH changes when 0.12 mol dm^{-3} NaOH is added to 25.0 cm^3 of a solution of a weak monoprotic acid, HA.

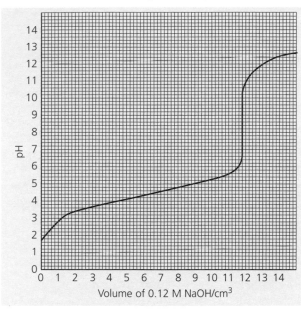

i) Use the graph to calculate the initial concentration of the acid HA. (2)

ii) Write an expression for the dissociation constant, K_a, of the weak acid HA. (1)

iii) Determine the volume of sodium hydroxide added when [HA] = [A$^-$] and use the graph to determine the pH at this point. (2)

iv) Use your answers to part **(a)(ii)** and part **(a)(iii)** to determine the value of K_a for the acid HA. (4)

b A buffer solution is formed, when approximately half of the original amount of the acid HA(aq) has been neutralised by the base NaOH(aq). Explain how this buffer solution is able to resist change in pH when

i) a small amount of NaOH(aq) is added, (2)

ii) a small amount of HCl(aq) is added. (2)

AQA CH04 March 2000 Q2

10 The graph below shows how the pH changes as 0.12 mol dm^{-3} HCl(aq) is added to 25.0 cm^3 of a solution of sodium carbonate. There are two end-points. The second end-point is at 30.0 cm^3.

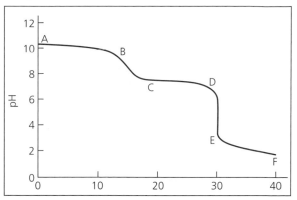

a Write equations for the reactions which occur in the solution between point A and point B on the graph and between point C and point D on the graph. (2)

b Estimate the minimum volume of hydrochloric acid needed in this experiment for carbon dioxide to be produced from a well-stirred solution of sodium carbonate. (1)

c Name an indicator which can be used to determine the end-point occurring between points D and E. Explain why this indicator does not change colour between points B and C. (2)

d Use the end-point occurring between points D and E to calculate the concentration of sodium carbonate in the given solution. (3)

e If the original solution had contained, in addition to sodium carbonate, an equal molar concentration of sodium hydrogencarbonate, at what volumes of hydrochloric acid would the two end-points have been detected? (2)

AQA CH04 March 2000 Q5

4 Naming organic compounds and isomerism

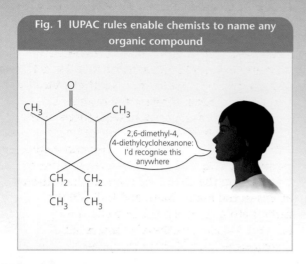

Fig. 1 IUPAC rules enable chemists to name any organic compound

2,6-dimethyl-4, 4-diethylcyclohexanone: I'd recognise this anywhere

You are unique, as described by your fingerprints or your DNA code. However, your name, individual physical characteristics and individual personality traits are not unique. Other people will have the same name; other people will also, for example have blue eyes and a friendly personality.

Different organic molecules are not quite so numerous as people, but each has it own unique characteristics, determined by its structure. For people, cases of mistaken identity occasionally have serious consequences, but it could be even more serious as far as organic molecules are concerned.

One type of thalidomide molecule is a very effective pain reliever but its mirror image caused many serious birth defects in the 1960s. Consequently it is essential to have a method of naming molecules that describes their structures and allows for even very subtle differences in these structures to be distinguished.

4.1 Introduction

Carbon atoms have the unique ability to bond together to make chains, branched chains and rings which form the skeleton of compounds that contain many carbon atoms. Hydrogen atoms are often bonded to these carbon skeletons, and other atoms (especially oxygen, nitrogen and halogens) can also attach in a variety of positions and combinations. This variety gives us millions of organic molecules, derived from just a few different types of atom.

We name organic compounds according to a set of rules devised by the International Union of Pure and Applied Chemistry (IUPAC). These rules give us a name for each and every molecule that describes the structure of the molecule unambiguously (Fig. 1). The complete set of IUPAC rules fills several books. Naming rules are given in *AS Chemistry*, page 163. Here, you will consider enough of the rules to allow you to name any molecule you may encounter in your study of A level organic chemistry, and many more!

The system easily distinguishes **isomers**, which are molecules with the same

molecular formula but different **structural formulas**. These may be **structural isomers** (including **chain**, **positional** and **functional group isomers**) or **stereoisomers** (including **geometrical isomers** and **optical isomers**) which are all discussed in detail later in this chapter. Fig. 2 shows a simple example of the importance of the IUPAC naming system; ethanol is drinkable, methoxymethane is not!

Fig. 2 Isomers of C_2H_6O

For the molecular formula C_2H_6O, the structural formula could be:

CH_3CH_2OH or CH_3OCH_3

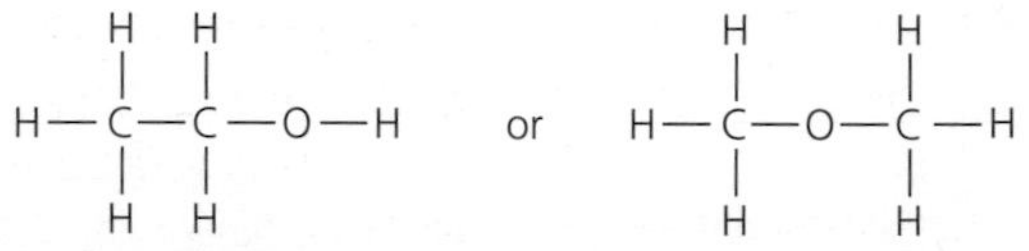

Using the IUPAC naming system, these are **ethanol** and **methoxymethane**.

They are *functional group isomers* of C_2H_6O.

4.2 Rules for naming organic compounds

The IUPAC system begins by considering three basic features common to all organic molecules. These features are each given names which we then combine to give the overall name. The features are:

- the size and shape of the carbon skeleton,
- the presence in the molecule of groups of atoms called **functional groups**,
- the position of these functional groups in the molecule.

Size and shape of the carbon skeleton

Table 1 Structures and names of simple alkanes		
Structure of alkane	Name	IUPAC code
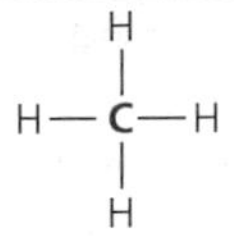 H—C—H or CH_4	methane	**meth** = 1 C in chain
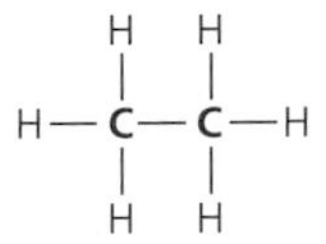 H—C—C—H or CH_3CH_3	ethane	**eth** = 2 C in chain
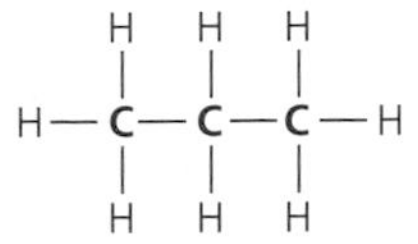 H—C—C—C—H or $CH_3CH_2CH_3$	propane	**prop** = 3 C in chain
$CH_3CH_2CH_2CH_3$	butane	**but** = 4 C in chain
$CH_3CH_2CH_2CH_2CH_3$	pentane	**pent** = 5 C in chain
$CH_3CH_2CH_2CH_2CH_2CH_3$	hexane	**hex** = 6 C in chain

Fig. 3 Naming alkanes

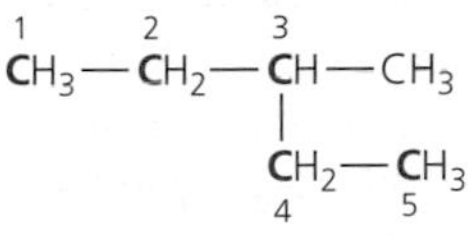

$$CH_3—CH_2—CH—CH_3$$
$$CH_2—CH_3$$

3-methylpentane

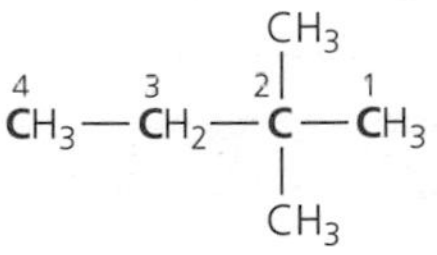

$$CH_3—CH_2—C—CH_3$$ with CH_3 above and below

2,2-dimethylbutane

$$CH_3—CH—CH—CH_3$$ with CH_3 and CH_3 below

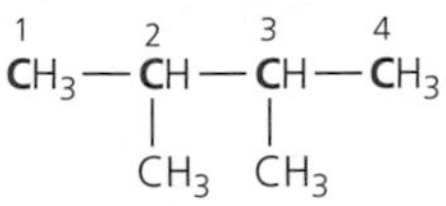

2,3-dimethylbutane

Table 1 shows the structures of methane, ethane, propane, butane, pentane and hexane, which are the first six members of the hydrocarbon family called the alkanes. We use these names to describe similar chains of carbon atoms in other molecules: a chain provides a name for the 'backbone' around which the rest of the molecule is constructed.

The carbon skeleton in ethanol is C–C (see Fig. 2 and Table 1). The name contains 'eth', which tells you there are two carbon atoms bonded together as in *eth*ane. However, in methoxymethane there are two separate carbon chains that each contain only one carbon atom. Each chain is named by 'meth' as in *meth*ane.

During most chemical reactions the carbon backbone remains unchanged and the same naming system occurs in both the name of the reactant and the product.

Each of the three general hydrocarbon groups, **alkanes**, **cycloalkanes** and **arenes**, has a particular type of carbon skeleton.

Alkanes
In alkanes the carbon atoms are bonded in continuous chains, though these may have branches. Table 1 shows how the number of carbon atoms in any chain is named.

1 Predict the IUPAC names for unbranched C_7H_{16}, C_8H_{18} and $C_{10}H_{22}$.

2 Using structural formulas showing *all* bonds, show the structures of
a pentane
b hexane

To name a molecule, first identify the longest chain, since the name of the **chain structural isomer** is based on the name for the longest continuous chain. This name comes at the end of the name for the whole molecule, as in the examples in Fig. 3. If the molecule contains branches to this chain, we name each branch using the relevant prefix from 'methyl-', 'ethyl-', 'propyl-', etc, depending on the number of carbon atoms in the branch. Repetitions of the same side-chain are shown using 'di-', 'tri-', 'tetra-', etc, as shown in the second and third examples in Fig. 3. To identify the positions of branches on the chain, we number the carbon atoms in the main chain, and indicate the carbon atom to which the side chain is bonded. Hence, **positional isomers** are distinguished by this naming system.

3 Give the name for $CH_3CH(CH_3)CH_3$.

4 Draw the structure of 2,4-dimethylpentane.

5 Name the straight chain alkane which has 3-methylpentane (see Fig. 3) as chain structural isomer.

6 Draw and name a positional isomer and a chain isomer of 3-methylpentane (see Fig. 3).

Cycloalkanes
Carbon atoms can bond together via single bonds to form rings, and we call such compounds **cycloalkanes**. The number of carbon atoms in the ring gives the name, using the same sequence that we used for alkanes (see Table 1), and the ring arrangement itself is indicated by the prefix 'cyclo-'. Two examples can be seen in Table 2.

Table 2 Cycloalkanes		
Structure	Abbreviation	Name
(ring structure)	△	cyclopropane
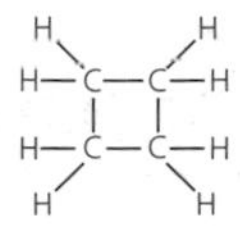 (ring structure)	▢	cyclobutane

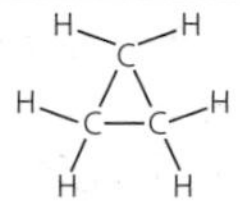

7 Other cycloalkanes are abbreviated by similar regular polygons. How would cyclopentane and cyclohexane be abbreviated?

Arenes

Arenes (also called *aromatics*) are compounds based on the structure of benzene, C_6H_6. The representation of the so-called 'delocalised' structure of benzene is shown in Fig. 4 (see also Chapter 6).

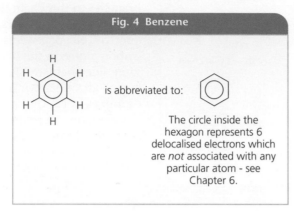

Fig. 4 Benzene

is abbreviated to:

The circle inside the hexagon represents 6 delocalised electrons which are *not* associated with any particular atom - see Chapter 6.

Many arene compounds exist and we can name these according to the different groups attached to the ring (see Fig. 5).

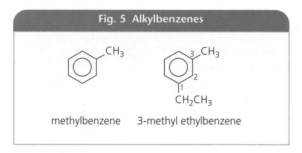

Fig. 5 Alkylbenzenes

methylbenzene   3-methyl ethylbenzene

8 Draw the structures of 1-phenylpropane and (1-methyl)ethylbenzene.

Functional groups

The groups of atoms that are responsible for the characteristic chemical reactions of a molecule are called **functional groups**. Each different functional group is represented by a unique IUPAC name in the overall name for the molecule.

In Fig. 2, C–O–H is the functional group in ethan*ol* and the '-ol' indicates this group is present. In meth*oxy*methane C–O–C is the functional group and compounds of this type have '-oxy-' in their name.

Each different functional group has its own IUPAC name (see Table 3) which is added to the name for the carbon skeleton (represented by * in the table). If the carbon atoms in the functional group are part of the longest chain, we must include these to get the correct name.

The functional group name is usually shown at the end of the overall name, but alternative prefix versions are sometimes available when more than one group needs to be identified. This enables us to distinguish **functional group isomers** where the molecules have the same molecular formula but contain different functional groups (e.g. butanal and butanone or ethanoic acid and methyl methanoate). **Positional isomers**, such as 1-chloropropane and 2-chloropropane, are indicated by prefixing the functional group's name by the number of the carbon atom in the main chain or ring to which it is bonded. If the functional group is bonded to other atoms such as nitrogen or oxygen, we indicate this by italicising the atomic symbol (e.g. N or O). When naming positional isomers we use the lowest possible numbers in the sequence. For example, 1,2-dichloropropane is used, rather than 2,3-dichloropropane. Table 3 gives a number of examples of these isomers and you should study it carefully.

9 Draw the structures of:

i bromoethane, **ii** propanoic acid, **iii** propanal, **iv** butan-2-one, **v** methylethanoate, **vi** 2-chloropropane, **vii** 4-methylnitrobenzene, **viii** pent-2-ene, **ix** methanoic acid, **x** methylpropene, **xi** butan-1-ol, **xii** 2-methylpropan-2-ol, **xiii** but-1,3-diene, **xiv** 2-amino-2-phenylethanoic acid, **xv** *N*-methylaminomethane, **xvi** methanoic ethanoic anhydride, **xvii** butanoyl chloride, **xviii** ethyl methanoate

10 Name each of the following molecules. Draw out the *full* structure before attempting to name the molecule.

i CH_3Br, **ii** C_6H_5Cl, **iii** $CH_3CH_2CH(OH)CH_3$, **iv** CH_3COOH, **v** $CH_3CH_2CH_2CH_2CH_2CHO$, **vi** $CH_3CH_2NH_2$, **vii** CH_3COCH_3, **viii** CH_2BrCH_2Br, **ix** CH_3CHBr_2, **x** $CH_3CH_2CH=CH_2$, **xi** C_6H_5COOH, **xii** $CH_3CH_2COOCH_2CH_3$, **xiii** $CH_3CH_2CH_2COCl$, **xiv** $(CH_3CH_2CH_2CO)_2O$, **xv** $CH_3COOCH_2CH_2CH_3$, **xvi** $CH_3CH_2CH_2COCH_2CH_3$, **xvii** $CH_3CH_2CH=CHCH(CH_3)CH_3$, **xviii** CH_3CH_2CN, **xix** $CH_3CH_2CH(CN)OH$

	Table 3 Code for functional groups				
Name	**Structure**	**Suffix code**	**Prefix code**	**Example**	**IUPAC name**
carboxylic acid	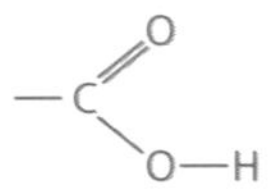 (structure)	(*)-oic acid	–	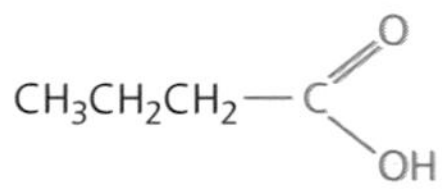 $CH_3CH_2CH_2$—C (=O)(OH)	butanoic acid
acyl chloride	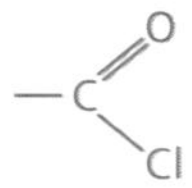 (structure)	(*)-oyl chloride	–	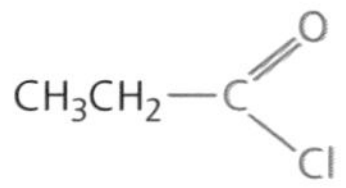 CH_3CH_2—C (=O)(Cl)	propanoyl chloride
acid anhydride	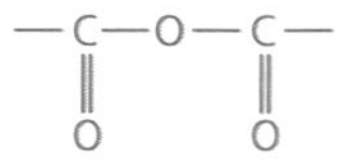 —C—O—C— (with O below each C)	(*)-oic anhydride	–	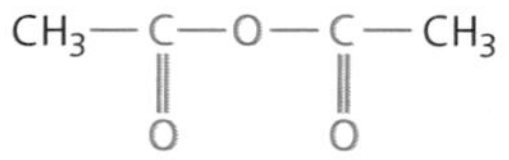 CH_3—C—O—C—CH_3 (O below)	ethanoic methanoic anhydride
alcohol	—O—H	(*)-ol	hydroxy-(*)	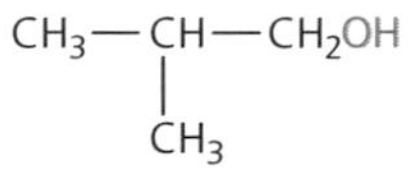 CH_3—CH—CH_2OH / CH_3	2-methylpropan-1-ol
aldehyde	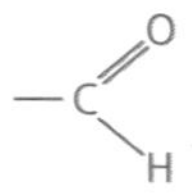 (structure)	(*)-al	–	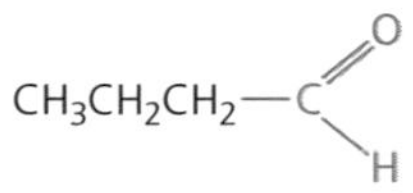 $CH_3CH_2CH_2$—C (=O)(H)	butanal
ketone	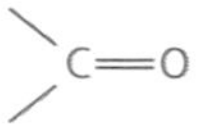 C=O	(*)-one	oxo-	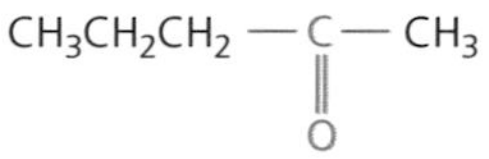 $CH_3CH_2CH_2$—C—CH_3 (=O)	pentan-2-one
alkene	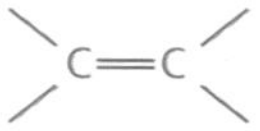 C=C	(*)-ene	–	CH_3CH_2—CH=CH_2	but-1-ene
nitro	—NO_2	–	nitro-(*)	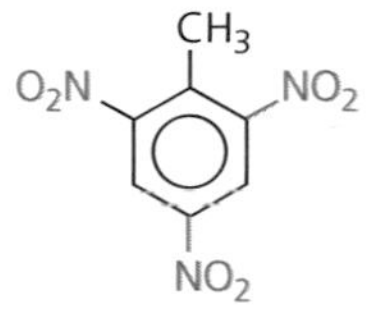 O_2N, CH_3, NO_2, NO_2 on benzene ring	methyl-2,4,6-trinitrobenzene
phenyl	(benzene ring)	–	phenyl-(*)	CH_3—CH—CH_3 with phenyl	2-phenylpropane
primary amine	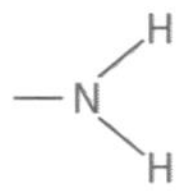 —N(H)(H)	–	amino-(*)	$CH_3CH_2NH_2$	aminoethane
secondary amine	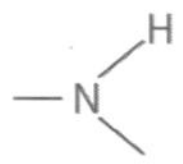 —N(H)	–	N-(*)-amino-(*)	$(CH_3CH_2)_2NH$	N-ethylaminoethane
tertiary amine	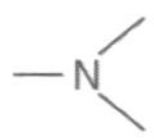 —N	–	N,N-di(*)-amino-(*)	$(CH_3CH_2)_3N$	N,N-diethylaminoethane

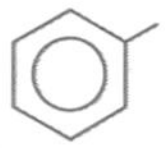

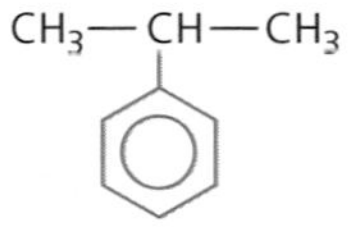

KEY FACTS

When naming an organic molecule :

■ Each different carbon skeleton has a unique IUPAC name.

■ Each functional group has a unique IUPAC name.

■ Positions of side chains and functional groups are shown by numbering the carbon atoms of the carbon skeleton.

4.3 Isomerism

Fig. 6 *Cis-* and *trans-* but-2-ene

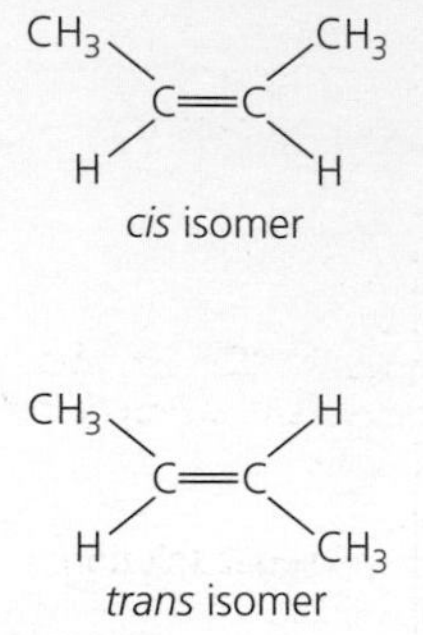

CH₃   CH₃
   C═C
H   H
cis isomer

CH₃   H
   C═C
H   CH₃
trans isomer

Fig. 7 Methylbut-2-ene

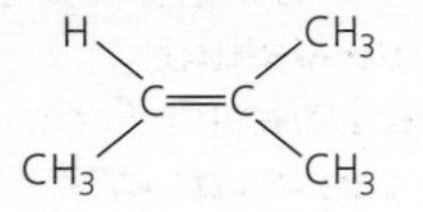

H   CH₃
   C═C
CH₃   CH₃

Fig. 8 Geometric isomers of 1,2-dichlorocyclobutane

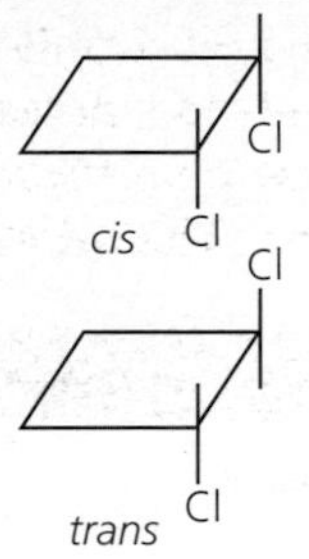

cis

trans

Isomers are molecules with the same molecular formula but different structures. There are two general types of isomerism: **structural isomerism** and **stereoisomerism**.

Structural isomers

Three types are:

- **Chain isomers** have the same total number of carbon atoms but different patterns of branching, e.g. pentane, methylbutane and dimethylpropane.

- **Positional isomers** have the same carbon skeleton but their functional groups are in different positions, e.g. 1-chloropropane and 2-chloropropane, 1,2-dinitrobenzene and 1,3-dinitrobenzene, and methylpropan-1-ol and methylpropan-2-ol.

- **Functional group isomers** have the same molecular formula but the carbon atoms are bonded differently to form different functional groups (see Table 3), e.g. alkenes and cycloalkanes, alcohols and ethers, carboxylic acids and esters, aldehydes and ketones, and primary, secondary and tertiary amines.

11 Draw and name the chain structural isomers of C_6H_{14}

12 Draw and name the positional and functional group isomers of:

a C_4H_8O

b C_3H_8O

c $C_4H_8O_2$

Stereoisomers

These have the same structural formula but have different arrangements of the atoms in space. There are two types of stereoisomerism: **geometrical** and **optical** isomerism.

Geometrical isomers

An alkene molecule cannot rotate about the C=C double bond and the four atoms bonded to the C=C bond lie in the same plane. This means that if two different groups are attached to each end of the C=C double bond, we can draw two different structures, called **geometrical isomers**. One isomer has the groups on the same side of the bond (the

cis isomer), and the other has them on opposite sides of the bond (the *trans* isomer). Fig. 6 shows the structures of *cis-* and *trans-* but-2-ene. Geometrical isomers cannot be formed if the two groups attached to one end of the C=C double bond are the same (Fig. 7).

Geometrical isomers can also occur in di-substituted cycloalkanes because the ring prevents free rotation of a C–C bond and therefore the two groups may be on the same side or on opposite sides of the ring (Fig. 8).

13 For the following compounds, decide which can have geometric isomers. Draw and name them.

a pent-1-ene

b pent-2-ene

c 2-methylpent-2-ene

d cyclohexene

e cyclohexan-1,4-diol

Optical isomers

A carbon atom that is bonded to four different atoms or groups is **asymmetric** (or **chiral**). **Optical isomerism** (also known as **enantiomerism**) occurs when a molecule contains an asymmetric carbon atom. As in Fig. 9, the molecule is without symmetry: like your left and right hands, it is possible to have two different structures that are mirror images of each other. Since the pair of mirror images are not superimposable, they must be isomers and are referred to as **optical isomers** or **enantiomers**. We can only show the difference between a pair of enantiomers properly by three-dimensional drawings.

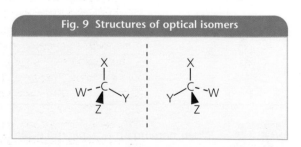

Fig. 9 Structures of optical isomers

A pair of enantiomers have identical physical and chemical properties with two exceptions. They interact differently with other asymmetric molecules, especially

14 Which of the following molecules can exist as optical isomers? For those molecules which can show this property, use three-dimensional structures to show the pair of optical isomers.

a chlorofluoromethane

b butan-2-ol

c 2-aminopropanoic acid (alanine)

d an isomer of heptane, C_7H_{16}

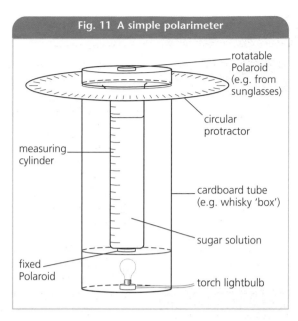
Fig. 11 A simple polarimeter

- rotatable Polaroid (e.g. from sunglasses)
- circular protractor
- measuring cylinder
- cardboard tube (e.g. whisky 'box')
- sugar solution
- fixed Polaroid
- torch lightbulb

enzymes. For example, one enantiomer of the amino acid alanine (see question 14) will be absorbed and converted to protein by the body but the other enantiomer will not.

Enantiomers also interact differently with plane-polarised light (hence the term *optical* isomers); this is referred to as a difference in **optical activity**.

Optical activity

A beam of light consists of waves that vibrate in all possible planes at right angles to the direction the beam is travelling. When passed through a **polariser**, all the waves are absorbed except for those vibrating in one particular plane (see Fig. 10). The light is then said to be **plane-polarised**.

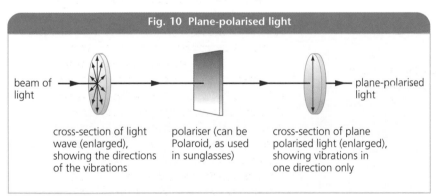
Fig. 10 Plane-polarised light

beam of light → plane-polarised light

cross-section of light wave (enlarged), showing the directions of the vibrations

polariser (can be Polaroid, as used in sunglasses)

cross-section of plane polarised light (enlarged), showing vibrations in one direction only

Optically active molecules rotate the plane of the plane-polarised light. One enantiomer rotates it clockwise and the other enantiomer rotates the plane by the same amount in the anti-clockwise direction. The former is called the **dextrorotatory** or **(+)** isomer and the latter is called the **laevorotatory** or **(–)** isomer. We can detect and measure this difference experimentally using a **polarimeter** as illustrated in Fig.11.

If we test a sample and find that it does not produce this rotation effect, then either it is not chiral or it consists of a 50:50 mixture of the two enantiomers in which the rotation effects have cancelled each other out. Such a

mixture is called a **racemate**. We often obtain racemic mixtures when we do addition reactions on a planar double bond (C=C or C=O), because the reaction can occur with equal probability from either side of the plane. For example (see Fig. 12), ethanal undergoes nucleophilic addition with hydrogen cyanide to form chiral 2-hydroxypropanenitrile. However, this product will be the racemate because there is a 50:50 chance of the cyanide nucleophile bonding to either side of the planar molecule. Attack at one side produces one enantiomer, attack at the other side produces the other.

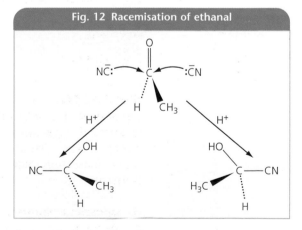
Fig. 12 Racemisation of ethanal

If you have studied this chapter thoroughly and worked through the examples, test questions, applications and examination questions you will have a sound foundation on which to base your further study of organic chemistry. The IUPAC system gives you a powerful means of international communication - it is like learning a new language!

APPLICATION A

Cracking petroleum

A petroleum cracking plant

The aim of cracking long-chain hydrocarbons is to produce branched structures in preference to straight chain structures. Branched structures burn more smoothly inside a car's engine. Heptane and 2,2,4-trimethylpentane are typical cracking products.

1 Draw the structure of heptane and 2,2,4-trimethylpentane.

2 Are these molecules isomeric? Explain.

3 Draw and name seven isomers of heptane.

4 Mark any chiral carbon atoms in these isomers with an asterisk.

When heptane reacts with bromine, any one of the 16 hydrogen atoms may be replaced (substituted) by a single bromine atom to produce four isomers of bromoheptane.

5 Draw and name all possible positional isomers of bromoheptane.

By removing HBr from 2-bromoheptane, hept-1-ene and hept-2-ene can be formed.

6 Draw the structures of these alkenes.

7 Show the geometrical isomers of the alkenes from question 6.

APPLICATION B

Diabetes

Diabetics cannot control their blood sugar levels because their pancreases do not produce enough insulin. One of the symptoms of diabetes, called ketosis, occurs when fats stored within the body are broken down to provide the energy that is normally supplied by glucose. This can result in the production of the ketone propanone, which is detectable as the smell of bad apples on the breath of a sufferer.

1 Draw the molecular structure of propanone.

2 Draw and name the functional group isomer of propanone.

3 The isomer from question 2 can be oxidised to the corresponding acid and reduced to the corresponding alcohol. Draw and name these molecules.

4 Propanoic acid reacts with ethanol to form the ester, ethyl propanoate, and propan-1-ol reacts with ethanoic acid to form another ester, propyl ethanoate. Draw the structures of these two esters and name the type of isomerism they depict.

5 Propanone can be converted directly or indirectly to each of the other molecules shown right. Name each of these molecules.

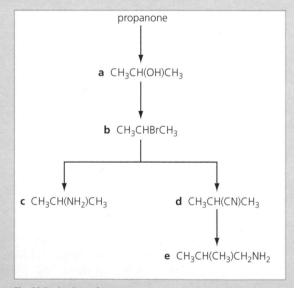

Fig. 13 Derivatives of propanone

Propanone is a problem to people suffering from ketosis but it is also a vital industrial chemical. Now you have experienced many of its structurally related compounds and how they can be clearly differentiated by the IUPAC naming system.

1 Give the structural formulas and names of the **four** isomers of C_3H_9N. (4)
NEAB CH06 Feb 1996 Q3a

2
a Name compound P.
b What structural feature in compound P results in optical isomerism?
c When compound P is made by reaction of propanal and hydrogen cyanide, the product is optically inactive. What name is given to such an inactive product and why is it inactive? (4)
NEAB CH03 June 1997 Q4b

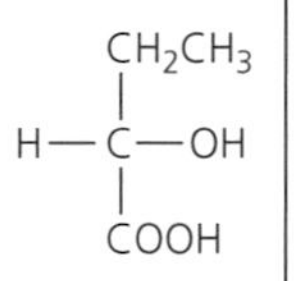

$$CH_2CH_3$$
$$|$$
$$H-C-OH$$
$$|$$
$$COOH$$

Compound **P**

3 Give the structures and names of two branched chain alcohols which are both isomers of butan-1-ol. (4)
NEAB CH03 Feb 1997 Q4c

4 3-Aminophenylethanone can be obtained from benzene in three steps:

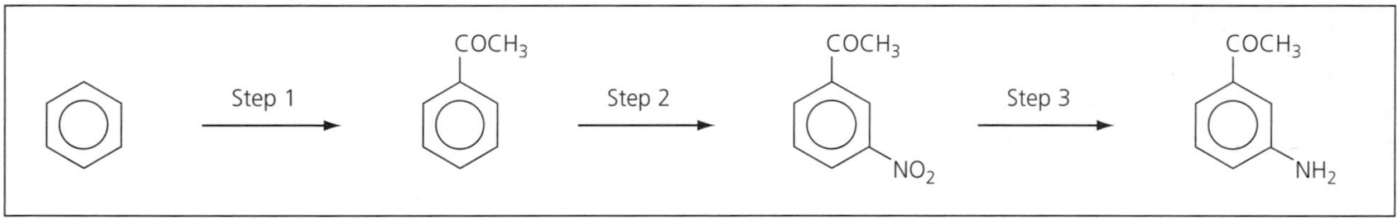

Name the molecules shown in the reaction scheme. (4)

5 The important monomer phenylethene (styrene) can be made from benzene in three steps:

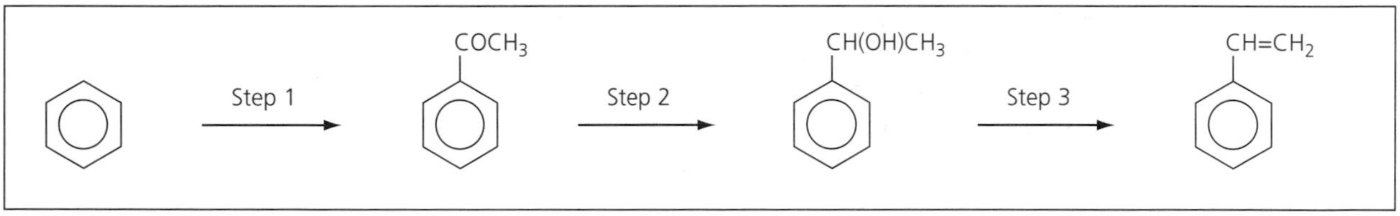

a Name the molecules shown in the reaction scheme.
b What type of isomerism is shown by the product of step 2? (5)
NEAB CH06 Feb 1996 Q2

6 The bromoalkane, $CH_3CH_2CHBrCH_3$, can be formed by reaction of $CH_3CH=CHCH_3$ with HBr. Name $CH_3CH_2CHBrCH_3$.
NEAB CH03 February 1997 Q3(a)

7 Name the ester, CH_3COOCH_3.
NEAB CH03 June 1997 Q2(b)

8 Give the structure of 3-bromopentane.
NEAB CH06 March 1998 Q4(a)(i)

9 Alcohol X has the structure $(CH_3)_2C(OH)CH(CH_3)_2$. Name alcohol X.
NEAB CH06 June 1999 Q3(a)(i)

5 Compounds containing the carbonyl group

However careful the tests during fermenting or brewing, carbonyl compounds occasionally form that make wine or beer undrinkable.

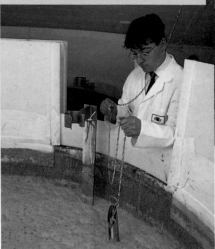

Ethanol is familiar as the substance commonly referred to as 'alcohol', found in wine, beer, cider and a wide variety of other alcoholic drinks. As you will know from your studies of AS level chemistry, ethanol is just one example of a large family of compounds to which the general term 'alcohols' is applied.

When the fermentation process for producing beer or wine goes wrong, the beverage tastes vinegary as a result of the ethanol being oxidised by oxygen in the air to the *carboxylic acid* called ethanoic acid.

As a part of the natural fermentation process, *aldehydes*, *ketones* and *esters* may also be produced which add to, or detract from, the overall flavour of the beverage. They may also add to the hangover effect when over-consumption of alcohol dehydrates the brain. These derivatives of alcohols – aldehydes, ketones, carboxylic acids and esters – all contain a C=O bond and are the central theme of this chapter.

5.1 Aldehydes and ketones

Aldehydes and ketones both contain a **carbonyl group** (>C=O, often abbreviated to –CO–), and so these two groups of compounds are closely related. The carbonyl group is situated at the end of a carbon chain in aldehydes and elsewhere in the carbon chain in ketones. Tables 1 and 2 show the names and structures of some simple aldehydes and ketones.

Aldehydes and ketones are both oxidation products of alcohols. We can make aldehydes by the *partial* oxidation of primary alcohols (which contain a CH_2–OH functional group). If we oxidise a secondary alcohol (which contains a –CH(OH)– group) we get a ketone. (Primary and secondary alcohols are described in *AS Chemistry*, Chapter 15.)

Table 1 Three aldehydes		
methanal	HCHO	
ethanal	CH_3CHO	
propanal	C_2H_5CHO	

Table 2 Three ketones		
propanone	CH_3COCH_3	
butanone	$CH_3COCH_2CH_3$	
phenylethanone	$C_6H_5COCH_3$	

Oxidation of primary alcohols

A solution of potassium dichromate(VI) acidified with dilute sulphuric acid oxidises primary alcohols ($R-CH_2-OH$) such as ethanol (where $R = CH_3$). The orange $Cr_2O_7^{2-}$ ion is reduced to the green chromium(III) cation (Cr^{3+}) during the reaction. This colour change indicates that oxidation has occurred. The oxidant $[Cr_2O_7^{2-} + H^+]$ is usually represented by [O], as shown in the following equation:

$$
\underset{\begin{matrix}|\\H\end{matrix}}{\overset{\begin{matrix}H\\|\end{matrix}}{R-C-O-H}} \xrightarrow{[O]} H_2O + R-\overset{O}{\underset{H}{C}} \xrightarrow{[O]} R-\overset{O}{\underset{O-H}{C}}
$$

R represents an alkyl or aryl (aromatic) group.

This equation shows how the aldehyde (e.g. ethanal, $R = CH_3$) can be oxidised further to form a carboxylic acid (e.g. ethanoic acid). Prolonged heating under reflux with excess acidified dichromate completely oxidises an alcohol to a carboxylic acid.

Metaldehyde, a cyclic polymer of four ethanal molecules, is the killer in slug pellets – good for the plants but it poisons the birds that eat the dead slugs and snails

In order to isolate an aldehyde we have to immediately distil it from the reaction mixture before it can be oxidised further to the acid. Such a distillation is possible because aldehydes have relatively lower boiling points because their molecules are bound by weak dipole-dipole intermolecular forces compared to the stronger hydrogen bonding between both alcohol and acid molecules.

1 Give the structures and names of the oxidation products of:
a methanol,
b propan–1–ol.

2 What is the change in the oxidation number of chromium when acidified dichromate is used to oxidise alcohols?

3 Write an equation to show the oxidation of butan-1-ol by potassium dichromate(VI), and name the organic products.

Oxidation of secondary alcohols

Oxidation of secondary alcohols by warm, acidified dichromate(VI) produces ketones. These have the general formula R_1COR_2 where R_1 and R_2 are alkyl or aryl groups. Again, a colour change from orange to green indicates that oxidation has occurred:

$$
\underset{\begin{matrix}|\\OH\end{matrix}}{\overset{\begin{matrix}H\\|\end{matrix}}{R_1-C-R_2}} \xrightarrow{[O]} R_1-\underset{\underset{O}{\|}}{C}-R_2 + H_2O
$$

secondary alcohol   ketone

$$
\text{e.g. } CH_3-CH(OH)-CH_2CH_3 \xrightarrow{[O]} CH_3COCH_2CH_3
$$

butan–2–ol   butan–2–one

The ketone propanone is used to make Perspex, a hard, clear plastic used in incubators for premature babies.

4 Give the name and structure of the alcohol that would be oxidised to:
a pentan-2-one,
b phenylethanone.

5 Write an equation to show the oxidation (use [O] to represent the oxidant) of cyclohexan-1-ol by acidified potassium dichromate(VI), and name the organic product.

Oxidation of tertiary alcohols

Tertiary alcohols are not easily oxidised. The standard oxidant, acidified dichromate(VI), remains orange when heated with a tertiary alcohol. There is no hydrogen atom to be removed from the carbon to which the –OH group is attached. Hence, the oxidant [O] cannot remove two hydrogen atoms to form water and the carbonyl group. However, a stronger oxidant (e.g. acidified potassium manganate(VII), $KMnO_4$) can break the C–C bonds and oxidise the alcohol. The products are a mixture of compounds, each with shorter carbon chains.

5.2 Reactions of aldehydes and ketones

The carbonyl group reacts in similar ways in both aldehydes and ketones, and undergoes **nucleophilic addition**. However, aldehydes are reducing agents but ketones are not.

Nucleophilic addition reactions

Addition of hydrogen cyanide to both aldehydes or ketones is an example of a nucleophilic addition reaction:

Hydrogen cyanide is a very weak acid. The degree of ionisation in solution, to produce hydrogen ions (H^+) and nucleophilic cyanide ions ($:CN^-$), is very small. In practice, the cyanide ion is derived from ionic potassium cyanide. This is acidified with sulphuric acid to provide the H^+ ions.

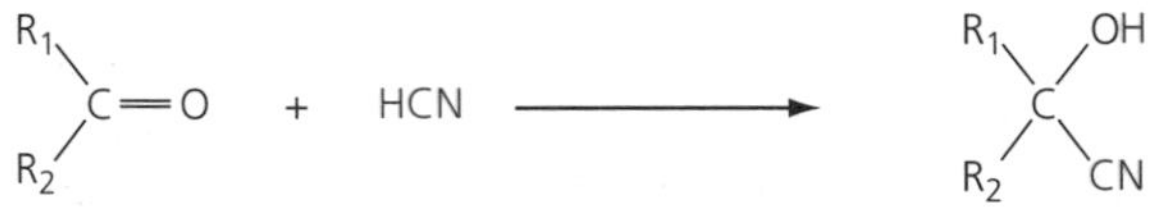

aldehyde: R_1 and/or R_2 = H
ketone: R_1 and R_2 = alkyl groups

a hydroxynitrile

2-hydroxypropanenitrile

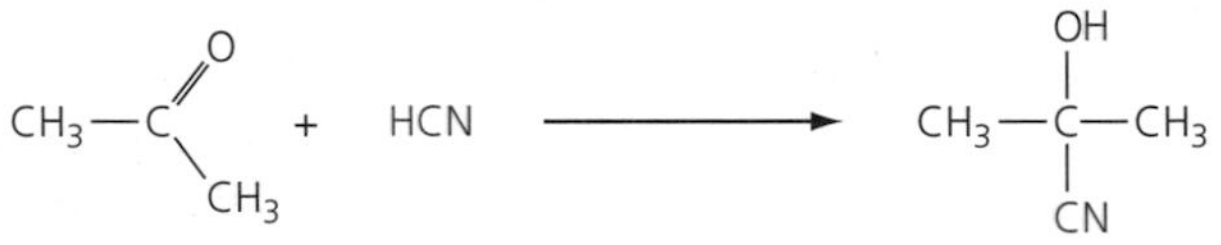

2-hydroxy-2-methylpropanenitrile

Fig. 1 shows how the high electronegativity of the oxygen atom in the C=O group results in this bond being polar, with the carbon atom electron-deficient. The carbon atom is therefore prone to attack by nucleophiles ('lovers of positive charge') which have lone pairs of electrons available for bonding. To accommodate this new bond to the carbonyl carbon, the π–bond of C=O bond breaks, and a saturated C–O bond is formed. Hence, an addition reaction results and, overall, the reaction mechanism is called a **nucleophilic addition across the C=O bond.**

Fig. 1 Nucleophilic addition of cyanide

The reaction happens in stages.

1 The nucleophile ($:CN^-$) attacks the partial positive charge on the carbon atom of the carbonyl group:

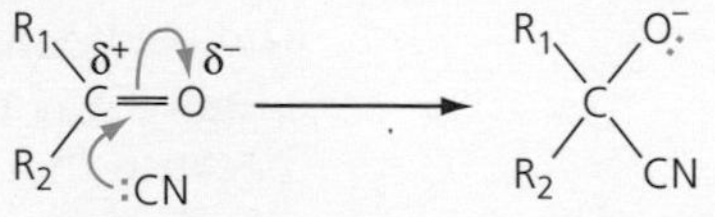

2 The reaction intermediate has an oxygen atom carrying a negative charge. An electron pair from this atom is then used to form a covalent bond with a hydrogen ion, forming a hydroxynitrile:

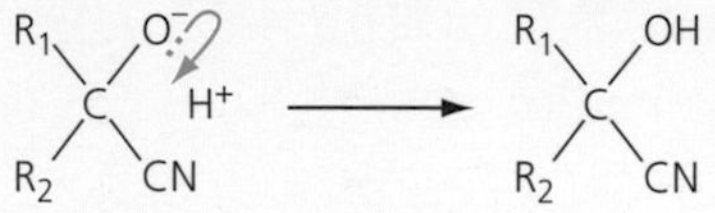

6 Write equations to show the reactions with hydrogen cyanide of:

a propanal,

b butan-2-one.

Name the organic products.

7 Both of the reactions of question 6 produce racemic products (see Chapter 4). Show the structures of the optical isomers concerned.

Redox reactions

Oxidation reactions

Aldehydes are easily oxidised to form carboxylic acids, but it is difficult to oxidise ketones. Ketones can be oxidised only by very strong oxidising agents, because there is no hydrogen atom attached to the carbon atom of their carbonyl group. This gives us a good way of distinguishing aldehydes and ketones: we can test an unknown sample with a very gentle oxidising agent such as **Fehling's solution**. Fehling's solution is blue (see photo) and contains copper(II) complex ions in alkaline solution. When the solution is warmed with an aldehyde, the aldehyde is oxidised to a carboxylic acid and the blue copper(II) is reduced to a brick red precipitate of copper(I) oxide:

$$R-C\overset{O}{\underset{H}{\big|}} + 2Cu^{2+}(aq) + 4OH^- \longrightarrow R-C\overset{O}{\underset{OH}{\big|}} + Cu_2O(s) + 2H_2O$$

blue   brick red

Tollens reagent, a solution made by adding dilute ammonia solution to silver(I) nitrate solution, can be used instead of Fehling's solution. It contains the silver(I) complex ion, diamminesilver(I) ion, $[Ag(NH_3)_2]^+$. When gently warmed with Tollens reagent, an aldehyde causes the colourless diamminesilver(I) ions to be reduced to metallic Ag(0). This causes a 'silver mirror' to coat the inside of the reaction vessel.

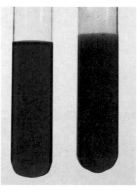 

The copper(II) complex ion gives Fehling's solution its blue colour. After an aldehyde is oxidised, the copper(II) ion is reduced to brick-red Cu_2O.

The silver(I) complex ion in Tollens reagent is reduced by an aldehyde to metallic silver.

Table 3 summarises the reactions of alcohols, aldehydes and ketones with the oxidising agents we have discussed above.

8 Briefly describe how separate samples of a primary, a secondary and a tertiary alcohol might be distinguished using the redox reactions discussed above.

Reduction reactions

The oxidation sequences to make aldehydes and ketones and described in section 5.1, can be reversed with a strong reducing agent such as lithium tetrahydridoaluminate (LiAlH$_4$, which is used in non-aqueous solvents) or sodium tetrahydridoborate (NaBH$_4$, which can be used in water). These agents reduce aldehydes to primary alcohols and ketones to secondary alcohols. Carboxylic acids can also be reduced to primary alcohols because they are first reduced to aldehydes.

Acids and aldehydes

$$R-C\overset{O}{\underset{OH}{\big|}} \xrightarrow{2[H]} R-C\overset{O}{\underset{H}{\big|}} \xrightarrow{2[H]} R-\overset{H}{\underset{H}{\overset{|}{C}}}-OH$$

acid   aldehyde   primary alcohol

Ketones

$$R_1-\overset{}{\underset{O}{\overset{}{C}}}-R_2 \xrightarrow{2[H]} R_1-\overset{H}{\underset{OH}{\overset{|}{C}}}-R_2$$

ketone   secondary alcohol

This equation shows that such reductions are also nucleophilic addition reactions. NaBH$_4$

Table 3 Oxidation of alcohols, aldehydes and ketones					
Oxidising agent	Primary alcohol	Secondary alcohol	Tertiary alcohol	Aldehyde	Ketone
$K_2Cr_2O_7/H^+$	✓	✓	✗	✓	✗
Fehling's reagent	✗	✗	✗	✓	✗
Tollens reagent	✗	✗	✗	✓	✗

Fig. 2 Nucleophilic addition of hydride

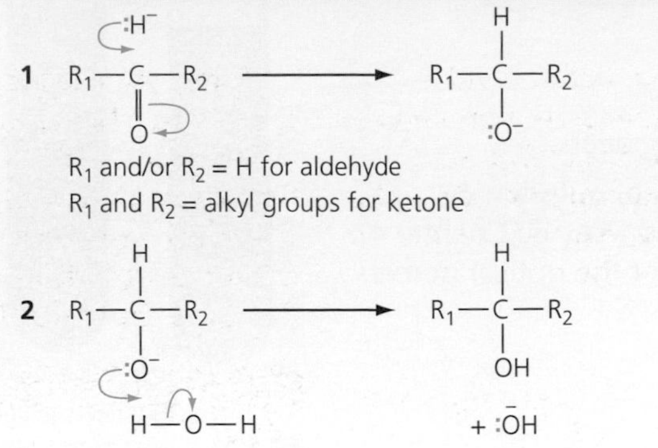

R_1 and/or R_2 = H for aldehyde
R_1 and R_2 = alkyl groups for ketone

provides hydride ions (:H$^-$), which act as the nucleophile. The mechanism for this process is shown in Fig. 2.

9 Naming the organic product, write equations for the reduction of:

a propanal,

b pentan-2-one,

c butanoic acid,

using sodium tetrahydridoborate (NaBH$_4$). Use [H] to represent the reductant.

5.3 Carboxylic acids

As described earlier, carboxylic acids are produced by the oxidation of primary alcohols or aldehydes. Ethanoic acid is the acid responsible for the vinegary taste of wine which has been oxidised by exposure to air (see the Opener to this chapter). The characteristic functional group of carboxylic acids is –COOH.

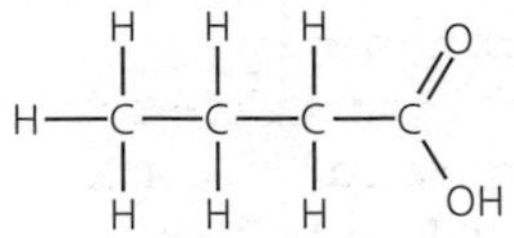

Carboxylic acids with fewer than six carbon atoms per molecule are water soluble. In aqueous solution they are only slightly ionised to give low concentrations of hydrogen ions and alkanoate ions:

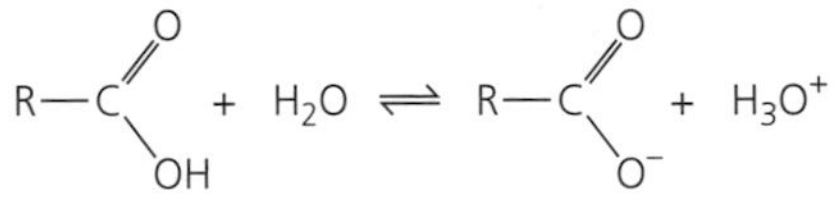

This partial ionisation in solution means that carboxylic acids are **weak acids**. Nevertheless, the concentration of hydrogen ions is sufficient to displace carbon dioxide gas from an aqueous solution of sodium carbonate. This gaseous product provides a useful test for the possible presence of a carboxylic acid:

$$2RCOOH + Na_2CO_3 \rightarrow 2RCOO^-Na^+ + CO_2 + H_2O$$

10 The values for the acid dissociation constant K_a (see Chapter 3) of some carboxylic acids are:

Ethanoic acid 2×10^{-5}/mol dm^{-3}

Benzoic acid 1×10^{-5}/mol dm^{-3}

Methanoic acid 1×10^{-4}/mol dm^{-3}

Use these to place 0.1 mol dm^{-3} solutions of these acids in order of increasing pH.

Table 4 The first four carboxylic acids

methanoic acid HCOOH

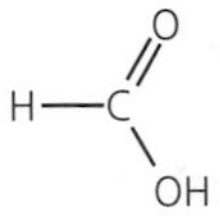

ethanoic acid CH$_3$COOH

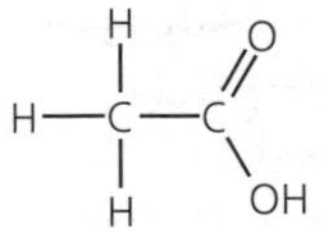

propanoic acid CH$_3$CH$_2$COOH

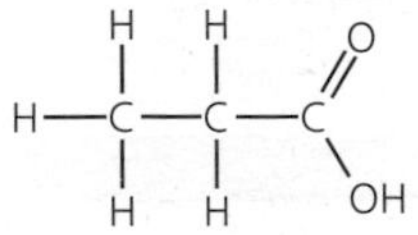

butanoic acid CH$_3$CH$_2$CH$_2$COOH

5.4 Esters

An ester is formed when an alcohol and a carboxylic acid react together in the presence of a concentrated, strong acid catalyst (e.g. sulphuric acid or conc. hydrochloric acid). The general equation for this type of reaction is:

$$R_1-\overset{\displaystyle O}{\underset{\displaystyle O-H}{C}} \quad H-O-R_2 \overset{\text{H}^+ \text{ catalyst}}{\rightleftharpoons} R_1-C-O-R_2 + H_2O$$
$$\text{acid} \qquad \text{alcohol} \qquad \qquad \text{ester}$$

Radioactive labelling of the starting materials shows that, in most cases, the C–O bond in the acid breaks rather than the C–O bond in the alcohol. Table 5 gives some examples of ester formation.

Table 5 Ester formation		
Acid	**+ methanol gives:**	**+ ethanol gives:**
ethanoic acid	methyl ethanoate	ethyl ethanoate
$CH_3-\overset{\displaystyle O}{\underset{\displaystyle OH}{C}}$	$CH_3-\overset{}{\underset{\displaystyle O}{C}}-O-CH_3$	$CH_3-\overset{}{\underset{\displaystyle O}{C}}-O-CH_2CH_3$
propanoic acid	methyl propanoate	ethyl propanoate
$CH_3CH_2-\overset{\displaystyle O}{\underset{\displaystyle OH}{C}}$	$CH_3CH_2-\overset{}{\underset{\displaystyle O}{C}}-O-CH_3$	$CH_3CH_2-\overset{}{\underset{\displaystyle O}{C}}-O-CH_2CH_3$

This reaction is often referred to as an **esterification**, but it is also known as a **condensation reaction** since a small molecule (e.g. H_2O) is eliminated from two carbon-based molecules that form a larger structure. Such reactions are vitally important in the production of natural and artificial polymers, as we will see in Chapter 9.

11 Write an equation for the reaction of butanoic acid with propan-1-ol, and name the organic product.

Ester formation reactions are reversible, and produce relatively low yields of esters. Higher yields of esters are obtained by reacting the alcohol with the acid chloride or acid anhydride instead of the carboxylic acid. These reactions are also called **acylations** and will be discussed in detail in Chapter 7.

Esters are the 'smells and flavours' of chemistry. They are responsible for many flower scents and fruit flavours. The artificial fruit flavours (e.g. cherry, banana, pear) used in some confectionery products are made by mixing synthetic esters together. Some of these are shown in Table 6.

Table 6 Ester fragrances	
Ester	**Essence**
ethyl methanoate	raspberry
ethyl butanoate	pineapple
ethyl ethanoate	pear drops
2-pentyl ethanoate	pear

Esters flavour the sweets in this Turkish shop

We can use the sweet or fruity smell of an ester as a test for the presence of an alcohol or carboxylic acid. If we suspect that the compound we are investigating is an alcohol, we can warm it with a carboxylic acid (e.g. ethanoic acid) in the presence of concentrated sulphuric acid as a catalyst. The excess acid, which has a pungent vinegary smell, is removed by adding warm, aqueous sodium carbonate solution. The sweet smell of an ester confirms the presence of an alcohol. As the ester is immiscible with water, because it cannot form hydrogen bonds to water, the other components of the reaction mixture will dissolve but the ester will float on the surface. The warmth of the solution then causes the ester to evaporate, and the sweet smell is easily detected.

Esters are also used as **plasticisers**, which are additives mixed into polymers to increase the flexibility of the polymer.

Poly(chloroethene), better known as PVC (see Chapter 9), is a strong and rigid polymer suitable for making drainpipes and guttering. However, when treated with up to 18% by mass of an appropriate plasticising ester it becomes clingfilm, used as a food wrapping material for non-fatty foods.

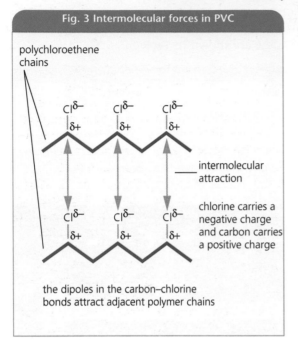

Fig. 3 Intermolecular forces in PVC

polychloroethene chains

intermolecular attraction

chlorine carries a negative charge and carbon carries a positive charge

the dipoles in the carbon–chlorine bonds attract adjacent polymer chains

Fig. 4

polychloroethene chains

plasticiser molecules

distance too great for intermolecular forces to be effective

PVC consists of very long polymer chains. The carbon–chlorine bond is strongly polar and there is considerable intermolecular attraction between the polymer chains (Fig. 3). PVC is therefore a strong and rigid plastic. Plasticiser molecules penetrate between the polymer chains and increase the distance between the chains. The polar effects of the carbon–chlorine bond are weakened and the rigidity of the three-dimensional structure is reduced (Fig. 4). As a result, the polymer chains can slide over each other and the resulting plastic is soft and pliable, as clingfilm must be.

The ester originally used as a plasticiser in clingfilm was di(2-ethylhexyl) adipate (commonly called DEHA). DEHA is an ester of 2-ethylhexan-1-ol and adipic acid (hexanedioic acid). Such placticised PVC is also used to make imitation leather for car seats, shoes, briefcases, etc. Over time the plasticiser evaporates, causing the 'leather' to lose its flexibility and crack.

The volatile esters used as solvents in nail varnish quickly evaporate in the air.

12

a Draw the structural formula of 2-ethylhexan-1-ol.

b Draw the structure of hexanedioic acid.

c Hence, write an equation for the formation of DEHA.

Unlike the acids and alcohols from which they are derived, esters have no –OH groups with which to form hydrogen bonds. They therefore have lower boiling points than their constituent acids and alcohols. This accounts for their 'smelly' nature: volatile compounds can reach our noses more easily than less volatile ones.

The lack of an –OH group also means that esters are not very soluble in water. The presence of carbon chains, however, makes them fat-soluble. When esters are used as plasticisers in clingfilm, their volatility and solubility in fat poses problems. Clingfilm containing DEHA has caused several health scares because this ester can dissolve from the film into food. Hence, such clingfilm is not used for foods such as cheese, which has a high fat content. Newer PVC clingfilms have been produced for general use, as well as for use in microwave ovens. The new types of clingfilm contain either a mixture of two plasticisers – a polymeric plasticiser (typically 1–11%) and DEHA (typically 10–13%) – or a polymeric plasticiser only. The polymeric plasticiser contains long molecules of polymerised esters, which cannot diffuse from the film to the food very easily, because of their high relative molecular mass.

Esters are also commonly used as solvents for organic compounds. Fragrances and nail varnishes, for example, often contain ethyl ethanoate. The ingredients are dissolved in the ester, which evaporates in air and leaves the scent or varnish behind.

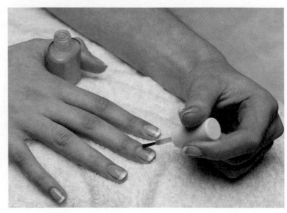

Hydrolysis of esters

The reverse of an esterification reaction is called **hydrolysis**. The ester is split ('lysis') by the action of water ('hydro') to re-form the acid and the alcohol. Such a hydrolysis requires heat and a concentrated sulphuric acid or sodium hydroxide catalyst. If the latter is used, it needs to be present in excess because the carboxylic acid produced from the ester would react to form a salt.

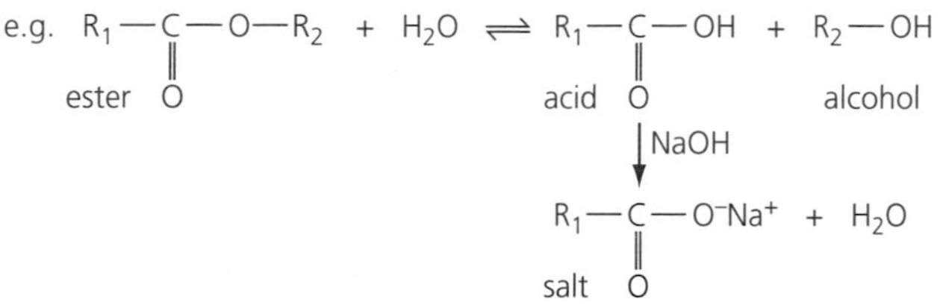

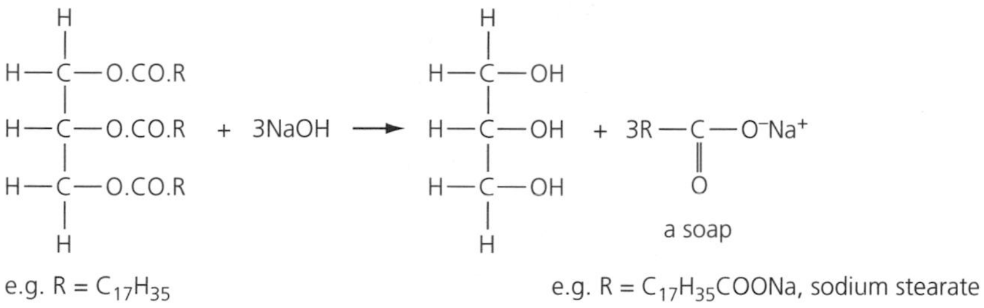

e.g. R_1—C—O—R_2 + H_2O ⇌ R_1—C—OH + R_2—OH
ester   O   acid   O   alcohol

↓ NaOH

R_1—C—O$^-$Na$^+$ + H_2O
salt   O

$$
\begin{array}{c}
\text{H} \\
| \\
\text{H—C—O.CO.R} \\
| \\
\text{H—C—O.CO.R} \quad + \ 3NaOH \longrightarrow \\
| \\
\text{H—C—O.CO.R} \\
| \\
\text{H}
\end{array}
\qquad
\begin{array}{c}
\text{H} \\
| \\
\text{H—C—OH} \\
| \\
\text{H—C—OH} \quad + \ 3R\text{—C—O}^-\text{Na}^+ \\
| \qquad\qquad\qquad\quad \| \\
\text{H—C—OH} \qquad\qquad\quad \text{O} \\
| \\
\text{H} \\
\text{a soap}
\end{array}
$$

e.g. R = $C_{17}H_{35}$ e.g. R = $C_{17}H_{35}$COONa, sodium stearate

13 Naming the products, write equations for the hydrolysis of:

a ethyl ethanoate by hot conc. sulphuric acid solution,

b methyl propanoate by hot sodium hydroxide solution.

Fig. 5 Reaction of glycerol and stearic acid

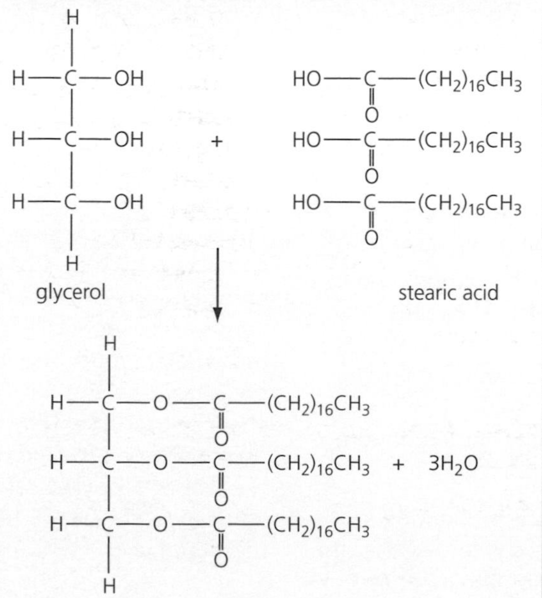

$$
\begin{array}{c}
\text{H} \\
| \\
\text{H—C—OH} \\
| \\
\text{H—C—OH} \qquad + \\
| \\
\text{H—C—OH} \\
| \\
\text{H} \\
\text{glycerol}
\end{array}
\qquad
\begin{array}{c}
\text{HO—C—}(CH_2)_{16}CH_3 \\
\|\\
\text{O} \\
\text{HO—C—}(CH_2)_{16}CH_3 \\
\|\\
\text{O} \\
\text{HO—C—}(CH_2)_{16}CH_3 \\
\|\\
\text{O} \\
\text{stearic acid}
\end{array}
$$

↓

$$
\begin{array}{c}
\text{H} \\
| \\
\text{H—C—O—C—}(CH_2)_{16}CH_3 \\
| \qquad\quad \|\\
\qquad\qquad \text{O} \\
\text{H—C—O—C—}(CH_2)_{16}CH_3 \quad + \ 3H_2O \\
| \qquad\quad \|\\
\qquad\qquad \text{O} \\
\text{H—C—O—C—}(CH_2)_{16}CH_3 \\
| \qquad\quad \|\\
\qquad\qquad \text{O} \\
\text{H}
\end{array}
$$

Soaps

Soaps are salts of long-chain carboxylic acids (also called **fatty acids**). These are derived from fats and oils which are naturally occurring **tri-esters** of the trihydric (i.e. containing three –OH groups) alcohol propane-1,2,3-triol (also known as 'glycerol'), and three fatty acids. These acids usually have 14, 16, 18 or 20 carbon atoms, one example being octadecanoic acid (stearic acid) as shown in Fig. 5.

Fats and oils are hydrolysed to produce **soaps** by boiling with aqueous sodium hydroxide solution in a process also called **saponification**. After the boiling is complete, common salt is added to precipitate the soap. Since the hydrolysis is done under alkaline conditions, the product is a mixture of glycerol and the salts of the fatty acids, soaps. Whenever soaps are manufactured, glycerol is a useful by-product that has extensive uses in pharmaceutical and cosmetic preparations.

Soaps are **anionic detergents**. As illustrated in Fig. 6, the negatively charged carboxyl group of a soap is attracted to water: it is hydrophilic. Conversely, the long hydrocarbon chains are attracted to oil rather than water: they are hydrophobic. When we add soap to an oil-water mixture and agitate the mixture, the oil is broken up into tiny droplets, in a process called emulsification, and dispersed throughout the water.

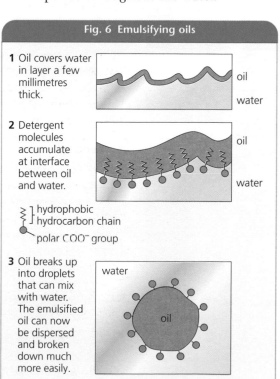

Fig. 6 Emulsifying oils

1 Oil covers water in layer a few millimetres thick.
oil
water

2 Detergent molecules accumulate at interface between oil and water.
oil
water

⟩ hydrophobic hydrocarbon chain
● polar COO$^-$ group

3 Oil breaks up into droplets that can mix with water. The emulsified oil can now be dispersed and broken down much more easily.
water
oil

The sodium salts (soaps) may be converted to free fatty acids by adding a strong acid such as hydrochloric acid:

Remember, carboxylic acids are weak acids, so addition of hydrogen ions moves the above equilibrium to the right, producing the insoluble fatty acid. This is why it can be a mistake to mix soap-based cosmetics with others based on acidic citric products – slimy fatty acids may precipitate!

$$CH_3-(CH_2)_{16}-C\overset{O}{\underset{O^-Na^+}{}} + H^+ \rightarrow CH_3-(CH_2)_{16}-C\overset{O}{\underset{OH}{}} + Na^+$$

sodium stearate

stearic acid

KEY FACTS

- Esters can be produced by the reaction between an alcohol and a carboxylic acid, an acid chloride or an acid anhydride.

- Esters are commonly used as solvents, flavourings and plasticisers.

- Fats and oils are naturally occurring tri-esters of glycerol with long chain fatty acids.

- Fats and oils are hydrolysed to soaps by boiling with aqueous sodium hydroxide.

APPLICATION A

Caught by the breathalyser

Look her in the eye. Then say a quick drink never hurt anybody. DRINKING AND DRIVING WRECKS LIVES.

'Blow into the bag please!' The original breathalyser of 1960.

Traffic police spend a lot of time trying to catch drivers who have had too much to drink. At one time, police officers had to judge whether drivers were sober or not by deciding if they could talk clearly, or could walk in a straight line.

The first breathalyser, invented in 1960, measured the amount of alcohol in a driver's breath. This chemical breathalyser relied on the reaction between ethanol vapour in the driver's breath and orange crystals of acidified potassium dichromate(VI), the oxidising agent contained in a glass phial. Ethanol reduces the crystals and they turn green. A police officer would check how much of the orange packing had changed to green. The extent of the change was then related to the amount of ethanol in the driver's breath.

1 Give the full molecular structure of ethanol.

2 Is ethanol a primary, secondary or tertiary alcohol? Explain your answer.

3 Write equations for the oxidations that occur inside the chemical breathalyser and name the organic products formed. You may use [O] to represent the oxidant.
a Give the formulas of the ions responsible for the original orange colour and the final green colour.
b What type of reaction does the chromium species of the oxidant undergo?

4 Diabetics tend to produce propanone in their breath and may be concerned that this could give a misleading breathalyser test.

Why is this fear unfounded?

If you eat a lot of sweets flavoured with esters, digestive hydrolysis may produce a varied mixture of alcohols and acids which are detectable on your breath. The mixture might include butan-1-ol, butan-2-ol, 2-methylpropan-2-ol, ethanoic acid and butanoic acid.

5 Draw the structures of these alcohols and acids.
a Which of these substances might cause a misleading breathalyser test? Explain.
b Which of these substances would not cause a misleading breathalyser test? Explain.
c Draw and name the ester responsible for the production of butan-1-ol and butanoic acid.

6 Write an equation for the digestive hydrolysis of the ester from question 5c.

The oxidant in the chemical breathalyser cannot be reused, misleading results can occur (although any positive result would be confirmed by blood or urine tests) and the orange-to-green colour change is not always easy to see, especially at night. The police now prefer to use the modern electronic breathalyser – the 'intoximeter' – which is cheaper than the 'blow in the bag' chemical device. The electronic breathalyser is about the size of a pocket calculator and coloured lights indicate clearly the level of ethanol in the breath.

The intoximeter works by measuring the amount of infrared (IR) radiation that the ethanol molecules absorb at different wavelengths. This analytical technique is discussed in Chapter 11.

The electronic breathalyser is very convenient and reliable for roadside tests.

APPLICATION B   Investigating simple sugars

Simple sugars (monosaccharides) are important biochemical molecules. Glucose (grape sugar) and fructose (fruit sugar) are just two examples (Figs. 7 and 8).

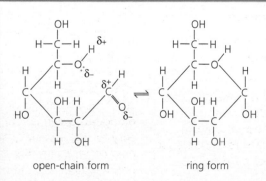

Fig. 7 Open-chain structures of glucose and fructose

glucose   fructose

1 Deduce the molecular formulas of glucose and fructose. What can be concluded from your answers?

2 Redraw the open-chain structures and mark each of the chiral carbon atoms (see Chapter 4) with an asterisk. Also, in glucose draw circles around any primary alcohol groups, squares around any secondary alcohol groups and triangles around any tertiary alcohol groups.

3 Are these sugars also aldehydes, ketones, acids or esters? Explain.

Fig. 8 Open and ring forms of glucose

open-chain form   ring form

4 Based on your answers to question 3, describe the procedure and results of a chemical test that *should* distinguish glucose from fructose. Explain the chemical basis of this test.

This test *does not* produce the expected results! Both glucose and fructose will produce the same visible effects. This is explained by the fact that, in the presence of an acid or alkaline catalyst, fructose will isomerise to glucose.

5 Explain this isomerisation in terms of redox reactions.

6 Draw the structure of the molecule derived from glucose during the test described in question 4.

7 Draw the structure of the molecule produced when fructose reacts with hydrogen cyanide.

8 Name the type of reaction involved in the reaction from the answer to question 7 and describe the mechanism.

9 Name the type of mechanism involved in the conversion of open-chain glucose to ring-form glucose.

10 Draw the structure of the first product resulting from the treatment of glucose with hot, acidified potassium dichromate(VI) solution.

11 Draw the structure of one of the products resulting from reacting fructose with ethanoic acid in the presence of concentrated sulphuric acid. What is the purpose of the sulphuric acid?

1

a Explain the meaning of the terms *empirical formula* and *molecular formula*. (3)

b Give the three molecular formulas for organic compounds which have the empirical formula CH_2O and relative molecular masses below 100. (9)
In fact, four compounds, **H**, **J**, **K** and **L**, fit this information.
H can be oxidised to a carboxylic acid, **M**, and also reduced to a primary alcohol, **N**.
M and **N** react together, when warmed in the presence of concentrated sulphuric acid, to form **J**.
L contains a carboxylic acid group and is a structural isomer of **J**.
K contains a carboxylic acid group and shows optical isomerism.
Draw the structures of the six compounds, **H**, **J**, **K**, **L**, **M** and **N**.
NEAB CH03 June 1998 Q7 a and b

2 Propenoic acid, $CH_2=CHCOOH$, can be converted into compound A, CH_3CH_2COOH.

a i) Name the compound A. (1)

ii) Give the reagent and conditions necessary for the conversion of propenoic acid into A. (2)

b Propenoic acid can be used to make a polymer.

i) Draw the repeating unit of the polymer. (1)

ii) Name the type of polymer formed. (1)

c Write an equation for the reaction of compound A with sodium hydroxide. (1)

d Compound A reacts with ethanol in the presence of a suitable catalyst.

i) Identify a suitable catalyst. (1)

ii) Name the type of compound formed. (1)

iii) Write an equation for the reaction of A with ethanol, showing clearly the structure of the organic product. (1)
NEAB CH03 June 1995 Q1

3

a Compound **W** can be converted into three different organic compounds as shown by the reaction sequence below. Give the structures of the new compounds **X**, **Y** and **Z**. (3)

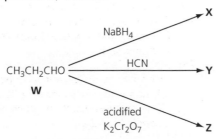

b Outline a mechanism for the formation of **Y**. (4)
NEAB CH03 Feb 1998 Q6a and b

4

a i) Write an equation for the reaction of butan-2-ol with ethanoic acid, showing clearly the structure of the organic product. (2)

ii) Name the type of organic compound formed in part **a i)** and suggest a use for this compound. (2)

iii) Give a homogeneous catalyst for the reaction in part **a i)** and state the meaning of the term homogeneous. (2)

b Write an equation for the complete combustion of butan-2-ol in an excess of oxygen. (1)
AQA CH03 March 1999 Q3 a and b

5 Consider the reaction sequence below.

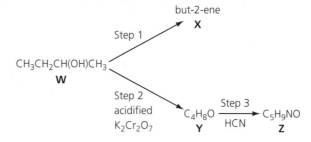

a Name **W**, deduce structures for **Y** and **Z** and state which of the four molecules **W**, **X**, **Y** and **Z** show optical isomerism.

b Give a reagent which could convert **Y** back into **W**.

c In step 3, **Y** undergoes nucleophilic addition. Explain the term *nucleophilic addition* and explain why compound **Y** undergoes this reaction.
NEAB CH03 Feb 1996 Q7

6 Butan-1-ol can be oxidised by acidified potassium dichromate(VI) using two different methods.

a In the first method, butan-1-ol is added dropwise to acidified potassium dichromate(VI) and the product is distilled off immediately.

i) Using the symbol [O] for the oxidising agent, write an equation for this oxidation of butan-1-ol, showing clearly the structure of the product. State what colour change you would observe. (3)

ii) Butan-1-ol and butan-2-ol give different products on oxidation by this first method. By stating a reagent and the observation with each compound, give a simple test to distinguish between these two oxidation products. (3)

b In a second method, the mixture of butan-1-ol and acidified potassium dichromate(VI) is heated under reflux. Identify the product which is obtained by this reaction. (1)

c Give the structures and names of two branched chain alcohols which are both isomers of butan-1-ol. Only isomer 1 is oxidised when warmed with acidified potassium dichromate(VI). (4)
NEAB CH03 Feb 1997 Q4

7 Consider the reaction scheme shown below.

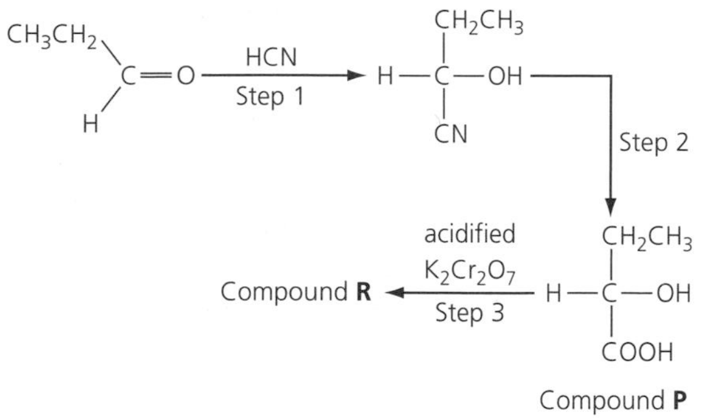

CH_3CH_2
C=O
H
HCN
Step 1
CH_2CH_3
H—C—OH
CN
Step 2
acidified
$K_2Cr_2O_7$
Step 3
Compound **R**
CH_2CH_3
H—C—OH
COOH
Compound **P**

a Name the type of reaction in Step 1 and outline a mechanism for the reaction. (5)

b i) Name compound **P**

ii) What structural feature in compound **P** results in optical isomerism?

iii) When compound **P** is made by the above method, the product is optically inactive. What name is given to such an inactive product and why is it inactive? (4)

c Draw the structure of compound **R** formed in Step 3. (1)

NEAB CH03 June 1997 Q4

8

a Compound **G**, shown below, is a tri-ester.

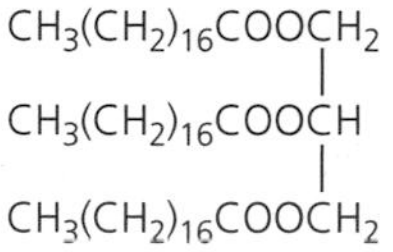

$CH_3(CH_2)_{16}COOCH_2$
$CH_3(CH_2)_{16}COOCH$
$CH_3(CH_2)_{16}COOCH_2$

i) Deduce the physical state of **G** at room temperature. (1)

ii) When completely hydrolysed by heating with aqueous sodium hydroxide, **G** forms an alcohol and the sodium salt of a carboxylic acid. Give the structural formula of the alcohol formed, write a formula for the sodium salt formed and state a use for this salt. (3)

b i) Complete and balance the equation below for the formation of di-ester **H**. (1)

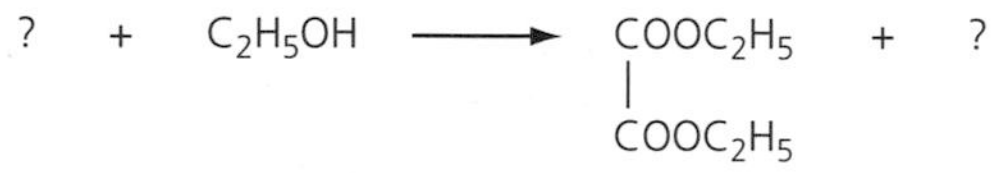

? + C_2H_5OH
$COOC_2H_5$
$COOC_2H_5$
+ ?

H

ii) Identify a substance which could catalyse this reaction to form **H**. (1)

iii) Draw a structural isomer of **H**, also a di-ester, which is formed by the reaction of ethane–1, 2-diol with one other compound. (2)

NEAB CH03 Feb 1998 Q5

9 The polymer poly(methyl 2-methylpropenoate) (Perspex) can be made by a process which involves the following reactions.

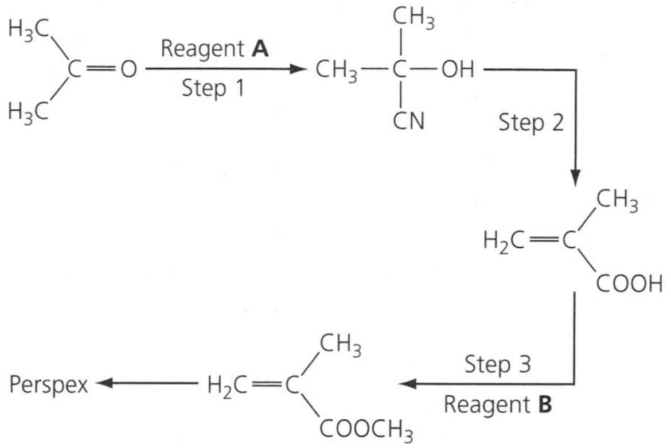

H_3C
C=O
H_3C
Reagent **A**
Step 1
CH_3
CH_3—C—OH
CN
Step 2
CH_3
H_2C=C
COOH
Step 3
Reagent **B**
CH_3
H_2C=C
$COOCH_3$
Perspex

a i) Identify reagent **A**. (1)

ii) Name and outline the mechanism for the reaction in Step 1. (5)

b i) Name reagent **B**. (1)

ii) Name the type of reaction occurring and give a substance which would act as a catalyst for the reaction in Step 3. (2)

c Write an equation for the reaction between the product of Step 2 with sodium hydroxide, showing clearly the structure of the new product. (1)

NEAB CH03 Feb 1997 Q5

10 An ester **A**, is used as a raspberry flavouring in some foods.

a Give **one** common use of esters other than as food flavourings.

b Ester **A** has molecular formula, $C_6H_{12}O_2$, and can be formed, in the presence of a homogeneous catalyst, by the reaction of acid **B** with alcohol **C** shown below.

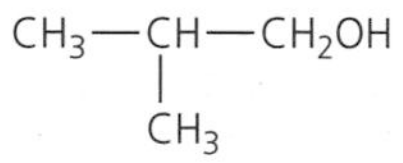

CH_3—CH—CH_2OH
CH_3

C

ii) Identify acid **B** and draw the structure of ester **A**.

iii) Give a suitable homogeneous catalyst for the formation of ester **A** and explain the term homogeneous.

iv) Alcohols **D** and **E** are isomers of alcohol **C**. Alcohol **D** exhibits optical isomerism, and alcohol **E** is unaffected by acidified potassium dichromate(VI) solution. Draw the structures of alcohols **D** and **E**.

NEAB CH03 June 1996 Q3

6 Aromatic chemistry

Victorian merchants made fortunes sailing to eastern countries for supplies of natural dyes such as blue indigo. This dye was mixed with others to make different coloured shades such as in the purple fabric that was much sought after by royalty and the rich. However, the bottom dropped out of the indigo trade when chemists developed methods for making a much wider variety of equally good, if not better, dyes on a large scale from oil-based materials.

Dye molecules are almost always *aromatic compounds*, and the essential feature of most of them is that they contain a benzene ring. This is also true of many pesticides, pharmaceutical drugs, polymers and explosives, which shows the importance of studying the chemistry of benzene specifically and aromatic chemistry in general.

Nowadays, most vivid dyes are synthesised from aromatic compounds and nitrogen compounds

6.1 The structure of benzene

Benzene was first isolated in 1825, but chemists only determined its precise structure in 1931. Despite this, as we can see from Fig. 1, a large number of chemicals based on benzene were made and used in the years between: all the products shown in Fig. 1 were developed before the structure of benzene was fully explained! In some cases the eventual main use of a particular chemical was totally different from that originally intended. For example, TNT, a powerful explosive, was initially developed as a yellow dye.

Until the structure of benzene was fully understood, the development of benzene derivatives and the processes for developing new products were very hit and miss. Nowadays, chemists use detailed structural information and theories to assess whether the production of a new material is feasible before ever trying to produce it in the laboratory.

The first structure for benzene was proposed by the German chemist Friedrich Kekulé. He suggested a ring structure that contained alternate single and double carbon–carbon bonds (Fig. 2). Kekulé's structure was accepted for several years, until experimental evidence was accumulated that could not be explained by his structure. These are some of the anomalies:

1 Kekulé predicted that there would be four isomers of dibromobenzene (Fig. 3), but only three isomers exist: 1,2-, 1,3- and 1,4-dibromobenzene. 1,6-dibromobenzene does not exist as a separate isomer.

1 What, according to Kekulé's original structure for benzene, is the difference between the 1,2- and 1,6-isomers of dibromobenzene?

2 Why didn't Kekulé think that a fifth isomer, 1,5-dibromobenzene, existed?

2 Ethene and other alkenes are reactive molecules, because of their C=C double bonds. If benzene contained three such C=C

74

Fig. 1 The discovery of important benzene derivatives

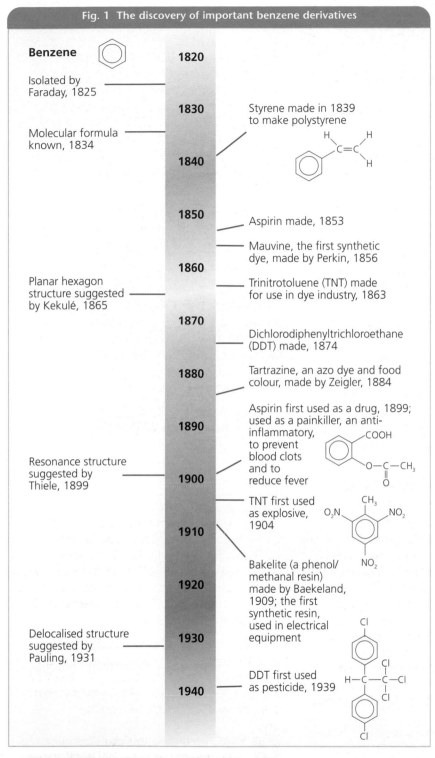

Benzene

Isolated by Faraday, 1825 — 1820

Styrene made in 1839 to make polystyrene — 1830

Molecular formula known, 1834 — 1840

Aspirin made, 1853 — 1850

Mauvine, the first synthetic dye, made by Perkin, 1856 — 1860

Trinitrotoluene (TNT) made for use in dye industry, 1863

Planar hexagon structure suggested by Kekulé, 1865 — 1870

Dichlorodiphenyltrichloroethane (DDT) made, 1874

Tartrazine, an azo dye and food colour, made by Zeigler, 1884 — 1880

Aspirin first used as a drug, 1899; used as a painkiller, an anti-inflammatory, to prevent blood clots and to reduce fever — 1890

Resonance structure suggested by Thiele, 1899 — 1900

COOH

O—C—CH₃

TNT first used as explosive, 1904 — 1910

Bakelite (a phenol/methanal resin) made by Baekeland, 1909; the first synthetic resin, used in electrical equipment — 1920

Delocalised structure suggested by Pauling, 1931 — 1930

DDT first used as pesticide, 1939 — 1940

Fig. 2 The Kekulé structure for benzene

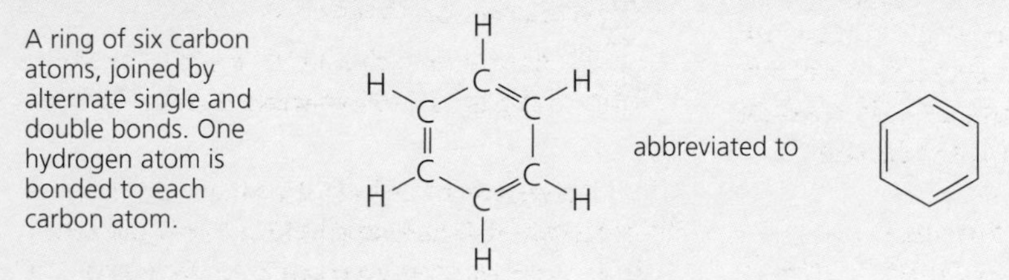

A ring of six carbon atoms, joined by alternate single and double bonds. One hydrogen atom is bonded to each carbon atom.

abbreviated to

Fig. 3 Kekulé's four isomers of dibromobenzene

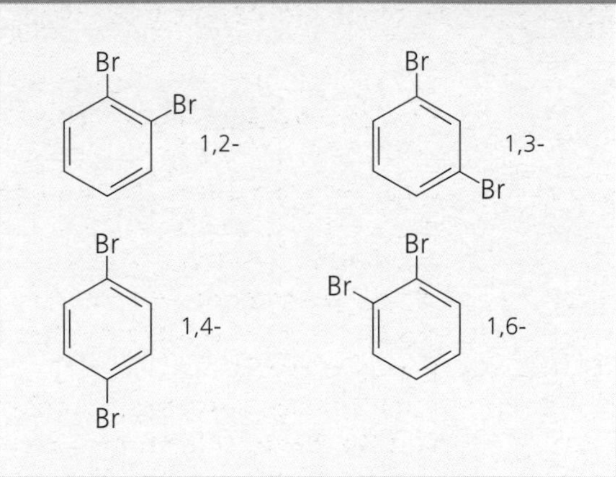

1,2-   1,3-   1,4-   1,6-

double bonds, you would expect it to be a very reactive molecule, but this is not the case. For example, unlike alkenes, benzene does not react readily with an aqueous solution of bromine. In fact, we will discover later in this chapter that benzene reacts only under quite vigorous conditions.

3 The enthalpy change when a C=C bond is hydrogenated (reacted with hydrogen) is -119 kJ mol^{-1} (see cyclohexene in Fig. 4), and therefore the Kekulé structure suggests that, in complete hydrogenation, $-119 \times 3 = -357$ kJ mol^{-1} should be evolved (Fig. 4). However, the hydrogenation of benzene releases only 208 kJ mol^{-1}. For some reason, benzene is 149 kJ mol^{-1} more stable than the Kekulé structure would suggest.

Fig. 4 The hydrogenation of benzene

cyclohexene + H₂ $\Delta H^{\oplus} = -119$ kJ mol^{-1} cyclohexane

benzene + 3H₂ $\Delta H^{\oplus} = -208$ kJ mol^{-1} cyclohexane

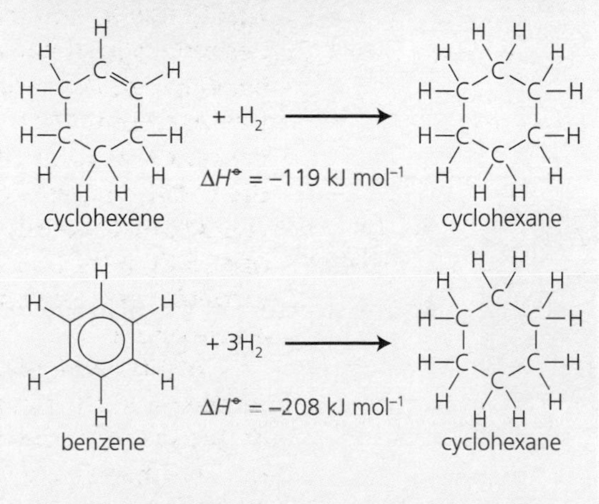

4 When the carbon–carbon bond lengths in benzene are measured by X-ray diffraction techniques they are *all* found to be 0.140 nm. This is intermediate between the lengths of longer C–C (0.154 nm) and shorter C=C bonds (0.134 nm) suggested by Kekulé's

structure (Fig. 5). This final piece of evidence provides a vital clue to a more acceptable structure for benzene.

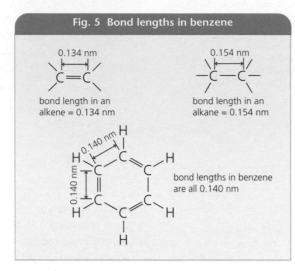

Fig. 5 Bond lengths in benzene

We can see that predictions based on Kekulé's structure do not match the actual properties of benzene. The structure for benzene currently accepted (Fig. 6) is the **delocalised structure** proposed in 1931 by Linus Pauling.

Fig. 6 The currently accepted structure of benzene

Each carbon atom has four bonding, or valence, electrons. Pauling proposed that one electron is used in the bond to the hydrogen atom and two are used in the bonds to the adjacent carbon atoms. The remaining bonding electron from each carbon atom (totalling six) form an electron cloud which is spread, or **delocalised**, evenly above and below the plane of the hexagon of carbons.

This structure for benzene helps to explain the four points which were the problems of Kekulé's structure:

1 The 1,6-dibromobenzene predicted by Kekulé is now identical to the 1,2-isomer.

2 Benzene does not contain C=C double bonds and so does not react in a similar manner to ethene.

3 The delocalised electron structure accounts for the extra stability of the benzene molecule (this has been shown theoretically by Pauling, using quantum mechanics calculations). This extra stability is called the **delocalisation energy**.

4 The delocalisation of electrons accounts for the observed intermediate carbon–carbon bond length. If the six delocalised electrons are 'shared' evenly between the six carbon–carbon bonds, each bond can be thought of (do not take this literally!) as having an extra 'half a bond'.

KEY FACTS

■ The benzene molecule (C_6H_6) is a planar regular hexagon.

■ The structure of the benzene molecule involves six delocalised electrons.

■ Delocalisation of electrons confers stability on the benzene molecule.

6.2 The reactions of benzene

The main reactions of benzene involve the replacement of one or more of the six hydrogen atoms by a functional group. The hydrogen atom is said to be 'substituted' by the reacting functional group. Hence, all such reactions are called **substitution reactions**. Because the delocalised electron system is a region of *high* electron density, such substitution reactions generally involve reaction with electron deficient (often positively charged) chemical species that have the potential to accept electrons to form new covalent bonds. These are called **electrophiles** ('electron lovers'), so this type of reaction is more fully described as **electrophilic substitution**, as opposed to electrophilic additions which are the characteristic reactions of alkenes studied in Chapter 13 of *AS Chemistry*.

One general reaction mechanism can be used to describe all the electrophilic substitution reactions of benzene and benzene derivatives (Fig. 7).

Step 1 The electrophile 'E+' is attracted to the delocalised electron cloud of the benzene ring structure.

Step 2 The electrophile bonds to the benzene ring via two of the six delocalised

electrons, leaving a partially delocalised system containing four delocalised electrons. This temporarily causes the loss of the natural stability of the six-electron delocalised system.

Step 3 A H^+ ion is lost by breaking the C–H bond. The electrons from this bond re-form the stable delocalised six-electron system.

Several different types of electrophiles (E^+) react with benzene (Fig. 8). These electrophiles cannot simply be added to benzene to give the required reaction. They have to be produced in the reaction mixture by mixing appropriate reagents. The reaction conditions are generally severe: we have to use heat, concentrated reagents and catalysts to make these reactions take place. This is because of the high stability (low reactivity) of the delocalised benzene structure.

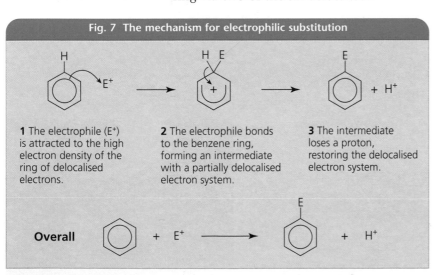

Fig. 7 The mechanism for electrophilic substitution

1 The electrophile (E^+) is attracted to the high electron density of the ring of delocalised electrons.

2 The electrophile bonds to the benzene ring, forming an intermediate with a partially delocalised electron system.

3 The intermediate loses a proton, restoring the delocalised electron system.

Overall

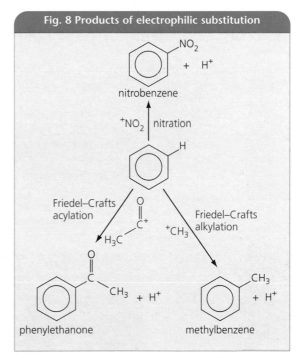

Fig. 8 Products of electrophilic substitution

nitrobenzene

$^+NO_2$ | nitration

Friedel–Crafts acylation

Friedel–Crafts alkylation

$^+CH_3$

phenylethanone

methylbenzene

- Delocalised electrons cause benzene to have a stable structure.

- The delocalised electrons cause benzene to react with electrophiles.

- Benzene usually undergoes substitution reactions.

- Nitration, alkylation and acylation are examples of electrophilic substitutions of benzene.

- Severe reaction conditions and catalysts are required to make these reactions occur.

6.3 Nitration of benzene (E⁺ = NO₂⁺)

When benzene is **nitrated** a nitro group, $-NO_2$, replaces one of the hydrogen atoms. The electrophile is the nitronium ion, NO_2^+, which is generated by mixing concentrated nitric acid and concentrated sulphuric acid, and heating to 50 °C. The overall reaction is represented by the equation in Fig. 9.

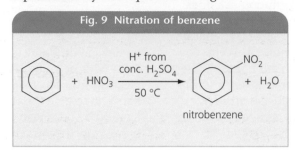

Fig. 9 Nitration of benzene

nitrobenzene

The concentrated sulphuric acid donates a proton to nitric acid. This produces an intermediate, $[H_2NO_3]^+$, which decomposes to yield the electrophilic nitronium ion as shown in Fig. 10.

Fig. 10 Production of nitronium ion

$$H_2SO_4 + HNO_3 \rightleftharpoons [H_2NO_3]^+ + HSO_4^-$$

$$\text{then} \quad [H_2NO_3]^+ \rightleftharpoons NO_2^+ + H_2O$$

The concentrated sulphuric acid also acts as a catalyst in the reaction as it is regenerated when the HSO_4^- ion reacts with the H^+ ion released in the third step of the reaction mechanism (see Fig. 7).

The nitration of benzene and other related compounds is an important industrial reaction. Nitrobenzene is often subsequently converted to phenylamine (aniline or aminobenzene) by reduction of the NO_2 group to an NH_2 group. In industry, iron is used, and in the lab, tin is used with moderately concentrated hydrochloric acid as the reductant. Alternatively, catalytic hydrogenation can be carried out with nickel and hydrogen gas.

3 Write an equation to describe the reduction of nitrobenzene to phenylamine. Use [H] to represent the reductant.

Phenylamine is important because it is widely used in the dye industry. As shown in Fig. 11, it can be converted to

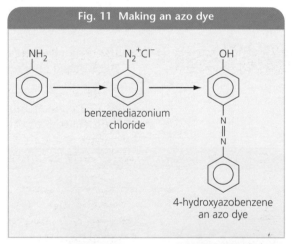

Fig. 11 Making an azo dye

benzenediazonium chloride

4-hydroxyazobenzene an azo dye

Indigo is a traditional dye derived from plants which gradually washes out; it is still popular for dyeing denim. Synthesised azo dyes are 'faster' dyes, less likely to wash out or fade, and more brightly coloured.

benzenediazonium chloride and then to an **azo dye**. These compounds contain two aromatic groups linked by the azo functional group,–N=N–, in a **coupling** reaction. The aromatic groups and the azo group are together called a **chromophore** or **chromophoric group**, and this gives rise to the colour of the dye.

Aromatic nitration is also important in the manufacture of explosives such as 2,4,6-trinitromethylbenzene, better known as 'trinitrotoluene' or TNT. As shown in Fig. 12, this involves substituting three nitro groups on to the benzene ring of methylbenzene (also called toluene).

Fig. 12 Nitro groups in explosives

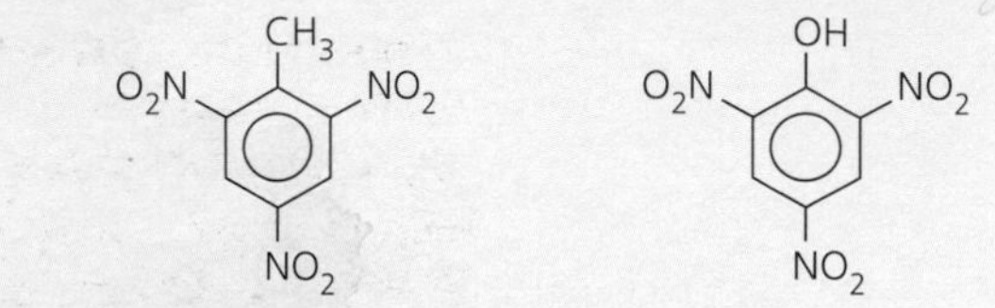

2,4,6-trinitromethylbenzene
(trinitrotoluene, TNT)

2,4,6-trinitrophenol
(picric acid)

Similarly, 2,4,6-trinitrophenol (also known as picric acid) is an explosive that was widely used during the First World War.

6.4 Friedel–Crafts reactions of benzene

The Friedel–Crafts reactions were developed in 1877 by the French chemist Charles Friedel (1832–99) and the American chemist James M. Crafts (1839–1917). They are electrophilic substitution reactions where an alkyl or an acyl group, from a chloroalkane or an acyl chloride respectively, is bonded directly on to a benzene ring. These reactions always use a catalyst such as aluminium chloride or iron(III) chloride. $AlCl_3$ and $FeCl_3$ are Lewis acids and accept a lone pair of electrons from the chloride of the chloroalkane or acyl chloride. As we can see from Fig. 13, this process generates alkyl and acyl carbocations,

which are the electrophiles required for reaction with benzene.

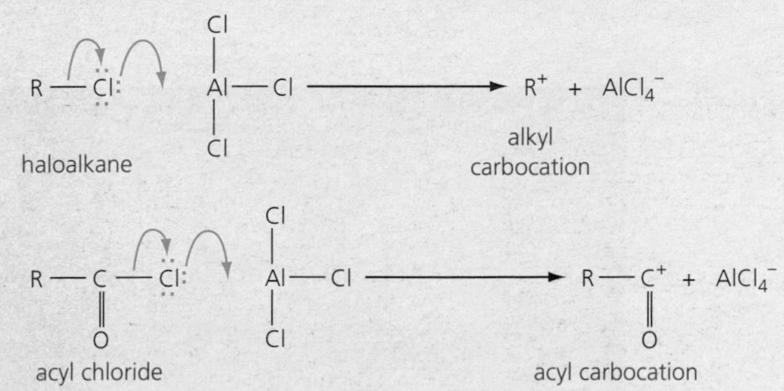

Friedel–Crafts alkylation reactions

A Friedel–Crafts **alkylation** reaction involves a benzene compound, a haloalkane and a catalyst ($AlCl_3$ or $FeCl_3$). As shown in Fig. 13, when $AlCl_3$ or $FeCl_3$ reacts with a haloalkane, an alkyl carbocation is formed, for example $C_2H_5^+$, an ethyl carbocation. It is this carbocation that is responsible for the electrophilic substitution reaction (Fig. 14). The substituted H^+ subsequently regenerates the catalyst.

Fig.15 shows the overall alkylation of benzene using chloroethane. As with the other electrophilic substitution reactions

Fig. 13 Forming alkyl and acyl carbocations

Fig. 14 The electrophilic alkylation of benzene

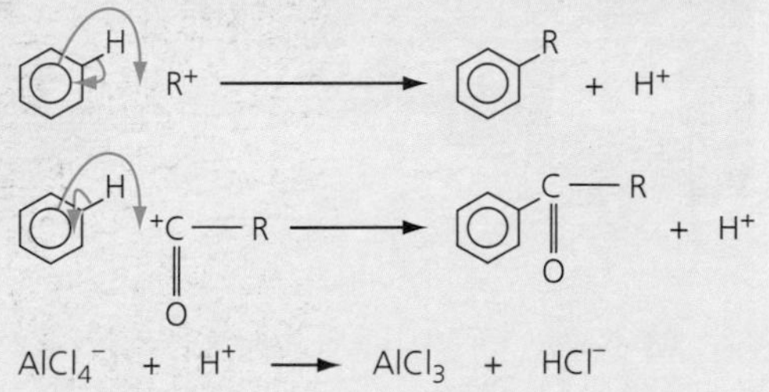

studied in this chapter, vigorous conditions are required, in this case, a catalyst and a temperature of about 80 °C. It is also essential to ensure the reaction mixture is totally free of water because $AlCl_3$ and $FeCl_3$ react vigorously with water.

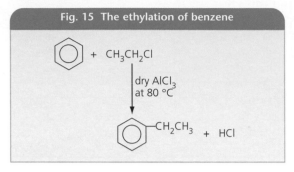

Fig. 15 The ethylation of benzene

Alkylation reactions are used extensively in the petrochemicals industry, and they are summarised in Fig. 16. For example, the reaction between chloromethane and benzene produces methylbenzene (also known as toluene). This can then be nitrated (see Section 5.3) to give 2,4,6-trinitromethylbenzene (trinitrotoluene or TNT, see the Opener and Fig. 12). Very large quantities of ethylbenzene are manufactured for conversion to phenylethene (also known as styrene), the monomer used to manufacture poly(phenylethene), better known as polystyrene.

Although benzene does react directly with chloroethane, commercial production of poly(phenylethene) uses the reaction between benzene and a mixture of ethene and hydrogen chloride with anhydrous $AlCl_3$ as a catalyst (Fig. 17). This approach is preferred for economic reasons because ethene and hydrogen chloride are readily available raw materials. Expanded polystyrene is probably one of the most common packaging and insulating materials in use today.

The alkylation of benzene by propene is known as the cumene process. The product of the reaction is 2-phenylpropane (cumene), which can be oxidised to form phenol and propanone (Figs 16 and 18). Propanone is an important solvent used in medical and cosmetic applications; phenol is used as a coupling agent in the dye-making industry.

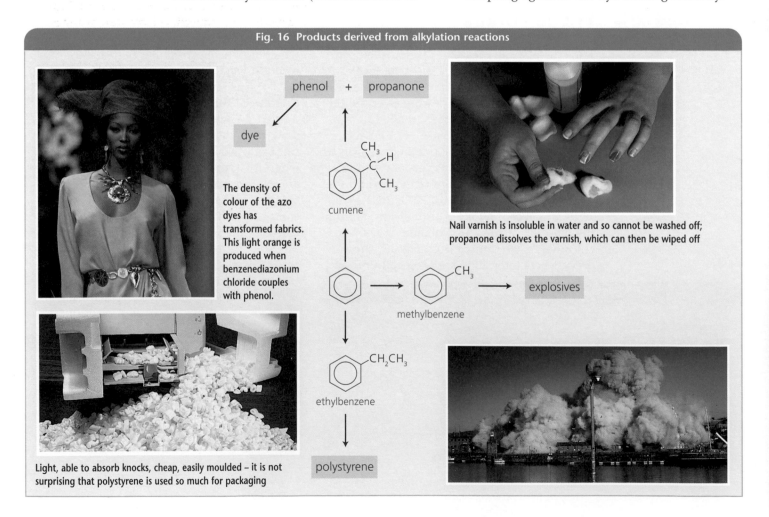

Fig. 16 Products derived from alkylation reactions

The density of colour of the azo dyes has transformed fabrics. This light orange is produced when benzenediazonium chloride couples with phenol.

Nail varnish is insoluble in water and so cannot be washed off; propanone dissolves the varnish, which can then be wiped off

Light, able to absorb knocks, cheap, easily moulded – it is not surprising that polystyrene is used so much for packaging

Fig. 17 Friedel–Crafts alkylation in the manufacture of polystyrene. Stages 1 to 3 are the Friedel-Crafts reactions

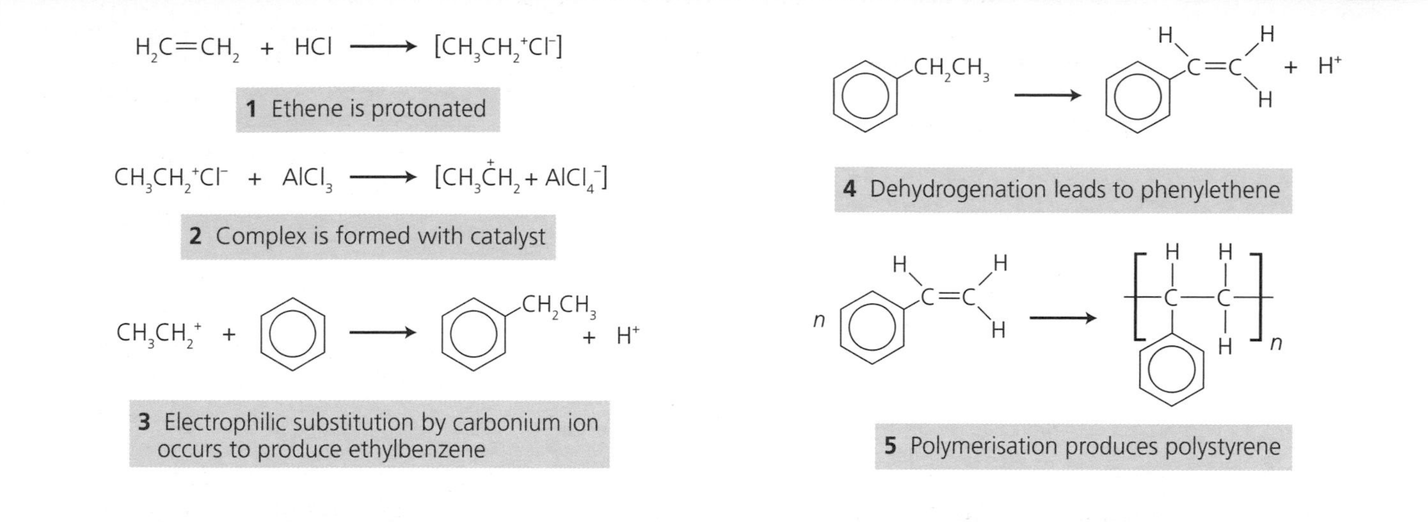

$H_2C{=}CH_2 + HCl \longrightarrow [CH_3CH_2^+Cl^-]$

1 Ethene is protonated

$CH_3CH_2^+Cl^- + AlCl_3 \longrightarrow [CH_3\overset{+}{C}H_2 + AlCl_4^-]$

2 Complex is formed with catalyst

$CH_3CH_2^+ + \bigcirc \longrightarrow$ (CH₂CH₃ benzene) $+ H^+$

3 Electrophilic substitution by carbonium ion occurs to produce ethylbenzene

4 Dehydrogenation leads to phenylethene

5 Polymerisation produces polystyrene

4 Overall, what type of reaction is involved in the cumene process?

5 Write a fully balanced equation for the oxidation of cumene to phenol and propanone.

Friedel–Crafts acylation

Friedel–Crafts **acylation** is similar to the alkylation reaction, but involves the use of acyl halides. Ethylbenzene, for example, can equally well be prepared using an acylation reaction. When benzene reacts with ethanoyl chloride, the product is phenylethanone (Fig. 19). This ketone is then reduced by hydrogen, using a nickel catalyst, to produce ethylbenzene.

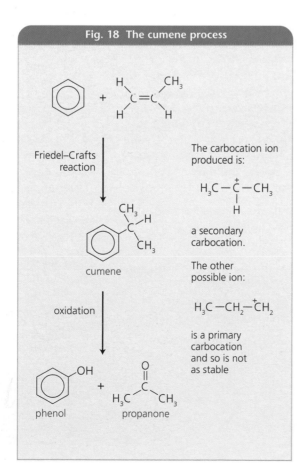

Fig. 18 The cumene process

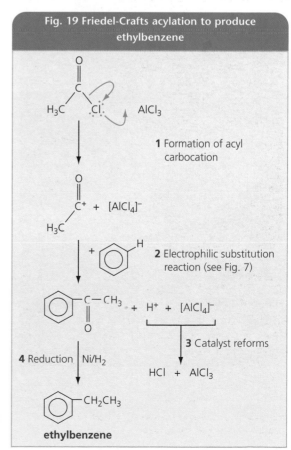

Fig. 19 Friedel-Crafts acylation to produce ethylbenzene

81

6.5 Summary

The reactions of benzene and its derivatives form the basis for the manufacture of many everyday materials. A thorough understanding of the structure of the benzene molecule and the nature of its electrophilic substitution reactions allows industry to create, control and develop new applications as the need arises. Many reaction sequences involve combinations of nitration, coupling, alkylation and acylation reactions, all of which are electrophilic substitution reactions. In this way, a great variety of complex structures containing benzene rings can be constructed.

KEY FACTS

■ The benzene molecule undergoes electrophilic substitution reactions.

■ Nitronium ions, alkyl carbocations and acyl carbocations act as electrophiles towards benzene.

■ Friedel–Crafts alkylation and acylation are catalysed by $AlCl_3$ and $FeCl_3$.

■ The Friedel–Crafts alkylation and acylation reactions use haloalkanes and acyl halides respectively.

■ New carbon–carbon bonds are formed during Friedel–Crafts reactions.

■ Ethylbenzene, used in the manufacture of polystyrene, can be made by both Friedel–Crafts alkylation or acylation with reduction.

APPLICATION A   **Explosives**

The well-known explosive TNT was widely used by armed forces, especially during the First World War, and is still used to demolish buildings and bridges and for blasting in mines and quarries.

The full IUPAC name for TNT is 2,4,6-trinitromethylbenzene. TNT is made by nitrating methylbenzene using 'fuming' (very concentrated) nitric and sulphuric acids at a temperature of over 120 °C. The process proceeds via mono- and dinitro-compounds.

1 Name and draw the structures of the three isomers of mononitromethylbenzene.

2 Name and draw the structures of the two isomers of dinitromethylbenzene which are precursors of TNT.

3 State the type of reaction involved in each of these nitration processes.

4 Give the reaction mechanism for the mononitration of methylbenzene.

5 The mononitration of methylbenzene occurs more easily than the mononitration of benzene. Does this suggest that the methyl group increases or decreases the electron density in the benzene ring?

6 The second and third nitrations of methylbenzene occur much less easily than the first nitration. Does this suggest that a nitro group increases or decreases the electron density in the benzene ring?

7 What will be the product of reacting 4-nitromethylbenzene with iron and hydrochloric acid? Give its name and structure.

8 What type of substance is produced after diazotisation and coupling of the molecule formed in question 7?

APPLICATION B

Alkylating and acylating arenes

In 1877, the chemists Charles Friedel and James Crafts discovered how to alkylate and acylate arenes. They used alkyl halides and acid halides respectively in the presence of so-called Friedel–Crafts catalysts such as aluminium(III) chloride and iron(III) chloride which act as a result of their Lewis acid character (p. 239). The ability to link chain structures to arene rings allowed the development of many new materials particularly useful for thermal insulation and packaging.

Fig. 20 Alkylation and acylation of benzene

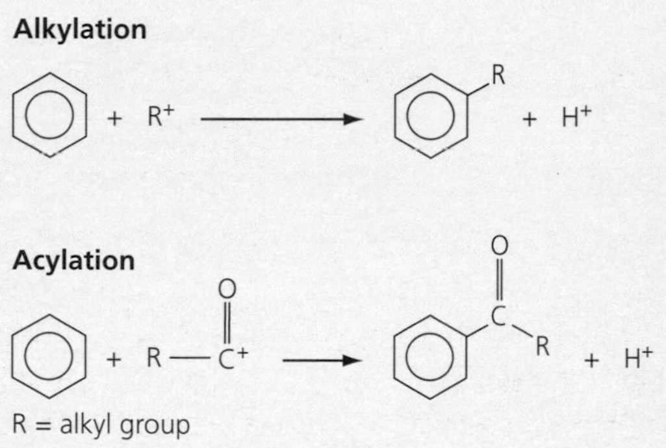

Alkylation

Acylation

R = alkyl group

1 What is a Lewis acid?

2 Draw the structure of one of the products of the reaction between methylbenzene and 1-chloropropane in the presence of anhydrous aluminium chloride.

3 Draw the structure of one of the products of the reaction between ethylbenzene and 2-chloropropane in the presence of anhydrous aluminium chloride.

4 Draw the structure of the product of the reaction between benzene and butanoyl chloride in the presence of anhydrous aluminium chloride.

5 Describe how methylbenzene and ethylbenzene (the reactants used in questions 2 and 3) can be made from benzene.

6 Describe the reaction mechanism for the conversion of benzene to methylbenzene

7 Ethylbenzene is now mostly made by another Friedel–Crafts reaction between benzene and ethene using concentrated phosphoric acid catalyst.
a Write a balanced equation for this reaction
b Ethene is readily available from which industrial process?

8 Ethylbenzene is often converted to phenylethene by heating at about 650 °C in the presence of a iron(III) oxide catalyst
a Write an equation for this reaction.
b Phenylethene is the precursor for which common polymer?
c Show the general repeat structure of this polymer.
d What are the common uses of this polymer?

1

a i) Estimate a value for the enthalpy of hydrogenation of the hypothetical molecule cyclohexa-1,3,5-triene, given that the enthalpy of hydrogenation of cyclohexene is −119.6 kJ mol^{-1}.

ii) The enthalpy of hydrogenation of benzene is −208 kJ mol^{-1}. Explain why this value differs from that you have obtained for cyclohexa-1,3,5-triene.

AQA CH06 Summer 1999 Q1 (part)

2 Consider the following reaction sequence:

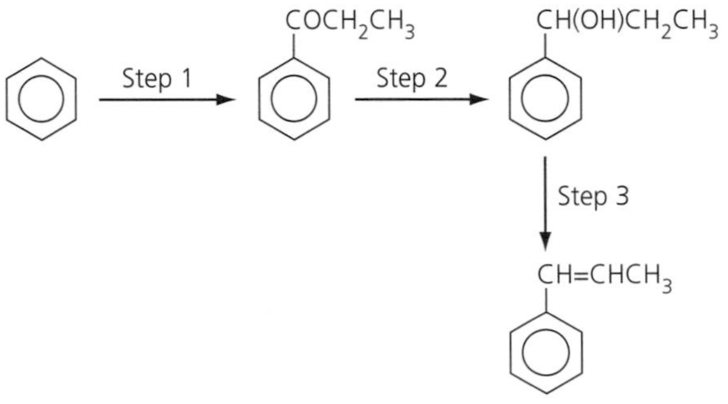

$$\bigcirc \xrightarrow{\text{Step 1}} \overset{COCH_2CH_3}{\bigcirc} \xrightarrow{\text{Step 2}} \overset{CH(OH)CH_2CH_3}{\bigcirc} \xrightarrow{\text{Step 3}} \overset{CH=CHCH_3}{\bigcirc}$$

a For each step, name the type of reaction taking place and suggest a suitable reagent or combination of reagents. (7)

b Outline a mechanism for Step 1. (4)

AQA CH06 Spring 1999 Q2 (part)

3 *N*-Phenylethanamide can be prepared from benzene in three steps:

$$\bigcirc \xrightarrow{\text{Step 1}} \overset{NO_2}{\bigcirc} \xrightarrow{\text{Step 2}} \overset{NH_2}{\bigcirc} \xrightarrow{\text{Step 3}} \overset{NHCOCH_3}{\bigcirc}$$

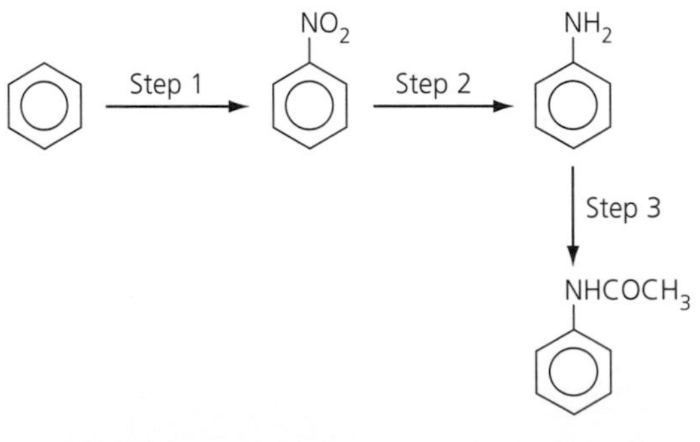

Give the reagents required to carry out Step 1 and write an equation for the formation of the reactive inorganic species present. Name and outline the mechanism for the reaction between this species and benzene. (7)

AQA CH06 Spring 1999 Q1 (part)

4 Ethylbenzene is obtained by the reaction of benzene, in the presence of aluminium chloride, either with chloroethane or with ethene and hydrogen chloride.

a i) Write an equation showing how a reactive species is generated from chloroethane and aluminium chloride.

ii) Write an equation showing how the same reactive species is generated from ethene, hydrogen chloride and aluminium chloride.

iii) Name the type of reaction between the reactive species formed and benzene. Outline a mechanism for this reaction. (7)

b Suggest two reasons why the ethane route is preferred for the industrial manufacture of ethylbenzene. (2)

c Give the structure of the monomer obtained by dehydrogenation of ethylbenzene. Name the polymer obtained from this monomer. (2)

NEAB CH06 Spring 1998 Q1

5 3-Aminophenylethanone can be obtained from benzene in three steps:

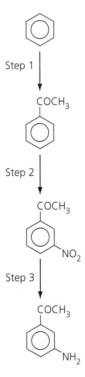

a For each step, name the type of reaction taking place and suggest a suitable reagent or a combination of reagents. (7)

b Write an equation showing how the electrophile is formed from the reagent(s) in Step 2. Outline a mechanism for the subsequent reaction between benzene and this electrophile. (5)

NEAB CH06 Summer 1997 Q1

6 (Phenylmethyl)amine, $C_6H_5CH_2NH_2$, can be prepared from (bromomethyl) benzene, $C_6H_5CH_2Br$, and also from benzenecarbonitrile, C_6H_5CN.

a Write an equation for the conversion of (bromomethyl) benzene into (phenylmethyl)amine. Name the type of reaction taking place.

b Name the type of reaction involved in the conversion of benzenecarbonitrile into (phenylmethyl)amine. Write an equation for this reaction and suggest a suitable reagent or a combination of reagent and catalyst.

NEAB CH06 June 1999 Q2a

7 Amines

Anaesthetising a patient during surgery involves more than just making the patient unconscious. An important aspect of the process is muscle relaxation, which enables the surgeon to operate more easily and helps the patient to breathe. In the past, large amounts of anaesthetic were used, sometimes not far short of a lethal dose. Nowadays, small amounts of specific muscle-relaxing drugs are used. This has made surgery much safer.

The first muscle relaxant used in surgery was curare. For centuries South American Indians have extracted curare from the bark of a tree and used it when hunting. The extract is applied to the tips of arrows and works by paralysing the prey. Chemists have extracted, purified and determined the structure of the main active compound in curare. They showed it to be derived from organic relatives of ammonia: the *amines* and *quaternary ammonium salts*. Many similar compounds have now been designed and synthesised for use in modern-day surgery.

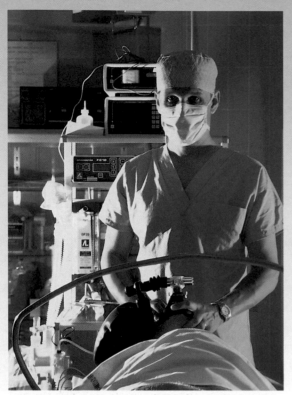
Modern anaesthesia uses a sophisticated range of chemicals and techniques to help the surgeon operate safely

A hunter using an arrow tipped with curare, a muscle relaxant

7.1 The structure of ammonia and amines

An ammonia molecule consists of a nitrogen atom bonded to three hydrogen atoms (Fig. 1). In addition to these three covalent bonds, the nitrogen atom also has a lone pair of electrons. This structure is the starting point for the study of the chemistry of ammonia and its derivatives, the amines. The lone pair on the nitrogen atom is responsible for the two main properties of ammonia: its tendency to act as a **nucleophile** and its action as a **Brønsted–Lowry base**.

There are three types of amines: **primary**, **secondary** and **tertiary** (Fig. 2). These are derived from ammonia when one, two or all three hydrogen atoms are replaced by alkyl or aryl groups. All amines have a lone pair of electrons and so, like ammonia, have nucleophilic and basic properties.

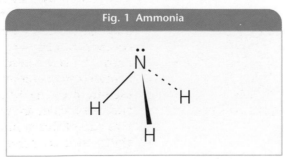
Fig. 1 Ammonia

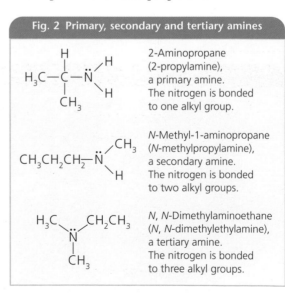

Fig. 2 Primary, secondary and tertiary amines

2-Aminopropane (2-propylamine), a primary amine. The nitrogen is bonded to one alkyl group.

N-Methyl-1-aminopropane (N-methylpropylamine), a secondary amine. The nitrogen is bonded to two alkyl groups.

N, N-Dimethylaminoethane (N, N-dimethylethylamine), a tertiary amine. The nitrogen is bonded to three alkyl groups.

Fig. 3 Propanolol

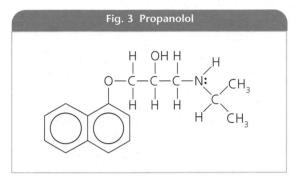

Fig. 4 Quaternary ammonium ions

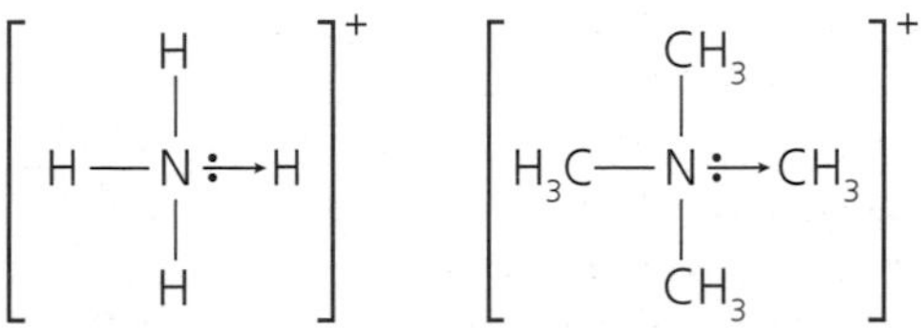

$$\left[\begin{array}{c} H \\ | \\ H-N\!\!\rightarrow\!\!H \\ | \\ H \end{array} \right]^{+} \qquad \left[\begin{array}{c} CH_3 \\ | \\ H_3C-N\!\!\rightarrow\!\!CH_3 \\ | \\ CH_3 \end{array} \right]^{+}$$

ammonium ion   tetramethylammonium ion, a quaternary ammonium ion

Fig. 5 Compounds that act on nerves

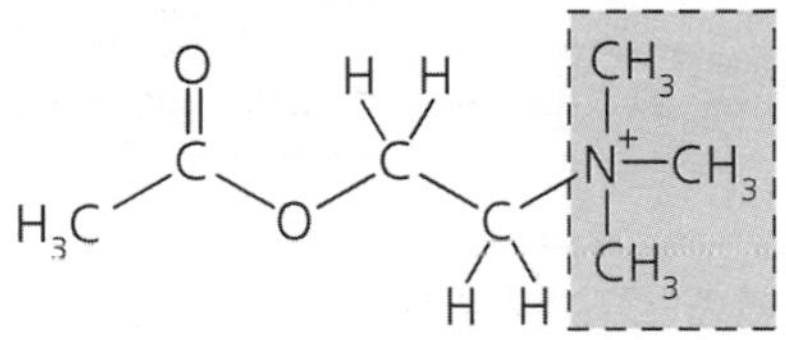

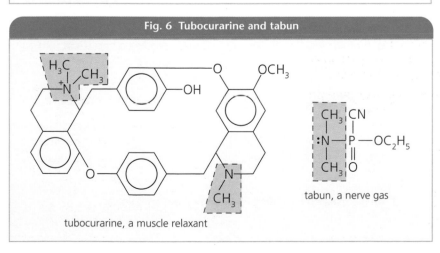

acetylcholine

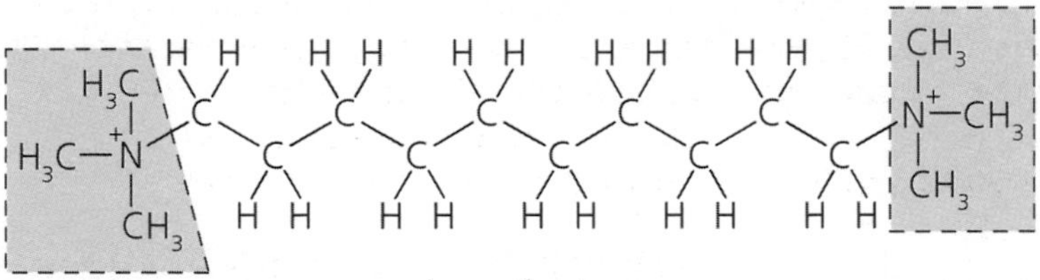

decamethonium

Fig. 6 Tubocurarine and tabun

tubocurarine, a muscle relaxant

tabun, a nerve gas

Naming secondary and tertiary amines can be a complicated business (see Chapter 4 for naming of organic compounds). First we must identify the longest hydrocarbon chain attached to the nitrogen. This, with the ending 'amine' or the prefix 'amino', forms the basis for the name. Then we identify the other smaller hydrocarbon groups attached to the nitrogen and list these at the beginning of the name. The prefix *N*– is used to show that the smaller groups are attached to the main chain via the nitrogen atom.

1 Draw the structures of:

a *N*–ethylbutylamine (or *N*–ethyl–1–aminobutane)

b *N*–ethyl–*N*–methylpropylamine (or *N*–ethyl–*N*–methyl–1–aminopropane)

c Classify each as primary, secondary or tertiary.

Many ammonium compounds are used to stimulate or suppress the action of nerves that control our muscles. The primary amine 2-aminopropane (see Fig. 2) is used to manufacture the drug propanolol (Fig. 3), a common heart stimulant used to control irregular heartbeat patterns.

Quaternary ammonium salts (Fig. 4) are produced from tertiary amines when the nitrogen atom's lone pair of electrons forms a dative (coordinate) bond to a fourth hydrocarbon group. Hence, a quaternary ammonium salt is like an ammonium ion (NH_4^+) where alkyl or aryl groups have replaced all four hydrogens. Like ammonium salts, they are crystalline, ionic solids.

Many compounds that affect the nervous system contain a quaternary ammonium group. Acetylcholine (Fig. 5), which helps in the transmission of nerve signals to muscle cells, contains one such group. Decamethonium (Fig. 5), a drug used as a muscle relaxant, contains two.

Curare is used by the inhabitants of South American forests on the tips of the arrows they use to hunt for food. The main component of curare, tubocurarine (Fig. 6), is both a tertiary amine and a quaternary ammonium compound and is still widely used in medicine. Some nerve gases (Fig. 6) and insecticides are also amines, and the presence of this group is fundamental to their ability to affect nerve action.

7.2 Nucleophilic properties of amines

Preparing aliphatic amines

Aliphatic primary amines can be made by reacting ammonia with an appropriate haloalkane. This must have the same number of carbon atoms as the amine we are trying to make, and the halogen atom(s) must be positioned on the carbon atoms to which the amino group(s) need to be bonded. Excess ammonia is dissolved in ethanol and heated with the haloalkane in a sealed vessel. Using a sealed vessel leads to an increased pressure, which promotes the reaction.

2 Based on the carbon chain of each of the following primary amines, which haloalkane could be used to make:

a ethylamine (or aminoethane)?

b 1–methylethylamine (or 2–aminopropane)?

c hexane-1,6-diamine (or 1,6-diaminohexane), a substance used to make nylon?

The formation of a primary amine from the reaction between a haloalkane and excess ammonia involves three stages (see Fig. 7):

1 The lone pair on the nitrogen atom of the ammonia molecule is attracted to the $\delta+$ charge on the carbon atom of the polar carbon–halogen bond.

2 The lone pair forms a covalent bond between the nitrogen atom and the carbon atom.

3 The amine is released from this salt by the removal of a proton. This can be caused by the excess of ammonia or by adding sodium hydroxide solution.

The overall reaction is a **nucleophilic substitution** of the halogen atom in the haloalkane by ammonia.

Chloro- and bromoalkanes are preferred as sources of the hydrocarbon groups because they are more readily available than iodoalkanes

Nylon is an extremely versatile material. Because of its high strength, a paradescender can use nylon ropes when climbing a mountain and a nylon parachute when returning to the ground

---

Fig. 7 Reaction between a haloalkane and ammonia

1 Attraction of ammonia's lone pair to $\delta+$ charge on carbon atom of the polar carbon–halogen bond (R is alkyl group, C_nH_{2n+1}).

$$H - \overset{\overset{\displaystyle H}{|}}{\underset{\underset{\displaystyle H}{|}}{N}}: \qquad \overset{\delta+}{R} - \overset{\delta-}{Br}$$

$$H - \overset{\overset{\displaystyle H}{|}}{\underset{\underset{\displaystyle H}{|}}{N}}: \overset{\delta+}{R} - \overset{\delta-}{Br} \longrightarrow \left[H - \overset{\overset{\displaystyle H}{|}}{\underset{\underset{\displaystyle H}{|}}{N}} - R \right]^+ Br^-$$

2 Covalent bond forms between nitrogen and carbon, with release of bromide ion. An alkylammonium salt is produced.

$$\left[H - \overset{\overset{\displaystyle H}{|}}{\underset{\underset{\displaystyle H}{|}}{N}} - R \right]^+ Br^- + NaOH \longrightarrow \overset{H}{\underset{H}{}}\ddot{N} - R + H_2O + NaBr$$

3 Amine released by adding alkali to alkyl ammonium salt.

---

and are more reactive than fluoroalkanes. For example, 2-phenylethylamine (or 1-amino-2-phenylethane), the primary amine responsible for the common yearning for chocolate, can be made by reacting 1-bromo-2-phenylethane with excess ethanolic ammonia (Fig. 8).

Fig. 8 The 'chocolate' amine

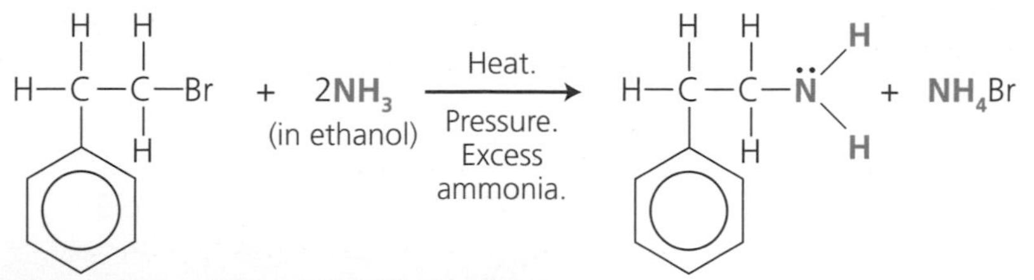

2-phenylethylamine
(2-phenylaminoethane)

It is not possible to guarantee that only one hydrogen on ammonia will be substituted in an alkylation reaction. Like ammonia, primary, secondary and tertiary amines have a lone pair of electrons. This allows repeated nucleophilic substitution by the haloalkane. Alkyl groups are electron-repelling, and so increase the electron density on the nitrogen they bond to. This in turn means that the product of each alkylation reaction is a better nucleophile than the starting material. As a result, mixtures of primary, secondary and tertiary amines and quaternary ammonium salts will always be produced (Fig. 9).

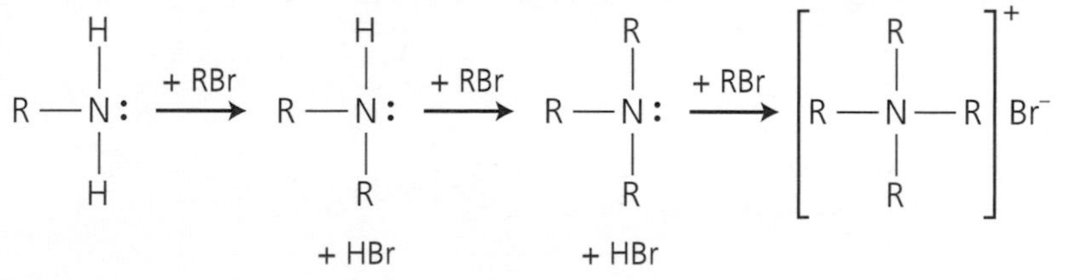

3 Give the names and structures of the other amines which could be produced when trying to make ethylamine (or aminoethane) using bromoethane and ammonia.

We can adjust the composition of the initial reaction mixture to favour a particular type of amine. Excess ammonia favours the production of primary amines because it is less likely that a second haloalkane molecule will react with a primary amine when there are a large number of unreacted ammonia molecules available. An excess of the haloalkane, to ensure that each ammonia molecule reacts with four haloalkane molecules, encourages the formation of quaternary ammonium salts.

Quaternary ammonium salts can be manufactured using long-chain (e.g. C_{12}) haloalkanes (Fig. 10). These compounds can be used as **cationic surfactants** because the positive charge on the nitrogen atom of the quaternary ammonium ion is attracted to negatively charged surfaces such as glass, hair, fibres, metals and plastics. This gives rise to applications in fabric conditioners, leather softeners, sewage flocculants (bringing particles to the surface), corrosion inhibitors, hair conditioners, anti-static agents and emulsifiers. Some quaternary ammonium salts with shorter carbon chains are also used in disinfectants where they combine surfactant action with germicidal properties.

Fig. 9 Successive substitutions of an amine

primary   secondary   tertiary   quaternary

Fig. 10 A cationic surfactant

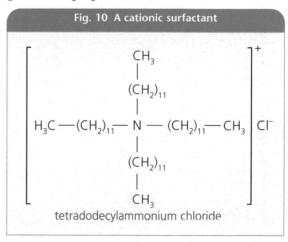

tetradodecylammonium chloride

7.3 Preparing aliphatic and aromatic primary amines

Preparing primary aliphatic amines

The production of primary amines from haloalkanes is not efficient because mixtures containing secondary and tertiary amines and quaternary ammonium salts are inevitably produced. The way around this problem is to introduce the $-NH_2$ group indirectly into a molecule.

The first step is to reflux an appropriate haloalkane, such as bromoethane, with a solution of potassium cyanide dissolved in a mixture of water and ethanol. The lone pair of the cyanide ion ($:CN^-$) allows nucleophilic substitution of the halogen atom to take place and introduces the cyanide group ($-CN$) into the organic molecule. Such organic cyanides are called **nitriles**. For example, refluxing bromoethane with potassium cyanide yields propanenitrile (Fig. 11). Note that this procedure allows the carbon chain to be extended by one carbon atom.

The nitrile group contains a carbon–nitrogen triple bond, and it is this bond that forms the basis of the amine group. The triple bond undergoes an addition reaction with two molecules of hydrogen, giving the primary amine. This addition of hydrogen (reduction) can be achieved by using lithium tetrahydridoaluminate ($LiAlH_4$) dissolved in dry ethoxyethane, followed by aqueous hydrolysis. An alternative reducing agent for this reaction is hydrogen gas with a nickel catalyst. For example, propanenitrile is reduced to propylamine (1-aminopropane) as shown in Fig. 12. Aliphatic primary amines can also be made by reducing nitro compounds and amides.

4

a Devise a reaction sequence to prepare 2-methylbutylamine (or 1-amino-2-methylbutane), stating which haloalkane you will use.

b Can 1-methylethylamine (2-aminopropane) be prepared by making and reducing a nitrile? If it can, give the equations; if it can't, explain why not.

Preparing primary aromatic amines

Primary aromatic amines are compounds that contain an $-NH_2$ functional group directly bonded to the benzene ring. The simplest example of a primary aromatic amine is phenylamine (or aminobenzene) as shown in Fig. 13. The methods we used to make aliphatic amines do not work for aromatic amines. For example, reacting bromobenzene with ethanolic ammonia does not produce a reasonable yield of phenylamine.

Fig. 11 Nucleophilic substitution of bromoethane by cyanide

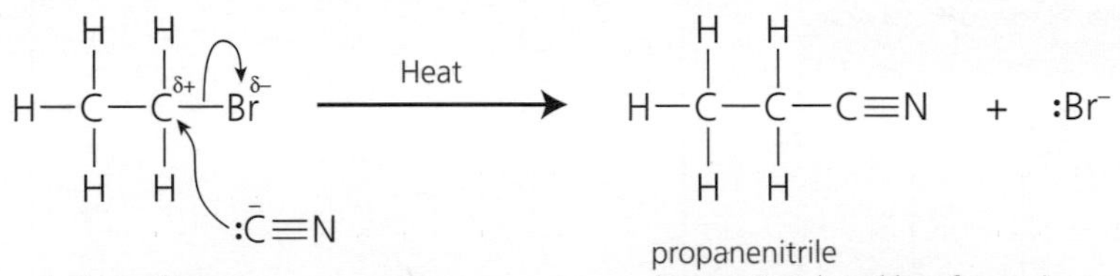

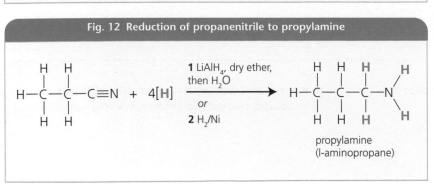

bromoethane

propanenitrile
(a new C–C bond has formed, extending the carbon chain)

Fig. 13 Phenylamine

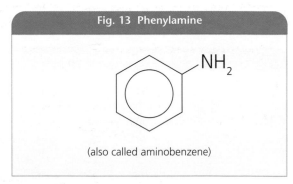

(also called aminobenzene)

There are two reasons why halobenzenes do not react strongly with ammonia:

1 The nitrogen atom in ammonia is nucleophilic, but benzene usually reacts with electrophiles (see Chapter 6) – the high electron density of the delocalised electron cloud of the benzene ring repels electron-rich nucleophiles such as ammonia.

2 The lone pairs of the halogen atom are delocalised towards the benzene ring (Fig. 14).

Fig. 12 Reduction of propanenitrile to propylamine

$$H-\overset{\overset{\displaystyle H}{|}}{\underset{\underset{\displaystyle H}{|}}{C}}-\overset{\overset{\displaystyle H}{|}}{\underset{\underset{\displaystyle H}{|}}{C}}-C\equiv N \;+\; 4[H] \quad \xrightarrow[\substack{or \\ \textbf{2}\ H_2/Ni}]{\substack{\textbf{1}\ LiAlH_4,\ dry\ ether, \\ then\ H_2O}} \quad H-\overset{\overset{\displaystyle H}{|}}{\underset{\underset{\displaystyle H}{|}}{C}}-\overset{\overset{\displaystyle H}{|}}{\underset{\underset{\displaystyle H}{|}}{C}}-\overset{\overset{\displaystyle H}{|}}{\underset{\underset{\displaystyle H}{|}}{C}}-N\overset{\displaystyle H}{\underset{\displaystyle H}{\big\langle}}$$

propylamine
(1-aminopropane)

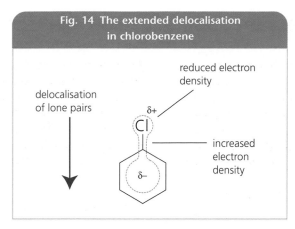

Fig. 14 The extended delocalisation in chlorobenzene

delocalisation of lone pairs

reduced electron density

$\delta+$

Cl

increased electron density

$\delta-$

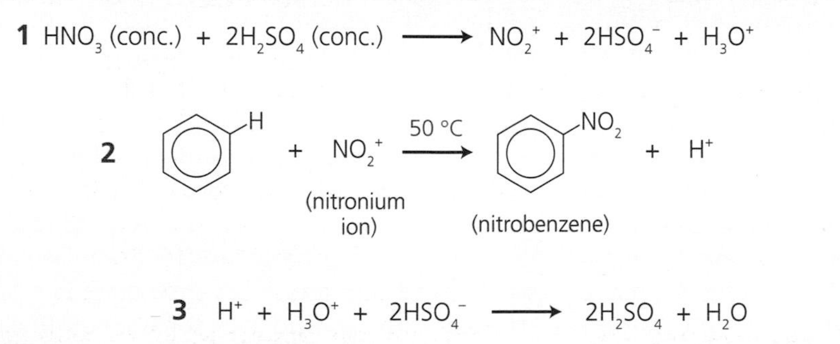

The nitro group is then reduced to a primary amine group using an acid together with a metal (Fig. 16). On an industrial scale, scrap iron and hydrochloric acid are used, but tin and hydrochloric acid is the preferred combination in the laboratory. Since excess acid is present, the product is the protonated form of phenylamine (or aminobenzene), $C_6H_5NH_3^+$. The amine is released from this ion by addition of a base such as sodium hydroxide; this removes the proton from the $C_6H_5NH_3^+$ ion.

The delocalisation of the halogen lone pairs has the effect of increasing the electron density of the carbon–halogen bond, which, in turn, reduces the polarity of the bond. Consequently, the carbon atom is less attractive to the nucleophilic lone pair in an ammonia molecule. The increased electron density of the bond also makes it shorter and stronger than the corresponding bond in a haloalkane. This means the bond is much more reluctant to break and, consequently, the substitution of the halogen by ammonia is much less likely.

As a result of these two factors, preparation of aromatic amines has to take a roundabout route. The nitrogen atom is substituted into an aromatic structure in the form of a nitro ($-NO_2$) group (Fig. 15; see also Chapter 6).

Fig. 16 Reduction of nitrobenzene

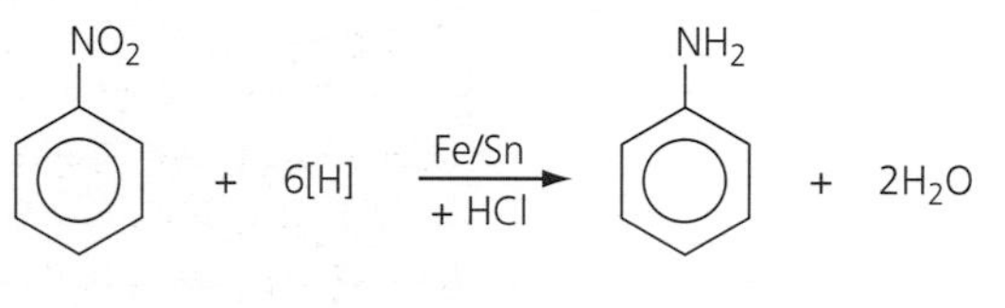

NO_2 + 6[H] $\xrightarrow[\text{+ HCl}]{\text{Fe/Sn}}$ NH_2 + $2H_2O$

The reduction can also be achieved using hydrogen gas and a nickel catalyst:

$$PhNO_2 + 3H_2 \xrightarrow{\text{Ni catalyst}} PhNH_2 + 2H_2O$$

where Ph stands for C_6H_5.

The conversion of aromatic nitro groups to aromatic amines is vitally important in the dye industry. It is used to introduce primary amine groups into a wide variety of aromatic molecules used for making dyes or into dye molecules themselves, causing changes in colour and ability to bond to fabrics.

Fig. 15 The nitration of benzene

1 HNO_3 (conc.) + $2H_2SO_4$ (conc.) $\longrightarrow$ NO_2^+ + $2HSO_4^-$ + H_3O^+

2 (benzene with H) + NO_2^+ $\xrightarrow{50\,°C}$ (nitrobenzene with NO_2) + H^+

(nitronium ion)   (nitrobenzene)

3 H^+ + H_3O^+ + $2HSO_4^-$ $\longrightarrow$ $2H_2SO_4$ + H_2O

KEY FACTS

■ Primary aliphatic amines are best produced by the reduction of nitriles.

■ Primary aromatic amines are prepared by the reduction of aromatic nitro compounds.

7.4 The base properties of ammonia and amines

Quaternary ammonium compounds such as tubocurarine (the main active component of curare, see Fig. 6) and other muscle relaxants (see Fig. 5) are ionic compounds. Animals killed using curare are safe to eat because the ionic quaternary ammonium salts are not absorbed from the gut into the body. As ammonia and amines are bases they can accept protons (H^+) from other molecules and also form positively charged ions.

In aqueous solution, ammonia and amine molecules can accept a proton from a water molecule, producing an ammonium ion (NH_4^+) or a substituted ammonium ion ($R-NH_3^+$) respectively, along with a hydroxide (OH^-) ion (Fig. 17). The presence of OH^- ions in the solution means that the solution is alkaline. Ammonia and amine molecules can accept protons because the lone pair of electrons on the nitrogen atom is available to form a dative (coordinate) bond with a proton. However, the reactions are equilibrium reactions (Fig. 17), in other words ammonia and amines are weak bases.

Fig. 17 Amines as weak bases

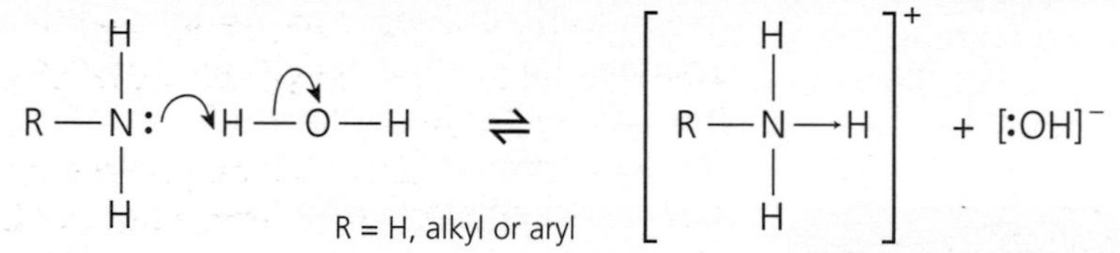

$$R-N:H-O-H \rightleftharpoons \left[R-N-H \right]^+ + [:OH]^-$$

R = H, alkyl or aryl

The degree of reaction of ammonia or an amine with water is an indication of the strength of the amine as a base. If we compare the basic properties of equimolar solutions of ammonia, primary aliphatic amines and a primary aromatic amine such as phenylamine (or aminobenzene) (see Table 1), we find that amines are all weak bases: they are not completely converted to ions by reaction with water. Table 1 also shows that an amine's base strength depends on the nature of the group attached to nitrogen. The base strength of primary amines is increased relative to ammonia by alkyl groups but decreased by aromatic groups. For comparison, Table 1 also contains data on sodium hydroxide, a strong base fully ionised in dilute solution.

Table 1 Comparison of different amines as bases		
Substance	pH of 1 mol dm^{-3} solution	% of molecules reacted in solution
ammonia	11.63	0.42
methylamine (aminomethane)	12.32	2.08
ethylamine (aminoethane)	12.35	2.25
phenylamine (aminobenzene)	9.32	2.08×10^{-3}
NaOH	14.00	100

In aromatic amines, the nitrogen lone pair is delocalised towards the benzene ring in much the same way as shown in Fig. 14 for chlorobenzene. Overlap occurs between the lone pair and the delocalised electron system of the benzene ring. The electron density on the nitrogen atom is lowered and this reduces the ability of nitrogen to accept a proton.

Alkyl groups, however, are electron repelling. They push electron density away from the alkyl group towards the amine group. Because of this, the electron density on the nitrogen atom increases, increasing its ability to accept a proton.

This difference in base strength accounts for the need to use dilute acid to dissolve aromatic amines in water. Acid forces the equilibrium shown in Fig. 17 to move to the right (by removing OH^-), causing the amine to dissolve because it is converted to its ionic salt. In contrast, provided the alkyl group is not too large, aliphatic amines are relatively soluble in water alone because there is a stronger natural tendency for the equilibrium (Fig.17) to move to the right.

KEY FACTS

■ Ammonia and amines act as bases because of the lone pair of electrons on the nitrogen atom.

■ Aliphatic primary amines are stronger bases than ammonia.

■ Aromatic primary amines are weaker bases than ammonia.

7.5 Acylation of amines and other molecules

Acyl groups (Fig. 18) can be introduced into many molecules by acyl chlorides (also known as acid chlorides) or acid anhydrides, which are known as **acylating agents**.

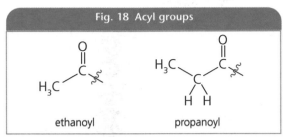

Fig. 18 Acyl groups

ethanoyl   propanoyl

Fig. 19 Ethanoic acid derivatives

carboxylic acid   acyl chloride   acid anhydride

ethanoic acid   ethanoyl chloride   ethanoic anhydride

Fig. 20 Nucleophilic addition-elimination of an acyl compound

addition

G = Cl or OCOCH₃

elimination

deprotonation

Acyl chlorides and acid anhydrides (Fig. 19) are derivatives of carboxylic acids. Acyl chlorides are derived from carboxylic acids by substitution of the –OH group by a chlorine atom. Acid anhydrides are derived by substitution of the –OH group by an alkanoate. Hence, the acyl chloride and acid anhydrides of ethanoic acid are ethanoyl chloride and ethanoic anhydride (Fig. 19) respectively.

 Draw the structures of benzoyl chloride and benzoic anhydride.

Reactions in which acyl groups are introduced into molecules are called **acylations**. Acylation reactions have many uses, for example in the pharmaceutical and textile industries. Acylation reactions are used to make drugs such as aspirin and textiles such as cellulose acetate. In general, acid anhydrides make better acylating agents than acyl chlorides. The anhydrides are cheaper to produce and, as they are less reactive, their reactions can be more easily controlled. They are also less corrosive, as acyl chlorides produce hydrogen chloride gas (HCl) when they react. For example, the industrial processes to produce aspirin and cellulose acetate use ethanoic anhydride rather than ethanoyl chloride.

Acyl chlorides and acid anhydrides are both extremely reactive towards nucleophiles. The electronegative oxygen atoms and/or chlorine atoms cause the carbon atom of an acyl group to be electron deficient and therefore open to attack by nucleophiles such as ammonia. This results in an **addition–elimination reaction** as shown in Fig. 20 for the ethanoylation of ammonia. The addition-elimination reaction occurs in four stages:

1 Attraction: the δ+ carbon atom of the polar C=O bond attracts the lone pair of a nucleophile such as ammonia.

2 Addition: the lone pair of the ammonia molecule forms a new bond to the carbon atom, producing a saturated ion which has both positive and negative charges.

3 Elimination: the C–Cl of the acyl chloride or the C–O of the anhydride breaks, liberating a chloride ion (Cl⁻) or ethanoate (CH₃COO⁻), as simultaneously, the C=O bond re-forms.

4 Deprotonation: the new compound is formed by removal of a proton by basic ammonia.

The reaction is a nucleophilic addition–elimination, but overall it looks like a nucleophilic substitution. The nucleophile is said to be acylated, i.e. an acyl group has replaced a hydrogen atom in the nucleophile. Ammonia and primary amines are suitable nucleophiles, as are water and alcohols.

All these nucleophiles have lone pairs, associated with an oxygen atom or a nitrogen atom, *and* a hydrogen atom directly bonded to that oxygen atom or nitrogen atom.

Acylation of ammonia (NH_3) or a primary amine ($R–NH_2$) produces an **amide** ($R–CO–NH_2$). Similarly, acylation of an alcohol ($R'–OH$) produces an **ester** ($R–CO–OR'$). These are shown in Figs. 21 and 22.

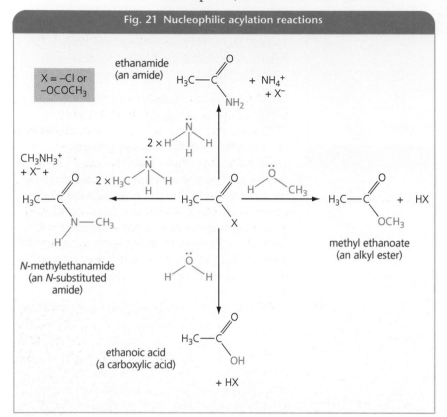

Fig. 21 Nucleophilic acylation reactions

X = –Cl or –OCOCH₃

ethanamide (an amide)

N-methylethanamide (an N-substituted amide)

methyl ethanoate (an alkyl ester)

ethanoic acid (a carboxylic acid)

6 Give the name and structure of the principal organic products from each of the following reactions:

a benzoic anhydride with water;

b propanoyl chloride with ammonia;

c butanoic anhydride with ethylamine (aminoethane).

Write an equation for each reaction.

Fig. 22 Amides and esters

Amides

Esters

Fig. 23 Synthesis of aspirin

phenol

CO₂ Kolbe process

2-hydroxybenzoic acid (salicylic acid)

ethanoic anhydride

aspirin

ethanoic acid

Extracts from willow bark or leaves were found to relieve pain and lower fever as long ago as 400 BC. The active ingredient in the extract was eventually shown to be 2-hydroxybenzoic acid (salicylic acid, see Fig. 23). Obtaining this directly from the willow tree is not practicable for several reasons: supplies are limited, the concentration of salicylic acid in the woody tissues varies with the season of the year, and any extract is inevitably impure.

The German chemist Adolf Kolbe (1818–84) discovered that phenol could be converted to salicylic acid by reaction with carbon dioxide. This provided the basis for a relatively easy and cheap manufacturing process, and the Bayer chemical company of Germany was soon producing large quantities of the 'new' drug. However, it soon became evident that salicylic acid caused unacceptable side-effects, including severe irritation of the mouth, throat and stomach.

Another German chemist, August Hofmann (1818–92), prepared several derivatives of salicylic acid. His aim was to find one with similar pain-relieving properties but no side-effects. Acylation of the phenolic –OH group with ethanoic anhydride finally proved to be the answer, producing the compound known as aspirin (Fig. 23). Aspirin was first used as a drug in 1899. Ironically, aspirin had first been made in 1853, but its value as a drug remained hidden for almost fifty years.

7 Give three reasons why the manufacture of aspirin uses ethanoic anhydride as the acylating agent, rather than ethanoyl chloride.

Morphine, a derivative of opium, is a powerful painkiller but it is also highly addictive. As shown in Fig. 24, it contains two –OH groups. If both of these groups are acylated using ethanoic anhydride, heroin is produced. Heroin is even more effective as a painkiller, but unfortunately it is also even more addictive. Codeine has been developed as a compromise; one of the –OH groups is methylated, rather than acylated, resulting in an effective painkiller which doesn't carry the same risk of addiction.

Fig. 24 Derivatives of morphine

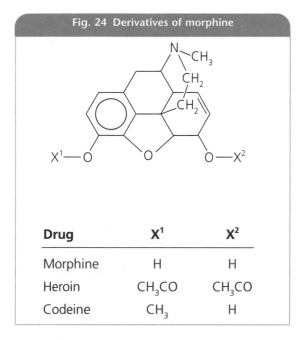

Drug	X^1	X^2
Morphine	H	H
Heroin	CH_3CO	CH_3CO
Codeine	CH_3	H

Codeine, a valuable painkiller, and heroin, a class A controlled substance, can both be made from opium poppies

The results of another victory in the continuing war against traffickers and the misery they cause

KEY FACTS

- Acylation reactions can be carried out using ethanoyl chloride or ethanoic anhydride.

- Industrial acylation reactions are usually carried out using ethanoic anhydride.

- Nucleophilic molecules containing :N–H or :O–H bonds are readily acylated.

- Acylation reactions are used in a wide variety of industrial processes, for example producing aspirin.

APPLICATION

Paracetamol and adrenalin

Making paracetamol

Paracetamol is a well-known analgesic (a pain-relieving medicine), usually sold in tablet form, often mixed with other analgesics. Using the IUPAC naming system paracetamol is known as 4-hydroxy-(N-ethanoyl-aminobenzene).

1 Draw the full structure of paracetamol.

One possible synthetic route for making paracetamol from benzene is summarised in the following flowchart:

benzene → nitrobenzene → aminobenzene → [warm with acidified sodium nitrate(III)] → phenol (or hydroxybenzene) → 4-nitro-hydroxybenzene → 4-amino-hydroxybenzene → paracetamol

2 Including all essential experimental details, write full equations (see Chapter 4) to represent the changes for each of the following steps in this flowchart
a benzene → nitrobenzene
b nitrobenzene → aminobenzene

3 Give detailed equations to show the stages of the mechanism for the conversion of benzene to nitrobenzene.

The reaction sequence:

hydroxybenzene → 4-nitro-hydroxybenzene → 4-amino-hydroxybenzene

can be achieved using the same sort of reactions as in question 2.

4 After protecting the OH from reaction, the final stage of the synthesis involves the ethanoylation of 4-amino-hydroxybenzene using ethanoic anhydride. The protecting group then needs to be removed. Ignoring the protecting group,
a write an equation for this reaction
b describe the mechanism for this reaction
c state two reasons for using ethanoic anhydride rather than ethanoyl chloride as the acylating agent.

Using more efficient methods than described here, thousands of tonnes of paracetamol are manufactured every year and, under controlled use, it has proved a very effective pain-reliever for many minor ailments. This is not to say it has no drawbacks: excessive use of paracetamol can lead to irreversible liver damage.

Relatives of adrenalin

The hormone adrenalin is secreted by the adrenal gland, which is found near the kidneys. Adrenalin is responsible for stimulating the nervous system into overdrive to allow rapid response to danger or unusual circumstances. The chemical structure of adrenalin is shown below.

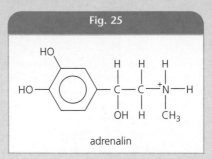

Fig. 25

adrenalin

5 Ignoring any effects on other functional groups, draw the structure of the amine (amine A) which would be produced by adding sodium hydroxide solution to adrenalin. How would this product be reconverted to adrenalin?

6 Explaining your answer, classify amine A as primary, secondary or tertiary.

7 Draw the structures of the two different products resulting from the reaction between amine A and chloromethane. How would reaction conditions be adjusted to favour the production of either of these possibilities?

8 Using R–NHCH$_3$ as an abbreviation for the amine A, describe the mechanism of the first reaction between it and chloromethane.

9
a Assuming a 1:1 reaction, draw the structure of the product resulting from the reaction of the aliphatic OH group in amine A with ethanoyl chloride. Classify the reaction according to its type.
b Which other groups/bonds in amine A would also react with ethanoyl chloride?

Adrenalin is a naturally occurring molecule which promotes a very important reflex action which allows us and other animals to react to, and escape from, danger. Moreover, its structure allows many conversions to related molecules. Perhaps these relatives will find equally useful artificial applications one day.

1 Outline a mechanism for the reaction of 1-bromopropane with an excess of ammonia.
NEAB CH03 June 1998 Q5 (a)

2
a Explain how methylamine can act as a Brønsted–Lowry base.
b Explain why phenylamine is a weaker base than ammonia.
c **i)** Name the type of mechanism involved when methylamine is formed from bromomethane and ammonia.
   ii) Give the structures of three organic compounds other than methylamine which can be obtained from the reaction between an excess of bromomethane and ammonia.
   iii) Name the type of compound formed in part **(c)(ii)** which can be used as a cationic surfactant.
AQA CH06 March 1999 Q3

3 *N*-Phenylethanamide can be prepared from benzene in three steps:

Write an equation for the reaction occurring in Step 3. Name and outline the mechanism for this reaction.
NEAB CH06 March 1999 Q1(c)

4
a Give the structural formulae of the **four** isomers of C_3H_9N.
b Show, by means of an equation, how one of the above isomers can be obtained from the reaction between 1-bromopropane and ammonia.
Name the type of reaction taking place and outline a mechanism.
c Explain why the reduction of propanenitrile is a more suitable method for preparing this isomer than the reaction you have given in part **(b)**.
NEAB CH06 February 1996 Q3

5
a Explain why phenylamine is a weaker base than (phenylmethyl)amine, $C_6H_5CH_2NH_2$.
b Write an equation for the formation of (phenylmethyl)amine from benzenecarbonitrile, C_6H_5CN, and name the type of reaction involved.
c (Phenylmethyl)amine can also be obtained from the reaction between (bromomethyl)benzene, $C_6H_5CH_2Br$, and ammonia. Name the type of reaction involved and explain why this method of synthesis is not as effective as that in part **(b)**.
d The secondary amine $(C_6H_5CH_2)NH(CH_2)_{11}CH_3$ can be converted into a cationic fabric-softening product by reaction with an excess of chloromethane. Name the type of product formed and give the structural formula of this compound.
NEAB CH06 June 1998 Q2

6
a Write an equation for the reaction between ethanoyl chloride and dimethylamine. Name and outline the mechanism of this reaction.
b Aspirin is manufactured by the reaction of 2-hydroxybenzenecarboxylic acid with ethanoic anhydride. Write an equation for this reaction and give two reasons why ethanoic anhydride, rather than ethanoyl chloride, is used.
NEAB CH06 March 1998 Q3

7
a Explain why propylamine is classed as a Brønsted–Lowry base.
b Explain why propylamine is a stronger base than ammonia.
c Propylamine is obtained from the reaction between 1-bromopropane and an excess of ammonia. Name the type of reaction taking place and outline a mechanism.
d Propylamine can also be prepared from propanenitrile, CH_3CH_2CN. Name the type of reaction involved and write an equation for this reaction. What advantage does this method of synthesis have over that in part **(c)**?
NEAB CH06 March 1998 Q2

8
a Explain briefly why methylamine is more basic than ammonia.
b Explain briefly why phenylamine is less basic than ammonia.
c Write an equation for the reaction between ethanoyl chloride and ammonia. Name the type of reaction taking place and outline a mechanism.
NEAB CH06 June 1997 Q2

9
a Write an equation for the formation of methyl ethanoate from ethanoyl chloride and methanol. Give the type of reaction taking place and outline a mechanism.
b Suggest why a base, such as triethylamine, is usually added to the above reaction mixture.
c Give **two** reasons why ethanoic anhydride, rather than ethanoyl chloride, is used in the manufacture of aspirin.
NEAB CH06 February 1996 Q4

8 Amino acids

It is said that we are what we eat. This is not strictly true: we are composed of similar materials to those we eat, but the food materials are broken down by digestive processes and then built up again in the way that our bodies require.

Every cell contains protein – our muscles are made up of a high percentage of protein – so it is not surprising that we need a daily balanced intake of protein. The protein molecules are broken down to provide a supply of *amino acids*. Our own characteristic proteins, vital for a healthy life, are then built up from these amino acids and their breakdown products.

Meat, dairy products, fish, peas, beans and lentils are all protein-rich foods.

8.1 Structure of amino acids

As the name suggests, amino acids are molecules that contain both amino ($-NH_2$) and carboxylic acid ($-COOH$) functional groups. Consequently they are sometimes called **bifunctional compounds**. Fig. 1 shows the structure of a general **α-amino acid** where the amino group is bonded to the carbon atom adjacent to the carboxylic acid group. Because this carbon atom is the second from the acid end of the molecule, α-amino acids may also be called 2-amino acids. Other amino acids are referred to as β-, γ-, δ-, etc. (or 3-, 4-, 5-, etc.) amino acids.

Only α-amino acids occur as part of proteins; other amino acids may occur naturally or can be synthesised. For example, γ-aminobutanoic acid (better known as GABA) is a molecule involved in nerve transmission in the brain, and 6-aminohexanoic acid can be a starting material for manufacturing one form of nylon.

1 Draw a general structure for:
a a β-amino acid,
b a 5-amino acid.

The R group represents the amino acid's side-chain. It may be a simple group such as $-H$ in the amino acid glycine, or $-CH_3$ in alanine. It can also be a much more complex group such as $-CH_2C_6H_5$ in phenylalanine, or $-CH_2COOH$ in aspartic acid. Table 1 shows the 20 different R groups that give rise to the 20 naturally occurring α-amino acids.

Apart from glycine (R = H), all α-amino acids have optical isomers (enantiomers), meaning that one form (dextrorotatory form) rotates plane-polarised light clockwise, while the second form (laevorotatory form) rotates it anticlockwise. The central carbon atom in these amino acids is bonded to four different groups: we describe the molecule as **chiral** (**asymmetric**). (This form of stereoisomerism is discussed in Chapter 4.) As shown in Fig. 2, the result is that each α-amino acid occurs as either of two non-identical mirror images which cannot be superimposed. However, only the L-isomers occur naturally in our body proteins.

2 Why does glycine (R = H) not have optical isomers?

Fig. 1 A general α-amino acid

General formula:
$RCH(NH_2)COOH$
R = any one of
20 possible
groups

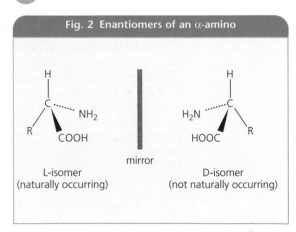

Fig. 2 Enantiomers of an α-amino

L-isomer
(naturally occurring)

mirror

D-isomer
(not naturally occurring)

Amino acid	Abbreviation	R group (side chain)	Amino acid	Abbreviation	R group (side chain)
glycine	Gly	–H	proline	Pro	
alanine	Ala	$-CH_3$			
valine	Val	$-CH(CH_3)_2$			
leucine	Leu	$-CH_2CH(CH_3)_2$			
isoleucine	Ile	$-CH(CH_3)CH_2CH_3$	phenylalanine	Phe	
serine	Ser	$-CH_2OH$			
threonine	Thr	$-CH(OH)CH_3$	tyrosine	Tyr	
lysine	Lys	$-CH_2CH_2CH_2CH_2NH_2$			
arginine	Arg	$-CH_2CH_2CH_2NHC(NH)NH_2$	tryptophan	Trp	
histidine	His				
aspartic acid	Asp	$-CH_2COOH$	methionine	Met	$-CH_2CH_2SCH_3$
asparagine	Asn	$-CH_2CONH_2$	cysteine	Cys	$-CH_2SH$
glutamic acid	Glu	$-CH_2CH_2COOH$			
glutamine	Gln	$-CH_2CH_2CONH_2$			

Table 1 The naturally occurring α-amino acids

8.2 Acid-base properties of amino acids

The –COOH group is weakly acidic and tends to donate its proton to water, while the $-NH_2$ group is weakly basic and tends to accept a proton from water via the nitrogen lone pair. These characteristics are discussed in Chapters 5 and 7 respectively.

An α-amino acid therefore has the ability to act as both a weak acid and a weak base. When dissolved in a highly acidic solution, the amino acid is transformed to its corresponding cation. Similarly, when dissolved in a highly alkaline solution, the amino acid exists as its corresponding anion (Fig. 3). Consequently, at some intermediate pH value, it exists simultaneously as both an anion and a cation. Such an ion is called the **zwitterion** or **amphion** of the α-amino acid.

In solution, the zwitterion exists at a unique pH value for each α-amino acid. This value is referred to as the acid's **isoelectric pH**. At the isoelectric pH the ion-dipole forces between the amino acid and the surrounding water molecules are at their weakest, and so this is the pH at which the α-amino acid is most likely to precipitate from solution. The isoelectric pH is not exactly 7 because, even though the proton donating power of the –COOH group and the proton accepting power of the $-NH_2$ group are approximately equal (though opposite), the overall effect is also influenced by the different electron-attracting powers of the different R groups.

Zwitterions also exist in the crystalline form of the amino acid. The electrostatic

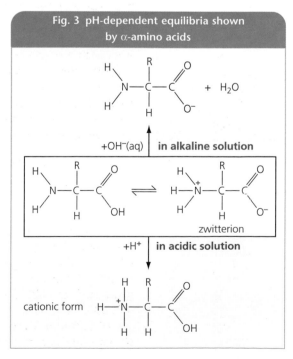

Fig. 3 pH-dependent equilibria shown by α-amino acids

attraction between the oppositely charged parts of the ion accounts for the relatively high melting point values of α-amino acids. For example, even the smallest amino acid molecule, glycine, is a crystalline solid at room temperature, whereas the corresponding amine ($CH_3CH_2CH_2NH_2$, m.p. 190 K) or acid (CH_3CH_2COOH, m.p. 252 K) *of similar molecular size* are both liquids.

3 State whether an amino acid ion subjected to an electric field will migrate towards a positive or a negative electrode if first dissolved in:
a dilute hydrochloric acid,
b dilute sodium hydroxide.

4 What would happen to a zwitterion under the influence of an electric field?

KEY FACTS

■ Amino acids have the same basic structure: $RCH(NH_2)COOH$. There are 20 alternatives for group R.

■ Amino acids have acidic properties due to the –COOH group and basic properties due to the –NH_2 group.

8.3 The peptide link

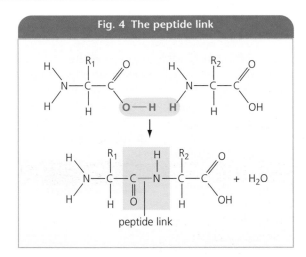

Fig. 4 The peptide link

peptide link

Fig. 5 Polyamide (polypeptide) formation

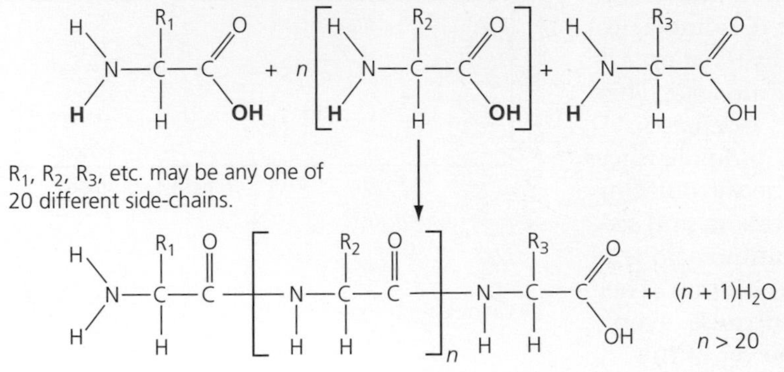

R_1, R_2, R_3, etc. may be any one of 20 different side-chains.

$n > 20$

Amino acids are the building units from which proteins are constructed. As shown in Fig. 4, the –NH_2 group of one amino acid can undergo a **condensation** reaction with the –COOH group of another amino acid, eliminating a water molecule and forming a **dipeptide**. The –CONH– group that links the two amino acids is called a **peptide link** and the C–N bond within this group is called the **peptide bond**. In general terms the arrangements of atoms correspond to the arrangement in N-substituted amides.

5 Draw structures of the two different dipeptides which may be produced by condensing 2-aminoethanoic acid (glycine) with 2-aminopropanoic acid (alanine).

The dipeptide we obtain from this reaction has an amino group at one end of the molecule and a carboxylic acid group at the other. Consequently, under the control of the body's enzyme system and predetermined by the genetic code, condensation reactions with other amino acids can occur repeatedly as shown in Fig. 5. The result is a **polyamide** or **polypeptide**. Its length ranges from 20 to more than 10 000 amino acids bonded together to form a continuous chain via repeated peptide links. This chain is the amino acid sequence of a protein and is called its **primary structure**.

The polyamide chain has many N–H and C=O groups along its length. Hydrogen bonds form between an N–H group from one acid and a C=O group from another acid further along the chain. As a result, the polyamide chain is coiled to form a helical shape held in place by a regular pattern of hydrogen bonds, as seen in Fig. 6. This helix is the **secondary structure** of the protein.

The helix is then folded into a characteristic three-dimensional shape (see Fig. 7) called its **tertiary structure**.

Various electrostatic or covalent bonds between the R side-chains of particular amino acids in the polyamide chain determine the pattern of folding. It is only in this form that the structure is capable of its characteristic biochemical functions and can properly be called a protein.

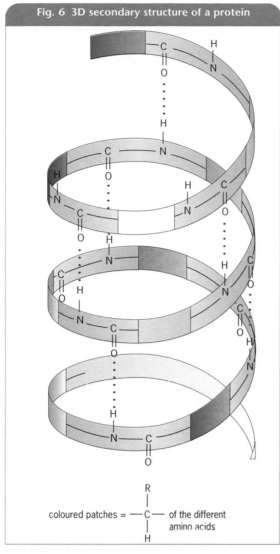

Fig. 6 3D secondary structure of a protein

coloured patches = —C— of the different amino acids

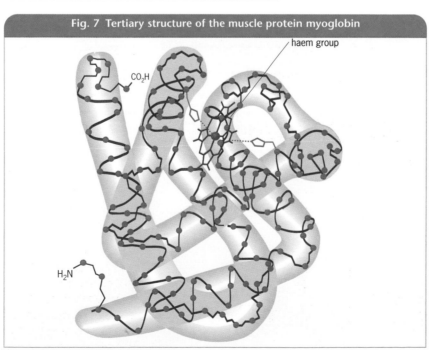

Fig. 7 Tertiary structure of the muscle protein myoglobin

haem group

CO$_2$H

H$_2$N

8.4 Hydrolysis of peptide bonds

The construction of a polyamide chain is a complex biochemical process, but its breakdown to liberate the component amino acids is relatively simple. Whenever we digest protein, our bodies break up these chains. This process can also be achieved by heating a protein with 5 mol dm^{-3} hydrochloric acid for about 24 hours. The secondary and tertiary structures of the protein are rapidly broken down, and this causes the polyamide chain to unravel. The chain is then **hydrolysed** (split by reaction with water) at each of the peptide links and liberates the component amino acids (Fig. 8).

6 What is the other purpose of the 5 mol dm^{-3} hydrochloric acid in this hydrolysis process?

We can use this hydrolysis as the first stage in determining the structure of a polyamide because it allows the types of 2-amino acids present in the polyamide chain to be identified and counted. Determining the sequence of the 2-amino acids and the three-dimensional structure of the chain is a much more difficult task and has earned some chemists Nobel prizes in the past!

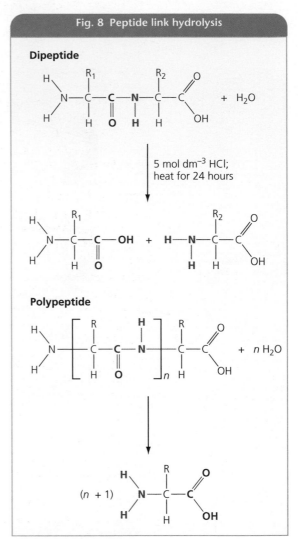

Fig. 8 Peptide link hydrolysis

Dipeptide

5 mol dm^{-3} HCl; heat for 24 hours

Polypeptide

$+ \; n \, H_2O$

$(n + 1)$

APPLICATION

Structure of proteins

The pain-killer leucine encephalin

The naturally occurring oligopeptide (a peptide with less than 20 amino acids) known as leucine encephalin contains five amino acids. This molecule was discovered in the 1970s during an investigation into how morphine and codeine's pain-killing properties worked. Research showed that these molecules fit into brain receptor sites, in a sense imitating the role of the brain's own pain-suppressing molecules, of which leucine encephalin is just one example.

When hydrolysed using 5 mol dm^{-3} hydrochloric acid for 24 hours, leucine encephalin releases four different amino acids which were shown to be glycine (Gly), tyrosine (Tyr), phenylalanine (Phe) and leucine (Leu) in the ratio 2:1:1:1. The structures of these amino acids are given in Table 1, page 99.

When partially hydrolysed using a shorter reaction time, apart from the individual amino acids, the following dipeptides and tripeptides are identified:

A   HOOC-Tyr-Gly-NH$_2$
B   HOOC-Phe-Leu-NH$_2$
C   HOOC-Gly-Gly-Phe-NH$_2$

1 Which of these are dipeptides and which are tripeptides? Explain.

2 Showing all atoms and all bonds, draw the structures of molecules **A** and **C**.

3 Deduce the order of the five amino acids in the leucine encephalin chain. Explain your reasoning.

4 Which of the amino acids released from leucine encephalin would not be optically active? Explain.

5 Write a balanced equation for the complete hydrolysis of molecule **B**.

Essential and non-essential amino acids

Alanine (2-aminopropanoic acid) and aspartic acid (1-carboxy-2-aminopropanoic acid) are non-essential amino acids. They are called 'non-essential' amino acids because it is not vital that we eat proteins containing them since our bodies can make them from other nutrients, provided we eat a properly balanced diet. However, lysine (2,6-diaminohexanoic acid) is one of the eight amino acids that are **essential**: we cannot synthesise them, so they must be eaten regularly.

6 Use the IUPAC names and the information from Table 1 to draw the full structures of alanine, aspartic acid and lysine.

7 All three of the amino acids mentioned here can exist as optical isomers (enantiomers). Draw three-dimensional diagrams to represent the optical isomers of lysine and outline the differences in their chemical and physical properties.

8 By comparing their structures, explain why equimolar solutions of lysine, alanine and aspartic acid would produce solutions of decreasing pH.

9 Draw the structures of the molecules or ions that would be present if:
a alanine were placed in a pH 1 buffer solution;
b alanine were placed in a pH 11 buffer solution;
c alanine were placed in a buffer solution of pH equal to the isoelectric point of alanine.

EXAMINATION QUESTIONS

1 The structures of aminoethanoic acid (*glycine*) and 2,6-diaminohexanoic acid (*lysine*) are shown below.

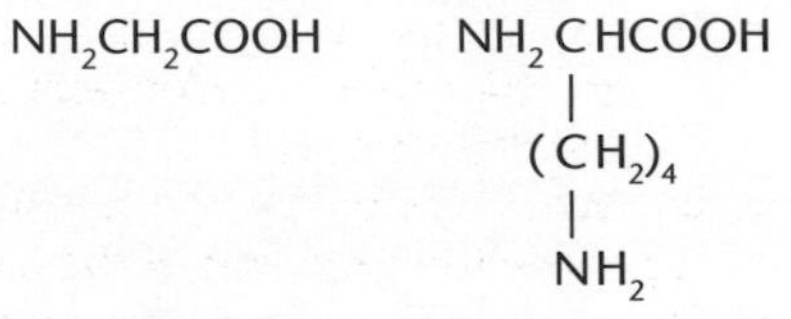
NH2CH2COOH
NH2 CHCOOH
(CH2)4
NH2

Glycine Lysine

Protein **B** contains a glycine residue next to a lysine residue. Draw these two amino acid residues bonded together as in this protein. (2)
NEAB CH09 Spring 1998, Q 4(a)

2 The structure of the D-isomer of 2,6-diaminohexanoic acid (lysine) is shown below:

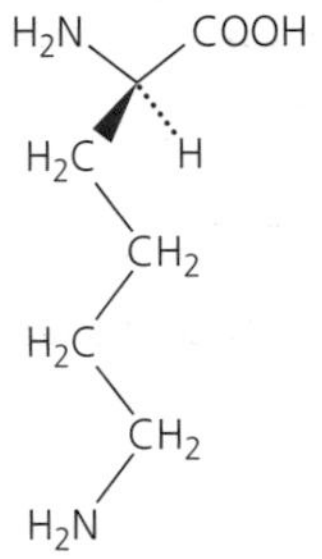
H2N
COOH
H2C
H
CH2
H2C
CH2
H2N

Draw the structure of L-lysine next to that of the D-isomer above. (1)
Copy and complete the structures below to show the main species of lysine present in aqueous solution at pH 2, pH 7 and pH 11. (6)

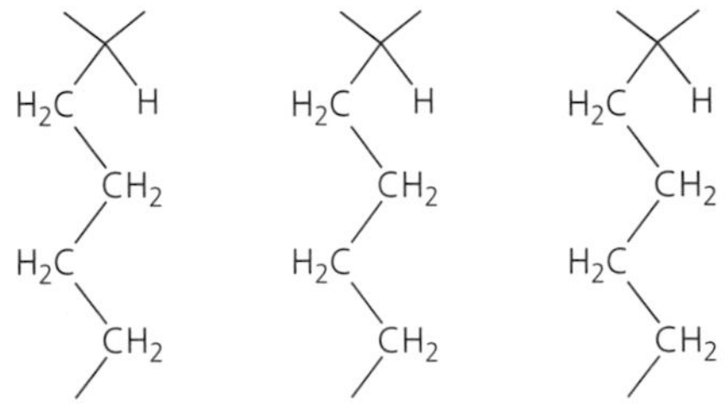
H2C
H
CH2
H2C
CH2

pH 2 pH 7 pH 11

Why is lysine of particular importance in human nutrition? (1)
NEAB CH09 Summer 1998, Q 4(a), (b) & (c)

3 The structure of 2-aminopentanedioic (*glutamic*) acid is given below:

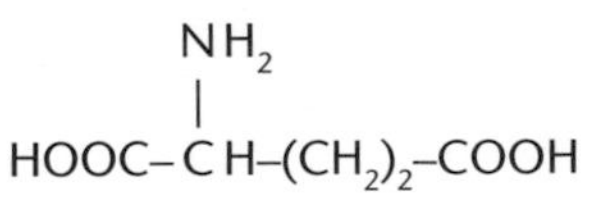
NH2
HOOC–CH–(CH2)2–COOH

Draw the principal ionic species formed from glutamic acid in aqueous solution at pH 1, pH 7 and pH 11. (3)
NEAB CH09 Summer 1997, Q1(a)

4 A typical amino acid molecule has the following displayed formula.

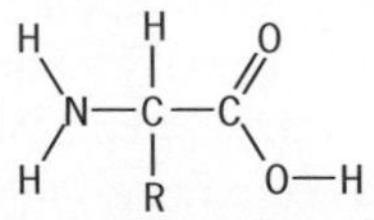
H
H
O
N
C
C
H
R
O—H

a i) Give the names of the two functional groups in this molecule.
ii) Give one possible chemical formula for R.
iii) Draw a displayed formula of the zwitterion form of the amino acid shown above.
b i) For the amino acid above, draw a diagram to show the peptide bond formed when two molecules react together.
ii) Name the type of reaction that takes place when amino acids react in this way.

5 Alanine and glycine are two important amino acids. Their structural formulas are given below.

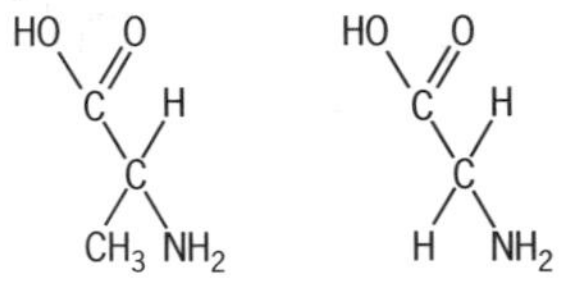
HO
O
C
H
C
CH3
NH2
HO
O
C
H
C
H
NH2

alanine glycine

a Which of these two amino acids can exist as optical isomers? Give a reason for your answer.
b Copy the amino acids above, and show how a peptide link is formed between them.

6 The structural formulae of three amino acids are shown below.

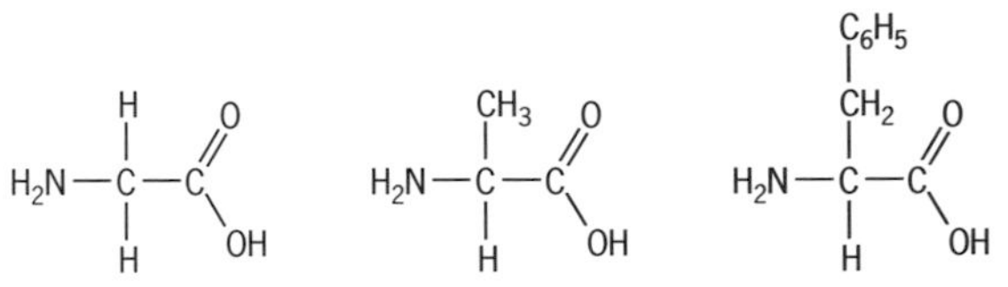
H
O
H2N
C
C
H
OH
CH3
O
H2N
C
C
H
OH
C6H5
CH2
O
H2N
C
C
H
OH

glycine alanine phenylalanine

a Explain why alanine and phenylamine have optical isomers, but glycine does not.
Amino acids can react both with acids and with bases and are also capable of forming zwitterions.
b Write equations for the reactions between:
i) phenylalanine and hydrochloric acid,
ii) alanine and sodium hydroxide.
The isoelectric point is described in terms of the pH at which an amino acid forms its zwitterion. The isoelectric point is different for each amino acid.
c i) The isoelectric point for glycine is at pH = 5.97. Draw the zwitterion formed by glycine at this point.
ii) Draw the ions that would be formed at pH = 5.75 by alanine and phenylalanine.

Glycine and alanine can react together to form the dipeptide shown below:

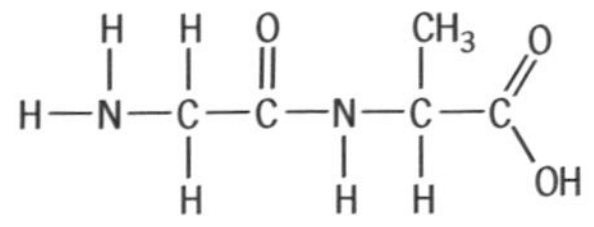

d A different dipeptide can be formed from the reaction between glycine and alanine. Draw its displayed formula.

7 The formula of lysine (2,6-diaminohexanoic acid) is $H_2N(CH_2)_4CH(NH_2)COOH$.
a Draw the displayed formula of the ionic species present in:
  i) a highly alkaline solution of lysine;
  ii) a highly acidic solution of lysine;
  iii) the zwitterion form of lysine.
b Poly(lysine) is the name of a polymer that containing only lysine residues. Write a possible structure of poly(lysine), showing two repeating units.

8 A tripeptide has the following formula:

$H_2N–CH(CH_3)–CO–NH–CH–(CH_3)–CO–NH–CH(CH_2SH)–COOH$

a Draw the displayed formula of this tripeptide.
b Put a circle round the peptide links contained in this structure.
c **i)** Explain the reactions that will take place when the tripeptide is boiled with aqueous hydrochloric acid.
  ii) What type of reaction are these reactions?
  iii) Draw the structures of the organic products.
d Draw and explain the structures of these organic products at: **i)** pH 1, **ii)** pH 7, **iii)** pH 12.
e **i)** Name the main intermolecular forces that exist between two molecules of the tripeptide.
  ii) Draw a diagram to represent their action.
  ii) What aspects of protein structure are associated with the types of intermolecular force illustrated in **e**?

9 The systematic name for alanine is 2-aminopropanoic acid. Its molecular formula is $CH_3CH(NH_2)COOH$.
a Draw the displayed formula of alanine:
  i) in its uncharged form;
  ii) when in a solution of pH 1.
b **i)** Write down the name given to an amino acid as a dipolar molecule.
  ii) Draw the displayed formula of alanine in this state.

9   Polymers

The word **polymer** comes from a combination of two Greek words: *poly* meaning 'many' and *meros* meaning 'part'. Polymers are large molecules built by the chemical combination of many small molecules. These building blocks are called **monomers**. For example, as we saw in Chapter 8, polypeptides are polymers derived from amino acid monomers. The number of combined monomer units may vary considerably, but can often be tens of thousands per molecule.

The discovery and development of polymers revolutionised twentieth-century living, and innovative polymers continue to be created. This is because *plastics*, as they are commonly known (because many are easily moulded to any desired shape) have provided materials with a vast variety of uses. Each polymer has its own specific properties that determine its applications. In general, plastics degrade slowly, have high strength-to-weight ratios, are easily shaped and are relatively cheap compared to wood, glass and metals.

The window frame, sponge, the woman's clothing, including her shoes, and the bucket, are all made of types of plastic.

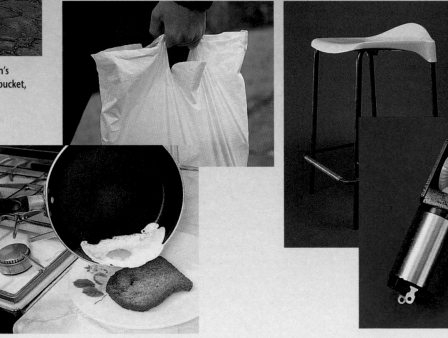

Plastics have very many advantages over traditional materials: Nylon gears have low friction; the plastic stool seat is rigid and hard wearing as well as being lightweight; the carrier bag is flexible and comfortable to hold; and the lining of the frying pan is temperature-resistant and non-stick.

Polymer science has undoubtedly provided many new and useful materials, but there are concerns over environmental issues. Many polymers are manufactured from chemicals that come from oil. As this precious, non-renewable resource will eventually be used up, chemists are searching for alternative ways to make polymers. One approach is to make use of natural polymers such as starch, cellulose and protein. Research is also aimed at using bacteria to 'grow' new and useful polymers. Another problem is that many polymers are non-biodegradable and are responsible for considerable environmental pollution and danger to wildlife. These problems are now being addressed as polymers that biodegrade when exposed to water or sunlight are being developed.

Polymers are today as essential a part of our lives as wood and metal were to people a hundred years ago. More chemists now work with polymers than with any other type of material, to provide us with the types of materials we demand in the goods we buy, with new polymers continually being created to satisfy particular needs.

9.1 Types of polymers

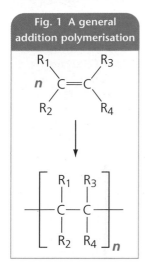

Poly(ethene) - or polythene as it is popularly known - and nylon are two well-known materials that illustrate the two general types of polymer. Poly(ethene) is an example of an **addition polymer** produced by repeated addition reactions in which many alkene monomers add to the end of the carbon chain to form very long chain molecules (see Fig. 1). Nylon is an example of a **condensation polymer** produced by repeated condensation reactions between appropriately designed monomers to form the condensation polymer *and* one small molecule such as water, methanol or hydrogen chloride for every two bonded monomers (see Fig. 2).

Fig. 2 A general condensation polymerisation

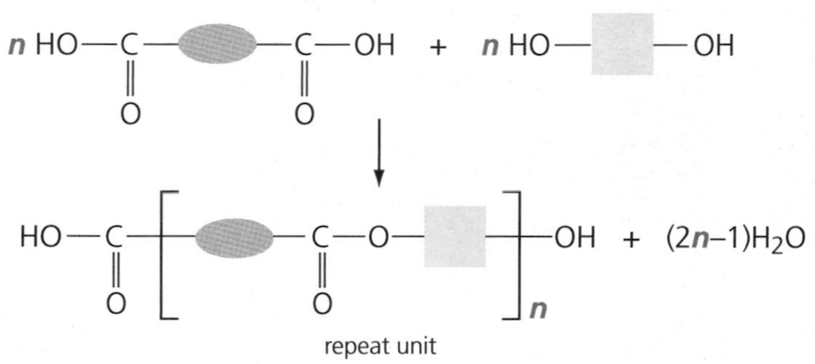

$$n \, HO-\underset{\underset{O}{\|}}{C}-\text{⬭}-\underset{\underset{O}{\|}}{C}-OH \; + \; n \, HO-\text{◻}-OH$$

$$HO-\underset{\underset{O}{\|}}{C}-\left[\text{⬭}-\underset{\underset{O}{\|}}{C}-O-\text{◻}\right]_n-OH \; + \; (2n-1)H_2O$$

repeat unit

9.2 Addition polymers

Alkenes (see *AS Chemistry*, Chapter 13) are molecules that contain C=C bonds. As we have seen in Chapter 4, alkenes undergo addition reactions. In these reactions the C=C double bond becomes a single bond, with two more single bonds created. Addition polymerisation works on the same general principle, except that the alkene molecules repeatedly add to themselves (see Fig. 3) many thousands of times, rather than just once.

Also, the reaction mechanism may involve free radicals, carbocations or carboanions, depending on the exact nature of the catalyst.

The first useful addition polymer was discovered by accident in 1933 by Fawcett and Gibson. These two British chemists worked for ICI and were experimenting with ethene under high pressure in an attempt to produce a ketone. However, a flaw in their process produced instead a small amount of white waxy solid that was later shown to be poly(ethene).

Poly(ethene) is called an addition polymer because many ethene monomer molecules bond together to form poly(ethene) as the *only* product (Fig. 4). Take particular note of the **repeat unit** which is shown in square brackets.

Poly(ethene) is manufactured in two forms: low-density poly(ethene) (LDPE) and high-density poly(ethene) (HDPE). LDPE is the more common form and, being softer and more pliable, is widely used for packaging and electrical insulation. Its molecules contain about 50 000 carbon atoms and there is considerable branching of the main chains, which makes the material softer and of melting points in the low range.

Fig. 3 Propene polymerisation

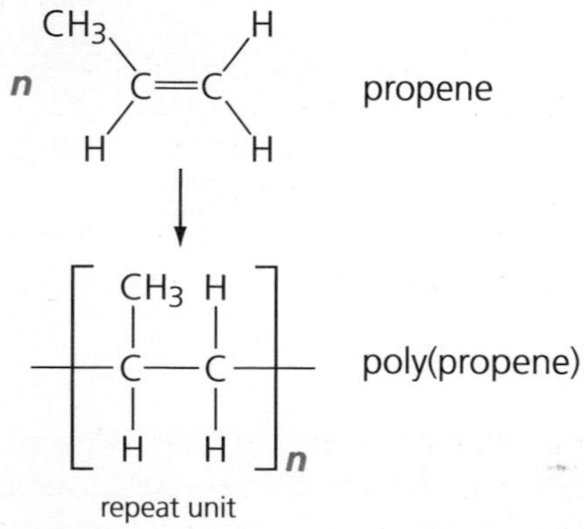

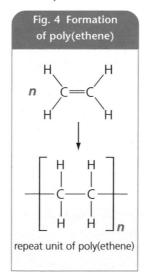

propene

poly(propene)

repeat unit

Fig. 4 Formation of poly(ethene)

repeat unit of poly(ethene)

HDPE molecules contain more than 50 000 carbon atoms, with little branching of the main chains and resulting in more rigid materials with a higher range of melting points.

1 In terms of appropriate intermolecular forces, explain why LDPE melts more easily than HDPE.

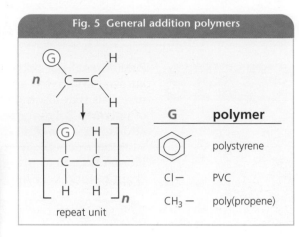

Fig. 5 General addition polymers

G	polymer
(benzene ring)	polystyrene
Cl—	PVC
CH_3—	poly(propene)

repeat unit

Hot low-density poly(ethene) can be blown to form 'polythene' bags.

The bottle on the left is made from HDPE, and the one on the right is made from LDPE.

The range of addition polymers has been vastly expanded by varying the structure of the alkene monomer. This is shown generally in Fig. 5 where G represents a general group which has been substituted for one of the hydrogen atoms in ethene. Table 1 shows some common addition polymers and their uses.

Although addition polymers are flammable, in other respects they are generally chemically inert. This lack of reactivity is due to the very strong C–C and C–H bonds and the lack of polarisation of the

Table 1 Common addition polymers and their uses

Alkene monomer	Polymer	Structure of polymer (monomer unit)	Uses
ethene	poly(ethene)		bags, insulation for wires, squeezy bottles
propene	poly(propene)		bottles, plastic plates, clothing, carpets, crates, ropes and twine
phenylethene	poly(phenylethene) (polystyrene)		insulation, packaging, food containers, model kits, flowerpots, housewares
chloroethene	poly(chloroethene) (PVC)		synthetic leather, water pipes, floor covering, guttering, window frames, curtain rails, wall cladding
methyl-2-methylpropenoate	poly(methyl-2-methylpropenoate)		light fittings, car lights, tap tops, lenses
tetrafluoroethene	poly(tetrafluoroethene) (Teflon)		non-stick pans, lubricant-free bearings

bonds within poly(alkenes). Consequently, once formed, it is very difficult to break down materials made from addition polymers. They are **non-biodegradable** and not easily recycled. This causes the environmental problem of how to get rid of worn out or non-fashionable items so that they can be replaced by newer, equally non-biodegradable merchandise! Using them as fuels in specially designed power stations is certainly one way forward here.

 Explain why poly(ethene) can be considered to be an alkane.

9.3 Condensation polymers

Manufactured condensation polymers (e.g. nylon) are artificial polymers with properties designed to mimic those of natural condensation polymers, such as silk, a polymer of amino acids, starch, which is a polysaccharide, and DNA, a polymer of nucleic acids. These polymers are biodegradable and combine structural strength, low density and resistance to some forms of chemical attack.

The name 'condensation' polymer derives from the type of chemical reaction used to link monomers. A condensation reaction links two organic molecules together by formation of an ester or amide group between the molecules with elimination of a small molecule such as water, methanol or hydrogen chloride (see Fig. 6).

Natural condensation polymers are all formed by elimination of water; the other possible elimination products are only formed in the synthesis of artificial condensation polymers.

In order to allow repeated condensation reactions, and so build up a polymeric structure, the monomers used for condensation polymers must be **bifunctional**. This means they must contain two functional groups. For example (see Chapter 8) amino acids, which contain both –COOH and –NH$_2$ functional groups, are the monomers for polyamides. However, for most synthetic polymers, it is more convenient to have two different monomers, one containing two acid (–COOH) functional groups and the other containing two alcohol (–OH) groups to make a **polyester** or two amine (–NH$_2$) groups to make a **polyamide**. Examples of bifunctional monomers are shown in Fig. 7.

The production of synthetic polyesters

A polyester is a polymer produced by linking together many small molecules (monomers) via ester linkages. When an alcohol with one –OH group reacts with an organic acid with

Fig. 6 Forming esters and amides by condensation reactions

ester

amide

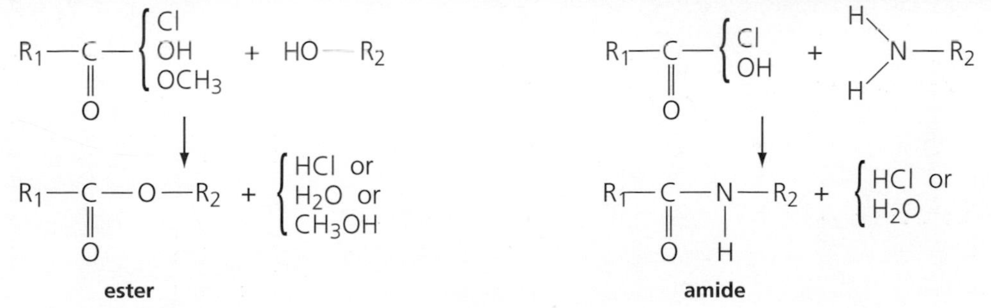

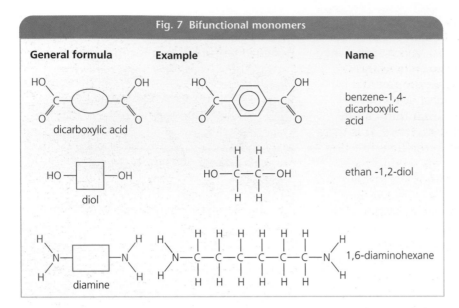

Fig. 7 Bifunctional monomers

General formula	Example	Name
dicarboxylic acid		benzene-1,4-dicarboxylic acid
diol		ethan-1,2-diol
diamine		1,6-diaminohexane

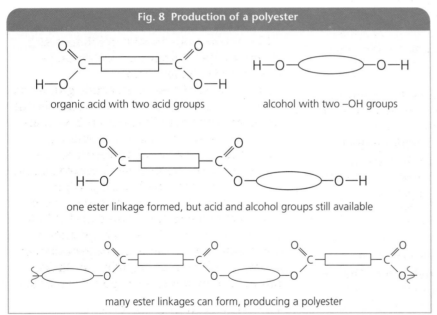

Fig. 8 Production of a polyester

organic acid with two acid groups   alcohol with two –OH groups

one ester linkage formed, but acid and alcohol groups still available

many ester linkages can form, producing a polyester

Fig. 9 Condensation reaction to form poly(ethylene)terephthalate (PET)

n HO–C... benzene-1,4-dicarboxylic acid   + n HO—CH$_2$—CH$_2$—OH   ethane-1,2-diol

1 heat and pressure
2 heat and antimony catalyst

HO[C...—O–CH$_2$—CH$_2$—O]$_n$H + (2n–1)H$_2$O

poly(ethane-1,2-diolbenzene-1,4-dicarboxylate) (PET; Terylene)

one –COOH group, an ester and water are formed and no further esterification reactions can take place. If the alcohol has two –OH groups and the organic acid has two –COOH groups, then the ester that is formed will still contain an –OH group at one end and a –COOH group at the other. Further esterification reactions are still possible, and, in theory, can continue to be possible indefinitely (Fig. 8). A polyester will have been formed. The chain length is limited because it becomes less likely, as the chain length grows, that further monomers will collide with the end of the growing chain.

The production of PET (Terylene)

Terylene is one of the trade names for the polyester made by linking 1,4-benzenedicarboxylic acid and the alcohol ethane-1,2-diol. Another name, p**oly(e**thylene)**t**erephthalate (or PET), is derived from terephthalic acid, which is the old name for 1,4-benzenedicarboxylic acid. Ethane-1,2-diol plays the part of the alcohol in the production of PET. Repeated condensation reactions occur between the –COOH group of 1,4-benzenedicarboxylic acid molecules and –OH groups of the ethane-1,2-diol molecules (Fig. 9).

Molten PET can be forced through a fine mesh (extruded) to form strands. These are water-cooled and chopped into small pellets for convenient storage and transportation. The

PET is an ideal material for carbonated drinks bottles as it is light, tough and won't shatter when dropped.

polyester is thermoplastic, which means it can be repeatedly heated to soften and melt it. This allows it to be extruded to form fine fibres for use in artificial fabrics, or to be blow moulded, into fizzy drink bottles and other containers.

A common process is to cast the molten polymer on a rotating drum to form a film which is then heated and stretched. It is then stretched at right angles to the first direction and 'set' at 220–240 °C. This produces a transparent, tough, dimensionally stable and chemically resistant film (sold by ICI as Melinex) which contains the polyester molecules partially aligned along both axes.

Melinex and similar products are used extensively for packaging, especially for food, helping to keep it fresh and free from contamination. Melinex is also used for making windsurfer sails, flexible printed circuit boards, cable insulation, touch pads (e.g. microwave controls) and, when coated with a thin layer of metal, it can be used for making things such as hot-air balloons.

Since its use as a substitute for silk in stockings, Nylon has found many other applications. It is used to make ropes, twines, Velcro, machinery parts and a wide range of clothing. In clothing and carpets, natural fibres such as cotton or wool are often mixed with Nylon to make them last longer.

The production of synthetic polyamides

Types of Nylon are the best-known examples of artificial polyamides. The brand name Nylon was created in the 1930s by the Du Pont chemical company in America. It followed the discovery of how to make the polymer by one of their employees, Wallace Carothers. Nylon proved to be a very acceptable, cheap substitute for the silk used to make women's stockings and became a huge commercial success for Du Pont.

One type of nylon is **nylon-6,6**. The 6,6 refers to the fact that both monomers (a diamine and a dicarboxylic acid – see Fig. 7) contain six carbon atoms. Hence, the diamine is 1,6-diaminohexane and the dicarboxylic

Windsurfer sails are commonly made of Melinex.

acid is 1,6-hexandioic acid and there are six carbon atoms between each of the amide links in the polymer chain.

The reaction between an amine group and a carboxylic acid group to form an amide link is not very efficient. The reaction is slow and the equilibrium established contains a significant proportion of unreacted amine and acid. Consequently, the dicarboxylic acid is usually first converted (see Fig. 10) to hexane-1,6-dioyl chloride by reaction with thionyl chloride ($SOCl_2$). This diacyl chloride then reacts (see Fig. 11) with the diamine much faster and, because of the evolution of hydrogen chloride, the equilibrium shifts almost entirely towards the polymer product.

3 Nylon-6,10 is also made from 1,6-diaminohexane and a dicarboxylic acid. Draw the structure of the dicarboxylic acid used to make this form of nylon and the structure of nylon-6,10 showing the repeat unit.

Unlike addition polymers, condensation polymers such as nylon-6,6 do not form side-chains because the reaction is confined to the ends of the monomer molecules. Consequently the molecules tend to be long, single chains which allows them to be drawn out at room temperature to form fibres. During this process the linear molecules align and become increasingly linked by hydrogen bonds between adjacent chains. This causes the strength of the fibre to increase during the drawing process.

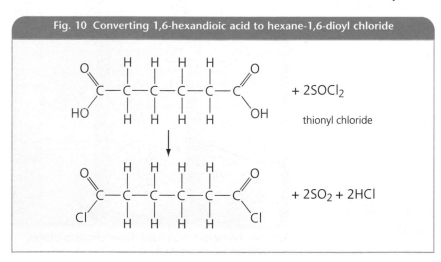

Fig. 10 Converting 1,6-hexandioic acid to hexane-1,6-dioyl chloride

Fig. 11 Forming Nylon-6,6

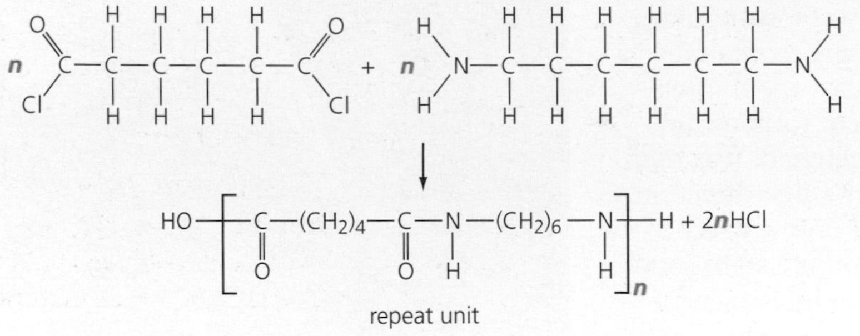

HO—[C—(CH₂)₄—C—N—(CH₂)₆—N]—H + 2nHCl

repeat unit

Fig. 12 Biodegradation of polyesters and polyamides

Polyester

Polyamide

hydrolysis

hydrolysis

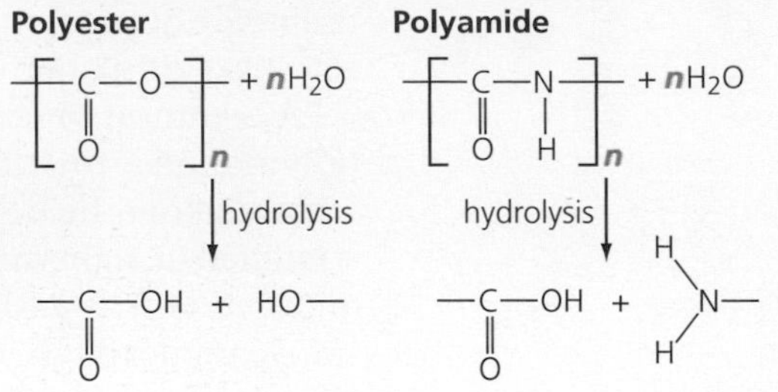

Biodegradability of polyesters and polyamides

As shown in Fig. 12, unlike the links between alkenes in poly(alkenes), the amide and ester links in polyamides and polyesters are hydrolysable. Under the influence of acid catalysts or bacterial enzymes, polyester and polyamide chains can be broken into smaller and smaller fragments. In other words, unlike poly(alkenes), such materials will biodegrade when placed in landfill sites.

4 Ester and amide bonds react with water in the presence of an acid catalyst.

a Write equations for such reactions for ethyl ethanoate and *N*-ethylethanamide.

b What type of reaction is involved in both of these reactions?

c Why is this reaction important from the point of view of biodegradability?

KEY FACTS

- Condensation polymers include polyesters and polyamides.

- Condensation polymers are biodegradable because the links are hydrolysable.

- Condensation polymers are derived from bifunctional compounds or two monomers.

- Proteins are naturally occurring condensation polymers.

9.4 Conclusion

Both addition and condensation polymers have contributed a lot to the way we live today, and will continue to do so as new polymers are developed and old ones are modified. The main problems facing the use of synthetic polymers in the future are the dwindling supplies of petroleum oil which is the source of all monomers, and the non-biodegradability of addition polymers.

Future developments may focus on the use of certain bacteria to 'grow' polymers as a natural part of their life-cycles. One example is the polyester Biopol, which was first grown using the bacterium *Alcaligenes eutrophus* in the early 1990s. The cost of the technology required is prohibitive at the moment, but this will be overcome in time.

Over time, bottles made from Biopol degrade far more quickly than non-biodegradable polymers.

APPLICATION A   Neoprene

Neoprene is a synthetic rubber with many applications in our modern world. For example, it is used for insulating wires, making drivebelts and hoses in car engines, coating a wide variety of materials to make them waterproof, and making wet suits. It is another example of an addition polymer but this one is made from a diene rather than a simple alkene. This means the molecule contains two C=C bonds.

The alkene used to make neoprene is 2-chloro-buta-1,3-diene which is a derivative of the hydrocarbon, buta-1,3-diene.

1 Draw the structures of these molecules.

First consider the polymerisation of buta-1,3-diene. This involves just one of the two C=C bonds. Consequently, one C=C bond reacts and one C=C bond is retained. This can happen in one of three ways.

A The addition polymerisation occurs using the C=C bond between C1 and C2, with C3 and C4 merely acting as a side-chain like CH_3 in poly(propene), C_6H_5 in polystyrene or Cl in PVC.

2 Show the structure of this addition polymer, and indicate the repeat unit.

B A new *trans* C=C bond is formed between C2 and C3, and one electron from the second C=C bond is used to bring about the polymerisation at the end carbon atoms.

3 Show the structure of this addition polymer, and indicate the repeat unit.

C This is like case **B**, except that a new *cis* C=C bond is formed between C2 and C3, and one electron from the second C=C bond is used for the polymerisation at the end carbon atoms.

4 Show the structure of this addition polymer, and indicate the repeat unit.
The addition polymer, Neoprene, is derived from 2-chloro-buta-1,3-diene in similar reactions.

5 With reference to **A** to **C** above, show the structures of the *four* possible polymer chains of Neoprene. Bear in mind that the Cl atom may or may not be part of the side-chain in type **C**.

In this raw state, diene polymers are soft and runny when warm but hard and brittle when cold. These are not desirable properties! To give a polymer the more durable rubber-like properties, it is vulcanized by heating it with sulphur, which reacts with some of the remaining C=C bonds to create C–S–C cross-links between polymer chains.

Vulcanisation was first applied to natural rubber by Goodyear in the late 1800s, and the process revolutionised the use of rubber. The application of this technique to synthetic addition polymers like neoprene opens up a wide range of practical uses.

APPLICATION B

Aramids

The condensation polymers discussed so far, including Terylene, nylon-6,6 and proteins, are all biodegradable because their ester or amide links are relatively easily hydrolysed. A group of polymers called **aramids** are also polyamides, but they are much tougher molecules to biodegrade. In fact, they are so tough and lightweight they are used to make bulletproof vests, fireproof suits, puncture-resistant bicycle tyres, reinforced concrete and Formula One racing cars.

Its fire resistance makes Nomex® ideal for lining the suits of racing drivers.

Being stronger and lighter than steel cable, Kevlar® is used to make sails.

Nomex® (made from 1,3-diaminobenzene and 1,3-benzenedicarboxylic acid) and Kevlar® are the two best-known examples of aramids. This general name derives from the fact that they involve benzene rings (from **ar**ene) linked via **amid**e bonds.

1

a Draw the structures of the Nomex® monomers.

b The monomers link via amide bonds to form chains. Showing the repeat unit, draw the structure of Nomex®.

2

a Kevlar® is made from 1,4-diaminobenzene and 1,4-benzenedicarboxylic acid. Draw the structures of these monomers.

b Again, the monomers link via amide bonds to form chains. Showing the repeat unit, draw the structure of Kevlar®

Kevlar® was discovered long before it was put to use. This is because it was so inert that processing proved very difficult. It does not dissolve in any of the usual solvents and it does not melt until 500 °C. It has these properties because the polymer chain is particularly regular in shape and has very strong and numerous intermolecular forces.

3 What type of intermolecular force will occur between Kevlar® polymer chains?

4 Draw a diagram to show this type of intermolecular force.

5 Which will have the higher melting point, Kevlar® or Nomex®? Explain your reasoning.

Having discovered such resilient materials, the chemists researching them set about finding ways of processing them into useful items. However, it was by chance that Stephanie Kwolek added calcium chloride as well as her favoured solvent (*N*-methyl pyrrolinidone). The ions from this salt are attracted to the C=O bonds in the polymer chain, and this breaks the intermolecular forces (see questions 3 and 4). Bingo, the polymer dissolved!

1 The polymer referred to as PET is formed when benzene-1,4-dicarboxylic acid and ethane-1,2-diol react together.

a Draw displayed formulas for benzene-1,4-dicarboxylic acid and ethane-1,2-diol.

b Describe the type of polymerisation occurring during the production of PET.

c Draw the structure of the repeat unit of this polymer.

2 Poly(phenylethene), also known as polystyrene, is formed when a monomer is polymerised.
In industrial production of poly(phenylethene), ethylbenzene is first produced in a reaction between ethene and benzene.

a Write a reaction for the formation of ethylbenzene, showing the structural formulae of the reactants and product.
Ethylbenzene then undergoes catalytic dehydrogenation to produce the monomer, which has a molecular formula of C_8H_8.

b i) What is meant by catalytic dehydrogenation?
  ii) Draw the structural formula of the monomer.

c Draw the structure of a poly(phenylethene) molecule, to include at least two monomer units.

d Name the type of polymerisation that has taken place.

3 Perspex is produced from the monomer shown in the diagram below.

To manufacture Perspex, the monomer undergoes an addition polymerisation reaction.

a What is meant by addition polymerisation?

b Draw the structure of the repeat unit for Perspex polymer.

4

a Compound **X** above can be polymerised. Draw a section of its polymer, showing two repeat units.
In a polymerisation reaction, phenylethene (styrene) can form poly(phenylethene), also known as polystyrene.

b What is the name of this type of polymerisation?

5 Describe the difference between *addition* polymerisation and *condensation* polymerisation using a suitable example of each type of polymerisation.

6
a Explain the term *biodegradable*.

b Describe two environmental benefits of using biodegradable plastics.

c Explain why polyesters and polyamides are biodegradable, while poly(alkenes) are not.

7 The polymer poly(chloroethene) is manufactured from chloroethene.

a Draw the displayed formula of the chloroethene.

b Draw the displayed structure of the polymer poly(chloroethene). Show three repeating units.

c What type of polymer is poly(chloroethene)?

8 Propane-1,2-diol, $CH_3CH(OH)CH_2OH$, is used to manufacture a polymer.

a What type of compound is propane-1,2-diol?

b Draw the structural formula of a suitable compound which could react with propane-1,2-diol to form a polymer.

c i) Draw the repeat unit of the polymer formed when these two compounds react.
  ii) Name the type of linkage formed between the molecules.

9 The polyamide Kevlar can be made from a diamine and a dicarboxylic acid. A section of Kevlar is shown below.

a What is meant by the term *polyamide*?

b Draw the structural formulas of the two monomers which react to make Kevlar.

c Name the type of polymerization that takes place when the monomers react to make Kevlar.

10 Organic synthesis and analysis

Chewing the bark or leaves from a willow tree has been known for centuries to give pain relief and reduce fevers. This was probably discovered by accident and the practice spread by word of mouth.

In the early nineteenth century the active ingredient was extracted, analysed and shown to be a substance called salicin, a compound closely related to 2-hydroxybenzoic acid. The tree population of the world could not be expected to supply enough of this material to satisfy growing demand. As a result, chemists, armed with their recent knowledge of its structure, set about the task of synthesising this or a closely related substance on a large scale.

In 1874, mass production of 2-hydroxybenzoic acid began, following a synthesis developed by the German chemist Kolbe, using readily available phenol (hydroxybenzene), sodium hydroxide and carbon dioxide. After prolonged use, this acid was shown to cause

In 1763, Edward Stone reported that an extract of willow bark was a 'Cure of Agues'. Nowadays, its derivative, aspirin, is used to relieve symptoms such as headaches. It also helps to avoid blood clots and so reduces the likelihood of heart attacks and strokes.

stomach irritations and its structure was modified again by esterification of the phenolic –OH group. This produced, in 1899, what we all know as aspirin, 2-ethanoyloxybenzoic acid, a drug used world-wide for all sorts of

treatments from headache to heart disease. This development is typical of the role played by analytical and synthetic chemists in a wide range of industries.

10.1 Introduction

Previous chapters have given you some insight into the vast variety of organic molecules that can occur both naturally and artificially. In this chapter you will be considering two problems that face organic chemists in research and industrial laboratories throughout the world: chemical **analysis** and **synthesis**.

Analysis tells us the chemical composition, the structure and the chemical characteristics of a particular compound. Synthesis tells us how to make a particular compound efficiently from available starting materials.

To understand both analysis and synthesis, you will need to apply the chemical reactions you have studied in previous chapters.

In analysis the absence or presence of a particular functional group can be determined from the observations made during chemical reactions. In synthesis, we choose reactions that will convert one material efficiently to the required product or to a material (called an **intermediate**) that can eventually, directly or indirectly, be converted to the required product.

10.2 Analysis

If you are presented with an organic molecule of unknown composition, you can gain detailed structural information about functional groups, relative molecular mass and distributions of hydrogen atoms by using the techniques of infra-red spectroscopy, mass spectrometry and nuclear magnetic resonance spectroscopy (see Chapter 11).

However, before using spectroscopic analysis, if we chemically analyse an organic compound we can find out a great deal of information about the molecule. This approach cuts down the number of possibilities considerably.

Combustion

Most organic compounds will burn under the right circumstances but how they burn in air, where the supply of oxygen is limited, often gives us useful general information about the molecule. Smaller molecules, particularly those containing only carbon, hydrogen and oxygen, tend to ignite more easily and burn more rapidly in air.

 1 Why does the presence of oxygen in the structure enhance the burning properties of an organic molecule?

However, it is the appearance of the flame that gives us clues about the nature of an organic molecule. Saturated molecules tend to burn with colourless, non-smoky flames, whereas molecules containing unsaturated carbon-to-carbon bonds produce sooty flames.

This is particularly true for benzene-related molecules where the percentage of carbon is very high. In these compounds there is a high degree of incomplete combustion, and large particles of soot (carbon) appear in the smoke.

Testing for alkenes

The electrophilic addition of bromine to alkenes (see *AS Chemistry*, pages 191–192) provides a test for the presence of C=C bonds in a molecule. As shown in Fig. 1, dilute, aqueous bromine solution (orange) is rapidly decolorised when reacted with an alkene.

 2 What is the predominant electrophile present in dilute, aqueous bromine solution?

 3 Identify the colourless organic product formed when but-1-ene decolorises aqueous bromine solution?

Testing for haloalkanes

The carbon–halogen bond in a haloalkane molecule is polar and susceptible to nucleophilic substitution (see *AS Chemistry*, Chapter 14). For example, hydrolysis of a haloalkane by warming it with dilute sodium hydroxide solution causes the covalently bonded halogen to be released in the form of a halide ion.

 4 Write an equation for the hydrolysis of a general haloalkane, RX (where X represents a halogen atom), by the hydroxide ion.

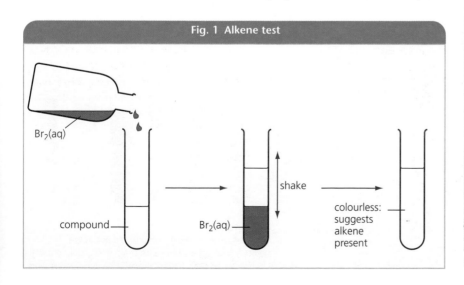

Fig. 1 Alkene test

Br₂(aq)

compound

Br₂(aq)

shake

colourless: suggests alkene present

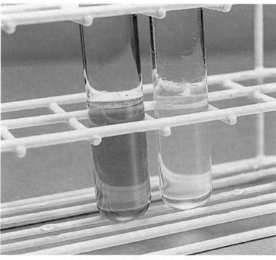

Bromine water before and after shaking with hex-1-ene.

Once the excess sodium hydroxide has been neutralised by dilute nitric acid, the halide ion can be identified using silver nitrate solution followed by ammonia solution (see *AS Chemistry*, Chapter 9). As shown in Table 1 and the photo, the visible effects tell you whether a chloro-, bromo- or iodoalkane is present or not and, indeed, which halogen it is. Fluoroalkanes are not detected because silver fluoride is water soluble.

At Step 2 (see Table 1), a chloroalkane (left) gives a white precipitate, a bromoalkane (centre) gives a cream precipitate and an iodoalkane (right) gives a pale yellow precipitate.

Table 1 Tests for haloalkanes

Procedure	For control e.g. water	For a chloroalkane	For a bromoalkane	For an iodoalkane
	Observations			
Step 1 Warm with dilute sodium hydroxide solution	no visible effect	aqueous and organic phases more miscible	aqueous and organic phases more miscible	aqueous and organic phases more miscible
Step 2 Add excess dilute nitric acid followed by silver nitrate solution	no visible effect	white precipitate of silver chloride	cream precipitate of silver bromide	yellow precipitate of silver iodide
Step 3 Add dilute ammonia solution then concentrated ammonia solution	no visible effect	white precipitate dissolves to form colourless solution when dilute ammonia added	cream precipitate dissolves to form colourless solution only when concentrated ammonia added	precipitate insoluble in both dilute and concentrated ammonia solution

Note: Fluoroalkanes are not included in this analytical scheme because Step 1 does not occur readily for fluoroalkanes and Step 2 does not produce a distinctive result because silver fluoride is water soluble.

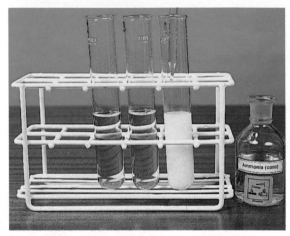

At Step 3, the precipitate from the chloride gives a colourless solution in dilute ammonia (left); the bromide precipitate dissolves only with concentrated ammonia (centre); and the iodide precipitate is not soluble in concentrated ammonia (right).

Testing for compounds containing oxygen

This group of organic compounds is very extensive and includes alcohols, aldehydes, ketones, acids and esters. As seen in Chapter 5 and summarised in Fig. 2, these molecules are closely related.

You could use infra-red spectroscopy (see Chapter 11) to find whether the C–O and C=O bonds that are characteristic of these molecules are present in a compound you are testing.

However, in order to distinguish the different types of compound, you have to use a combination of chemical tests based on acidity, redox reactions and condensation. It is important to apply these tests in the correct order so that eliminations can be made as efficiently as possible. This sequence is summarised in Table 2 opposite. Some results are illustrated in the photos below and on page 119.

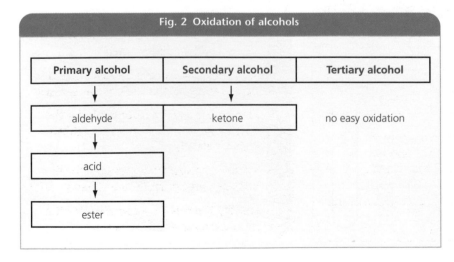

Fig. 2 Oxidation of alcohols

Primary alcohol	Secondary alcohol	Tertiary alcohol
↓	↓	
aldehyde	ketone	no easy oxidation
↓		
acid		
↓		
ester		

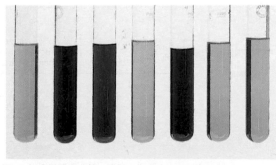

Results at Test 4, after warming with acidified potassium dichromate(VI) solution, are shown for (from left) water, primary alcohol, secondary alcohol, tertiary alcohol, aldehyde, ketone, acid.

Test	1 Add water and test the pH of the solution	2 Add sodium hydrogen carbonate solution	3 Add sodium metal	4 Warm with acidified potassium dichromate(VI) solution	5 Warm with ethanoic acid and conc. sulphuric acid catalyst	6 Warm with ethanol and conc. sulphuric acid catalyst	7 Reaction with Brady's reagent (2,4-dinitro-phenylhydra-zine in acid)	8 Fehling's test (see Chapter 5): warm with alkaline Cu^{2+}	9 Tollens test (see Chapter 5): warm with ammoniacal Ag^+
primary alcohol	neutral	no effect	colourless H_2 gas evolved quickly	orange to green	sweet smelling ester produced	no reaction	remains as orange solution	remains blue	remains colourless
secondary alcohol	neutral	no effect	colourless H_2 gas evolved slowly	orange to green	sweet smelling ester produced	no reaction	remains as orange solution	remains blue	remains colourless
tertiary alcohol	neutral	no effect	colourless H_2 gas evolved very slowly	remains orange	sweet smelling ester produced	no reaction	remains as orange solution	remains blue	remains colourless
aldehyde	neutral	no effect	no reaction	orange to green	no reaction	no reaction	bright orange-red precipitate	blue solution to brick red precipitate	colourless solution forms silver mirror
ketone	neutral	no effect	no reaction	remains orange	no reaction	no reaction	bright orange-red precipitate	remains blue	remains colourless
carboxylic acid	red with indicator: pH about 3	colourless CO_2 evolved	colourless H_2 gas evolved quickly	remains orange	no reaction	sweet smelling ester produced	remains orange	remains blue	remains colourless

Table 2 Testing for molecules containing oxygen

For Test 8, with Fehling's solution, an aldehyde gives a brick-red precipitate, while a ketone remains blue.

For Test 9, with Tollens reagent, an aldehyde (left) gives a silver coating, while a ketone (right) remains colourless.

Most of the chemistry involved in these tests is discussed in *AS Chemistry* or earlier chapters in this book. The reactions with sodium metal and Brady's reagent are exceptions but are included because they easily give specific information.

Moreover, we can also adapt these tests to characterise an ester, acyl halide or acid anhydride. If we hydrolyse esters, acyl halides or acid anhydrides (see Fig. 3), we produce other compounds that can be identified using the tests described above. Acyl halides and anhydrides react particularly vigorously during the initial hydrolysis.

5 Why does the acid in Table 2 give a weakly acidic pH?

6 Which group of atoms is responsible for the production of hydrogen gas when reacted with sodium metal?

7 Why must the sample be absolutely dry before applying the sodium metal test?

8 Identify the species responsible for the green colour in the test with acidified dichromate(VI).

9 Identify the species responsible for the brick-red precipitate in the Fehling's test?

10 Identify the species responsible for the silver mirror in the Tollens test?

Fig. 3 How to characterise esters, acyl halides and acid anhydrides

	Ester	Acyl halide	Acid anhydride
Hydrolyse using warm NaOH(aq)	alcohol + sodium salt of carboxylic acid	sodium salt of carboxylic acid + sodium halide	sodium salt of carboxylic acid
+ excess HNO₃(aq)	alcohol + carboxylic acid	carboxylic acid + sodium halide	carboxylic acid
Apply Test 5 from Table 2 *and* Test 6 from Table 2 *and* silver nitrate-ammonia tests from Table 1	Tests 5 and 6 positive	Test 6 and silver nitrate tests positive	Test 5 only positive

Testing with sodium metal

As you will know from Module 1 studies, sodium metal reacts vigorously with water to produce sodium hydroxide solution and hydrogen gas:

$$Na + H_2O \rightarrow Na^+(aq) + OH^-(aq) + \tfrac{1}{2}H_2(g)$$

This reaction depends on the reducing properties of sodium. Electrons are donated to the water molecule and this causes one of the O–H bonds to break: molecular hydrogen gas is released and a hydroxide ion is left behind.

Primary alcohols, secondary alcohols, tertiary alcohols and acids all react in a similar fashion because they too contain the necessary O–H bond:

$$Na + ROH \rightarrow Na^+(aq) + RO^-(aq) + \tfrac{1}{2}H_2(g)$$
$$Na + RCOOH \rightarrow Na^+(aq) + RCOO^-(aq) + \tfrac{1}{2}H_2(g)$$

The organic products are alkoxides (RO⁻) and alkanoates (RCOO⁻) respectively. Acids react faster because the O–H bond in these molecules is more strongly polarised. Primary, secondary and tertiary alcohols react increasingly slowly because the added hydrocarbon groups 'get in the way' and slow down the reduction process.

Testing with Brady's reagent

For Test 7 in Table 2, Brady's reagent is a solution of 2,4-dinitrophenylhydrazine dissolved in sulphuric acid. It is a bright orange solution which reacts to form dense, orange-red precipitates called hydrazones when mixed with any carbonyl compound. This is shown in Fig. 4. Hence, we can use this solution to confirm the presence or absence of aldehydes and ketones.

11 If a positive Brady's test is observed, what should be the next step in your analysis?

12 What type of reaction is involved in a positive Brady's test?

Testing for amines and amino acids

Amines and amino acids are both related to ammonia. Ammonia displaces water ligands to form a distinctively dark blue complex when added to aqueous copper(II) ions:

$$[Cu(H_2O)_6]^{2+}(aq) + 4NH_3(aq) \rightarrow$$
$$[Cu(NH_3)_4(H_2O)_2]^{2+}(aq) + 4H_2O\ (aq)$$
dark blue ammonia complex

With aqueous copper(II) ions and excess ammonia, amines and amino acids give a dark blue soluble complex.

Amines and amino acids can be detected in a similar manner. It is worth noting that aromatic amines tend to form green, insoluble complexes, so giving further useful information on the identity of an organic compound.

13 Which structural feature common to ammonia, amines and amino acids allow this test to work?

Once we have eliminated the possibility that the compound is ammonia by testing its combustibility in air, we can then distinguish amines from amino acids by adding Full Range indicator, followed by drops of very dilute sodium hydroxide solution.

The initial pH of an amine is considerably higher than that of the amino acid. If we have an amine, when we add the hydroxide, the indicator colour changes rapidly to the highly alkaline pH region. But this change requires significantly more alkali if the compound is an amino acid.

Fig. 4 Brady's test

2,4-dinitrophenylhydrazine + **aldehyde or ketone** ⟶ carbonyl-2,4-dinitrophenylhydrazone + **water**

14 Why do amines and amino acids react differently with sodium hydroxide solution?

15 What chemical tests would you apply to identify the functional group(s) present in each of the following molecules?
  i $CH_3CH_2CH_2COOH$ ii $CH_3CH_2CH_2OH$
  iii $C_6H_5CH_2Br$ iv $CH_3CH_2CH_2CH_2CHO$
  v $CH_3CH(OH)CH_2COOH$ vi CH_3COCl
  vii $CH_3CH=CHCH_2CH_2CH_3$
  viii $CH_3C(CH_3)(OH)CH_2CH_3$
  ix $CH_3CH_2NH_2$

16 What chemical tests would you apply to distinguish between the following pairs of molecules?
  a $CH_3CH_2CH_2CH_2CH_2OH$ and $CH_3C(CH_3)(OH)CH_2CH_3$
  b $CH_3CH_2CH_2CH_2Cl$ and $CH_3CH_2CH_2CH_2Br$
  c $CH_3CH_2CH_2CH_3$ and $CH_3CH=CHCH_3$
  d $CH_3CH_2COCH_3$ and $CH_3CH_2CH_2CHO$
  e $CH_3CH_2CH_2COOCH_3$ and $CH_3CH(OH)CH_2COCH_3$

10.3 Conclusion

As part of the analysis of an unknown organic compound, identifying functional groups by chemical testing can be very fruitful. But remember that it is very easy to get misleading results because of contamination and incorrect procedures. Furthermore, negative tests are often more useful than positive tests; negative tests allow us to eliminate the possibility that certain functional groups are present.

10.4 Organic compound synthesis

Designing the synthesis of one organic molecule from another is a true test of your knowledge of organic chemistry. Furthermore, it is a major activity in industrial research laboratories because it allows new molecules, which may have useful properties, to be made from other, readily available molecules. Penicillin, quinine, aspirin, vitamins and insect repellents are just a few examples of useful molecules which are not available in adequate quantities in nature and have to be synthesised. Synthesis is also the final stage in the analysis of any new compound. Once a chemist has deduced a structure using chemical testing and spectroscopy, that structure must be confirmed by showing that it can be reproduced identically using synthesis.

You can use just the reactions you have studied in AS Module 3 and A2 Module 4 to devise a vast number of different syntheses, both for well-known molecules and, possibly, totally unknown molecules. When designing such a synthesis, many factors will influence your final choice of pathway. These include:

- The availability of a suitable starting material from which the target molecule can be synthesised. The suitable starting material is usually arrived at by working backwards from the target and considering what carbon groups need to be 'joined' and what functional groups need to be created.

- The number of stages needed to convert the starting material to the target molecule; these will include both chemical reactions and subsequent separation and purification processes.

- The percentage yield that can be expected from each of the stages.

- Whether competing reactions can occur at any of the stages of the synthesis, resulting in lower yields and contamination.

- Are there any additional problems involved in scaling the synthesis to manufacturing proportions, particularly relative costs of starting materials, special designs for chemical plant and safety considerations?

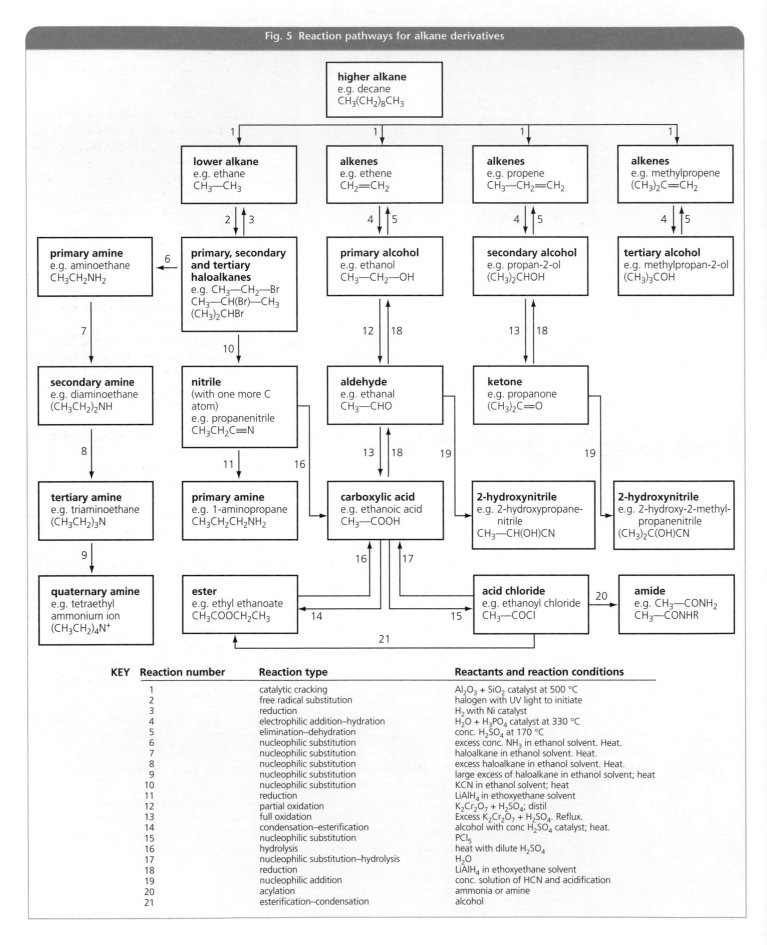

Fig. 5 Reaction pathways for alkane derivatives

KEY	Reaction number	Reaction type	Reactants and reaction conditions
	1	catalytic cracking	$Al_2O_3 + SiO_2$ catalyst at 500 °C
	2	free radical substitution	halogen with UV light to initiate
	3	reduction	H_2 with Ni catalyst
	4	electrophilic addition–hydration	$H_2O + H_3PO_4$ catalyst at 330 °C
	5	elimination–dehydration	conc. H_2SO_4 at 170 °C
	6	nucleophilic substitution	excess conc. NH_3 in ethanol solvent. Heat.
	7	nucleophilic substitution	haloalkane in ethanol solvent. Heat.
	8	nucleophilic substitution	excess haloalkane in ethanol solvent. Heat.
	9	nucleophilic substitution	large excess of haloalkane in ethanol solvent; heat
	10	nucleophilic substitution	KCN in ethanol solvent; heat
	11	reduction	$LiAlH_4$ in ethoxyethane solvent
	12	partial oxidation	$K_2Cr_2O_7 + H_2SO_4$; distil
	13	full oxidation	Excess $K_2Cr_2O_7 + H_2SO_4$. Reflux.
	14	condensation–esterification	alcohol with conc H_2SO_4 catalyst; heat.
	15	nucleophilic substitution	PCl_5
	16	hydrolysis	heat with dilute H_2SO_4
	17	nucleophilic substitution–hydrolysis	H_2O
	18	reduction	$LiAlH_4$ in ethoxyethane solvent
	19	nucleophilic addition	conc. solution of HCN and acidification
	20	acylation	ammonia or amine
	21	esterification–condensation	alcohol

10 ORGANIC SYNTHESIS AND ANALYSIS

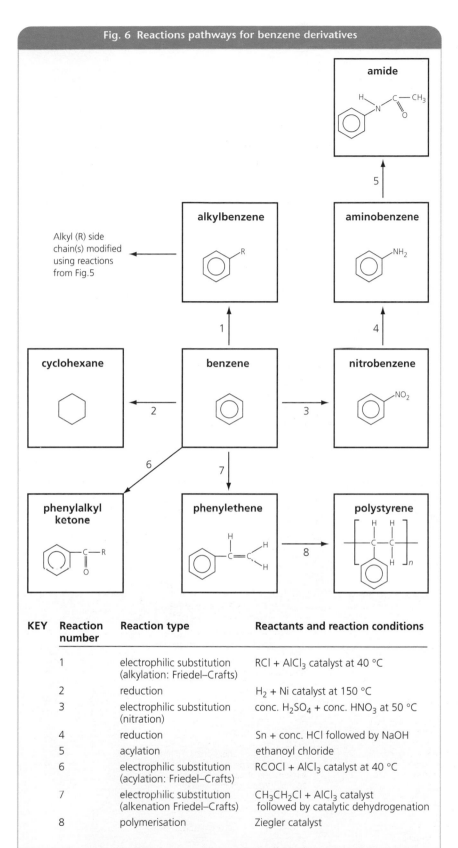

Fig. 6 Reactions pathways for benzene derivatives

Alkyl (R) side chain(s) modified using reactions from Fig.5

KEY	Reaction number	Reaction type	Reactants and reaction conditions
	1	electrophilic substitution (alkylation: Friedel–Crafts)	$RCl + AlCl_3$ catalyst at 40 °C
	2	reduction	H_2 + Ni catalyst at 150 °C
	3	electrophilic substitution (nitration)	conc. H_2SO_4 + conc. HNO_3 at 50 °C
	4	reduction	Sn + conc. HCl followed by NaOH
	5	acylation	ethanoyl chloride
	6	electrophilic substitution (acylation: Friedel–Crafts)	$RCOCl + AlCl_3$ catalyst at 40 °C
	7	electrophilic substitution (alkenation Friedel–Crafts)	$CH_3CH_2Cl + AlCl_3$ catalyst followed by catalytic dehydrogenation
	8	polymerisation	Ziegler catalyst

17 In a three-stage synthesis, the molar yields of the individual stages are 50%, 40% and 75%. What is the overall molar percentage yield?

The interconnections between the more significant synthetic reactions are shown in Figs 5 and 6. These emphasise the fact that, because of delocalisation effects, functional groups bonded directly to benzene ring structures are likely to behave differently to those bonded to saturated structures. Hence, this is often an important consideration when selecting reactions to achieve a particular synthesis.

18 State the reagent(s) and conditions needed to achieve each of the following conversions. Write equations for the reactions that occur.
a $CH_3CH_2CH(OH)CH_3$ to $CH_3CH_2COCH_3$
b $CH_3CH_2CH_2OH$ to $CH_3CH_2CH_2OCOCH_3$
c $CH_3CH_2CH_2CH_2OH$ to $CH_3CH_2CH=CH_2$
d $CH_3CH_2CH_2COOH$ to $CH_3CH_2CH_2CH_2OH$
e CH_3CH_2Br to $CH_3CH_2NH_2$

19 Devise reaction schemes for each of the following syntheses:
a Ethanoic acid from chloroethane
b Propanoic acid from bromoethane
c Propanone from propene
d Ethyl ethanoate from ethanol
e Aminobenzene from benzene

The pathways described allow you to select possible synthetic routes to your target molecule from different starting materials via different numbers of steps, each with differing yield expectations. However, in industry, only by experimentation and with a thorough costing of the alternative routes, is it possible to assess the relative merits of the synthetic processes.

Biologically active materials

Insects attract their mates by releasing organic molecules called pheromones. Muscalure is the pheromone produced by the housefly. It can be used to attract all female flies in an area and so prevent the production of the next generation.

Muscalure

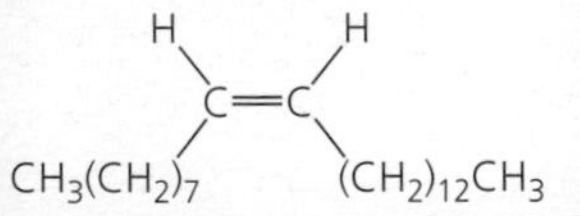

Thyroxine is a hormone which is partly responsible for the overall metabolic rate of the body. It is produced by the thyroid gland in the neck. A deficiency of thyroxine leads to low metabolic rate and goitre. An excess of thyroxine leads to an over-active metabolism with weight loss and irritability.

Thyroxine

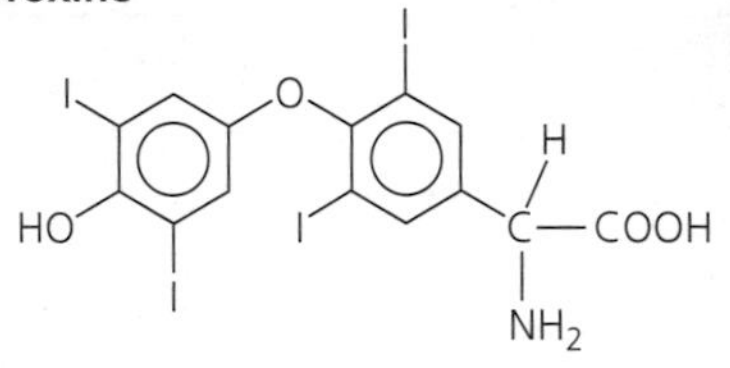

Better known as amphetamine, benzedrine is a stimulant which has direct effects on the central nervous system. It has been used to alleviate depression, fatigue, Parkinsonism, obesity and many other clinical conditions. However, because of the effects it has on mood, benzedrine is often used as a substance of abuse.

Benzedrine

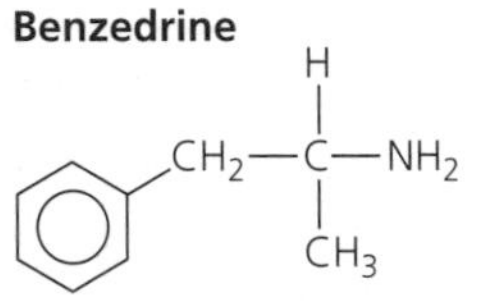

Grandisol is another insect pheromone which is found naturally in the male cotton boll weevil. This insect is capable of destroying entire cotton crops and causing extreme economic losses. Use of this pheromone to trap and so control the population of the insect is of vital economic importance.

Grandisol

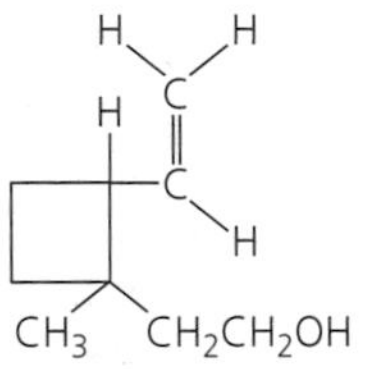

Extracts from willow bark have been used for pain relief for many centuries. They have since been shown to contain derivatives of 2-hydroxysalicylic acid. This is a precursor of aspirin: the conversion is completed on an industrial scale using an acylation reaction with ethanoic anhydride. Apart from its common use for pain relief, aspirin is used to help prevent heart disease.

Aspirin

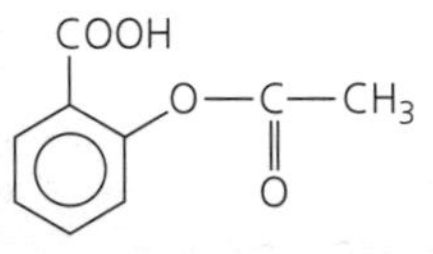

The molecules described in the table contain a wide variety of functional groups.

1 Identify these groups for each of the molecules.
2 Devise a series of chemical tests which could be used to distinguish five unlabelled samples which are known to contain these five materials.

APPLICATION B

How can benzocaine be synthesised?

Benzocaine is chemically related to cocaine. Cocaine is extracted from the leaves of the coca plant which is grown in South America. Both cocaine and benzocaine have medical uses, particularly as anaesthetics; but cocaine is dangerously addictive and has become a substance of abuse, associated with world-wide crime.

Fig. 8 Benzocaine

1 Given the following additional information, devise a scheme for making benzocaine from benzene.

2 Explain your choice of reaction sequence and write a balanced equation for each step of the process.

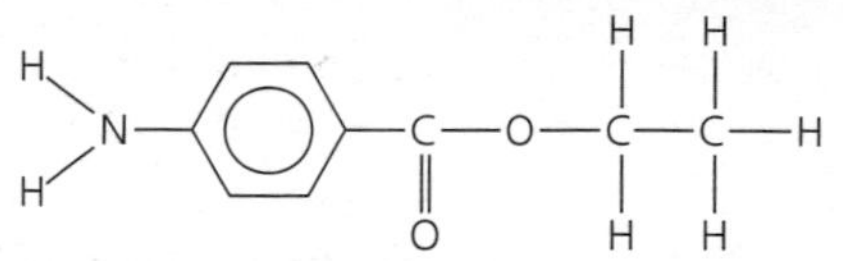

Additional information

- The presence of a nitro group on a benzene ring causes the next substitution to occur preferentially at the 3-position.

- The presence of an alkyl group on a benzene ring causes the next substitution to occur preferentially at the 2 and 4 positions.

- Any alkyl group on a benzene ring can be oxidised to a –COOH group by heating with alkaline potassium manganate(VII). In an equation, this can be represented by [O].

QUESTIONS

1 Suggest the simple tests you could use to distinguish between the following pairs of compounds:
a CH_3COOH and $HCOOCH_3$
b $CH_3CH_2CH_2Br$ and $CH_3CH_2CH_2OH$
c CH_3CH_2COCl and $CH_3CH_2CH_2Cl$
d $CH_3CH_2CH_2CH_2OH$ and $(CH_3)_3COH$

2 A compound has the structural formula shown in the diagram.
a Briefly describe the simple chemical tests that identify the functional groups present in the compound above.
b Giving appropriate reagents and reaction conditions, outline a reaction scheme to convert the above compound into:

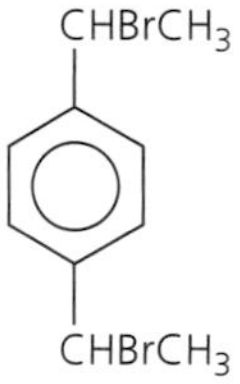

3 Aqueous acidified potassium dichromate(VI), was reacted with samples of butan-1-ol, butan-2-ol and 2-methylpropan-2-ol.
a Copy and complete the table below for each alcohol, stating what you would observe and identifying the organic product of each reaction.

Alcohol	Observation	Organic product (if any)
Butan-1-ol		
Butan-1-ol		
2-methylpropan-2-ol		

b For butan-1-ol and butan-2-ol, describe a simple chemical test to distinguish between the organic products you have given in **a**.

4 State the reagents and conditions needed to carry out each of the following conversions.
a 4-ethyl-nitrobenzene to 4-ethyl-aminobenzene
b propan-2-ol to propene
c propan-1-ol to propanoic acid
d propanone to 2-bromopropane [2 steps]
e benzene to 1-phenylpropane

5 Design outline schemes for each of the following preparations.
a butanoic acid from 1-bromopropane [2 steps]
b ethyl ethanoate from ethanol [2 steps]
c ethanoic acid from bromoethane [3 steps]
d 4-methyl-aminobenzene from benzene [3 steps]
e N-phenylethanamide from benzene [3 steps]

6 Describe simple chemical tests (more than one stage may be needed), including the expected observations, which may be used to differentiate between samples of:
a hex-1-ene and hexane
b propan-1-ol and propanoic acid
c propan-1-ol and propan-2-ol
d 1-aminobutane and butanenitrile
e ethanal and ethanoic acid

11 Structure determination

MRI (magnetic resonance imaging) scanners are powerful tools for medical diagnosis. The first commercial MRI scanners were available in 1983, only seven years after the scanner was first developed, and are now used in major hospitals world-wide. These scanners provide fast and accurate diagnosis of a very wide range of illnesses.

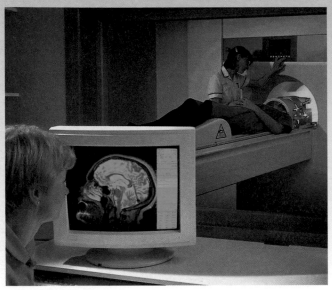

Scanning a patient by n.m.r. His body is lying along the central axis of a superconducting electronic magnet.

MRI has its origins in a technique called nuclear magnetic resonance spectroscopy (n.m.r.), that chemists have been using regularly since the 1950s. The technique helps chemists to deduce molecular structures by pinpointing and counting the positions of atoms such as ^1H atoms in molecules. This led the way to MRI scanners because ^1H occurs in all organic and biochemical molecules and is particularly abundant in the body in the form of water: water makes up about 70% of living tissues. This makes it possible for MRI to investigate living tissues by looking at the different effects caused by their varying water content.

MRI easily shows the distribution of different tissues. Mainly because of their different water content, bone is easily distinguished from muscle, and muscle from fat, giving clearly defined images. More importantly, diseased tissue is distinguished from healthy tissue!

This technique and others are used in this chapter to elucidate structures.

How do chemists know the structure of any particular molecule? By the time you reach this chapter, you will be aware of the vast variety of structures for organic compounds that occur both naturally and artificially. Variations in numbers of carbon atoms per molecule, arrangements of those carbon atoms, functional groups and isomers all add to the almost unlimited number of organic structures, each of which is unique. Consequently, structure determination involves some fascinating detective work, where experimental results from a variety of sources are each analysed to provide a piece of the jigsaw, finally arriving at the overall picture.

What experimental sources are used? Apart from chemical analysis, which can give general information such as the empirical formula of the compound and indicate the presence of particular functional groups in the compound, spectroscopic methods are the most generally useful sources of evidence for molecular structure.

1 An organic compound is found to contain carbon, hydrogen and oxygen. 0.44 g of the compound reacts to produce 0.88 g of carbon dioxide and 0.36 g of water. Calculate the empirical formula of the compound.

2 The relative molecular mass of the compound from Q1 is 88. Deduce the molecular formula of the compound.

3 Given that the compound from Q1 and Q2 is weakly acidic, deduce the possible structures for the compound.

In general, spectroscopic methods involve making observations of how molecules react when subjected to various physical stimuli such as infra-red radiation, microwave radiation, ultra-violet radiation, magnetic fields and radio-frequency radiation. The most generally useful are **infra-red spectroscopy** (section 11.1) and **nuclear magnetic resonance spectroscopy** (section 11.3). These are the main topics of this chapter.

Infra-red spectroscopy

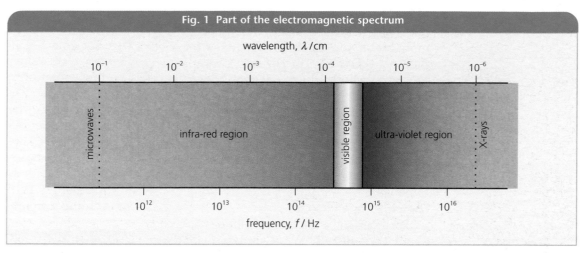

Fig. 1 Part of the electromagnetic spectrum

wavelength, λ /cm

10^{-1} 10^{-2} 10^{-3} 10^{-4} 10^{-5} 10^{-6}

microwaves | infra-red region | visible region | ultra-violet region | X-rays

10^{12} 10^{13} 10^{14} 10^{15} 10^{16}

frequency, f / Hz

What is infra-red spectroscopy? First, we need to consider what it is that is measured in infra-red spectroscopy. Light from the sun consists of a wide range of **electromagnetic radiation** of different frequencies (Fig. 1). There is infra-red radiation, which is responsible for the warmness we all feel on a summer's day, there is 'visible' light, which we can see as different colours, and there is ultra-violet radiation, which is associated with getting a suntan. These three different types of radiation differ in wavelength and in the amount of energy that they carry.

The wavelength λ (cm) of an electromagnetic wave is the distance between adjacent peaks (Fig. 2).

Infra-red radiation corresponds approximately to wavelengths between 1.0×10^{-3} and 0.7×10^{-6} m (1 mm and 0.7 μm) (see Fig. 1). The different regions of the electromagnetic spectrum can also be described by frequency f (Hz or s^{-1}), energy per mole E (J mol^{-1}) or wavenumber v (cm^{-1}). These are interrelated and each can be calculated from one or more of the others using the following universal constants:

- the speed of light ($c = 3.00 \times 10^8$ m s^{-1}),
- Planck's constant ($h = 6.63 \times 10^{-31}$ kJ s),

The interrelationships are described by the following equations:

- $f = c/\lambda$ (Hz or s^{-1})
- $E = hf = hc/\lambda$ (J mol^{-1})
- $v = 1/\lambda$ (cm^{-1})

For historical reasons, wavenumbers are most often used in infra-red spectroscopy. The wavenumber equals the number of wavelengths that will fit into 1 centimetre: $v = 1/\lambda$. Hence, radiations with shorter wavelengths, higher energies and higher frequencies correspond to larger wavenumbers, and vice versa.

The full range of infra-red radiation is not used in infra-red spectroscopy. Most infra-red spectrophotometers (the instruments for carrying out infra-red spectroscopy) operate at wavenumbers from 600 to 4000 cm^{-1}. This provides a range of radiation suitable for detecting all types of covalent bond.

4 Calculate the range of wavelengths (λ) used in infra-red spectroscopy.

5 Calculate the range of frequencies (f) used in infra-red spectroscopy.

6 Calculate the energies (E) corresponding to the infra-red radiation used in infra-red spectroscopy.

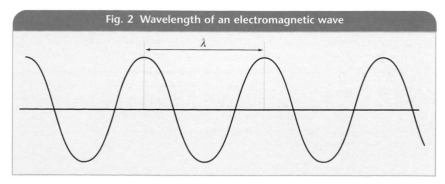

Fig. 2 Wavelength of an electromagnetic wave

λ

Absorption of infra-red radiation

Why does measuring infra-red radiation give information about the structure of a molecule? It is the covalent bonds in molecules that are actually responsible for molecules absorbing different parts of the infra-red spectrum. The covalent bonds in molecules act like springs in continuous, high-frequency vibration (Fig. 3). The natural vibrational frequencies of covalent bonds are in the same range as the frequencies of infra-red radiation, but they vary slightly from bond to bond. They are characteristic of any particular bond and depend on the type of bond (single, double, etc.) and the atoms combined by the bond.

For example, a C–Cl bond vibrates at about 2.0×10^{13} s^{-1}, but a C–Br bond vibrates at about 1.7×10^{13} s^{-1}.

 7 Why does the bond in C–Br vibrate more slowly than the bond in C–Cl?

When a molecule is exposed to infra-red radiation, a vibrating bond will absorb infra-red energy when the infra-red frequency and the natural vibrational frequency of the bond coincide. This is referred to as a **resonance** effect. Since different bonds have different natural vibrational frequencies, we can identify different bonds by analysing which infra-red frequencies are absorbed.

There are several types of bond vibration. They are generally either 'stretching' vibrations, i.e. along the bond (Fig. 4) or 'bending' vibrations, i.e. across the bond (Fig. 5). Both the stretching and bending vibrations can be 'symmetric' (in step) or 'asymmetric' (out of step), with each having its own characteristic vibrational frequency. These different types of bond vibration occur simultaneously for any particular group of atoms, making the overall vibrational pattern for even the simplest molecule very complex.

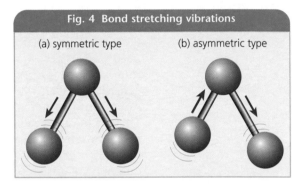

Fig. 4 Bond stretching vibrations
(a) symmetric type   (b) asymmetric type

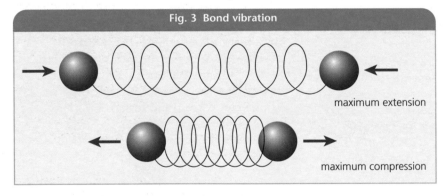

Fig. 3 Bond vibration

maximum extension

maximum compression

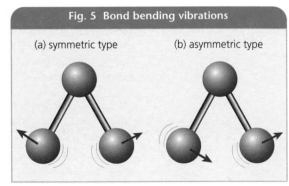

Fig. 5 Bond bending vibrations
(a) symmetric type   (b) asymmetric type

KEY FACTS

■ Covalent bonds are in continuous vibration.

■ Different bonds absorb different infra-red frequencies.

■ Absorption of infra-red frequency depends on the frequency of the radiation matching the vibrational frequency of the bond.

Infra-red spectra

Almost all molecules contain several different types of bond. Each bond will have several simultaneous stretching and bending types of vibration, with each type of vibration having a characteristic vibrational frequency. Consequently, molecules will cause many absorptions (some strong, some weak, some sharp, some broad) in the infra-red region. Together, these absorptions are called the 'infra-red spectrum' of the molecule.

Recording an infra-red spectrum

The infra-red absorptions are measured using an infra-red spectrophotometer (Fig. 6). The infra-red (IR) beam, produced by a heating element, is split into two identical beams – the sample beam and the reference beam. By comparing the frequencies remaining in the sample beam after passing through the sample with those present in the reference beam, the wavenumber of each absorption and the degree of each absorption can be measured. The use of two beams enables strong, misleading absorptions caused by water vapour and carbon dioxide gas in the air inside the spectrophotometer to be cancelled out because they are common to both beams.

Liquid, dissolved or suspended samples are usually 'sandwiched' as a thin film between polished discs made from sodium chloride or potassium bromide. NaCl and KBr are used because they are both transparent to infra-red radiation. This 'sandwich' is positioned in the sample beam. Gaseous samples are examined in specially designed sealed cells with a much larger thickness (path length) of sample to allow adequate absorption.

8 Can you think of any practical problems which might result from using a sample in 'solution' or 'suspension' form?

Identifying functional groups

By comparing structures and spectra, absorptions of a particular wavenumber can be associated with a particular covalent bond. This is shown in Fig. 7a–c for an alkane, an alkene and an alkyne.

The spectrum of 2,3-dimethylbutane (an alkane; Fig. 7a) contains two main groups of absorptions. These can be shown to be due to stretching of the C–H bond (the absorptions at 2960 cm⁻¹ and 2876 cm⁻¹) and bending of the C–H bond (the absorptions at 1464 cm⁻¹ and 1380 cm⁻¹).

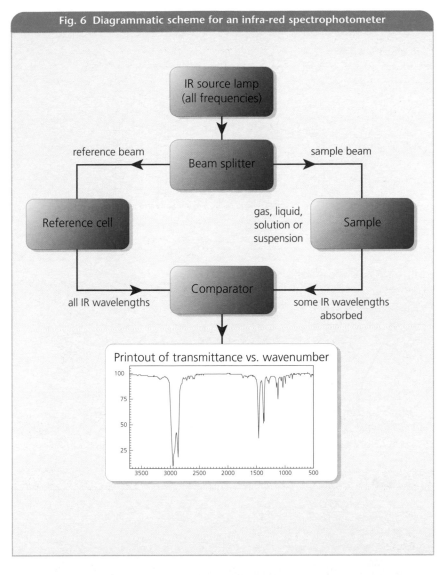

Fig. 6 Diagrammatic scheme for an infra-red spectrophotometer

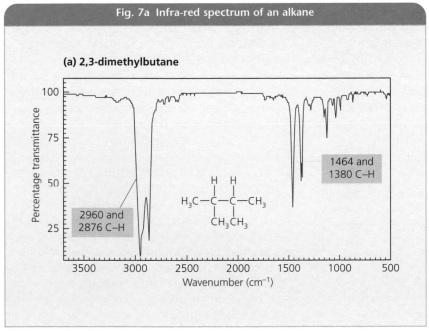

Fig. 7a Infra-red spectrum of an alkane

129

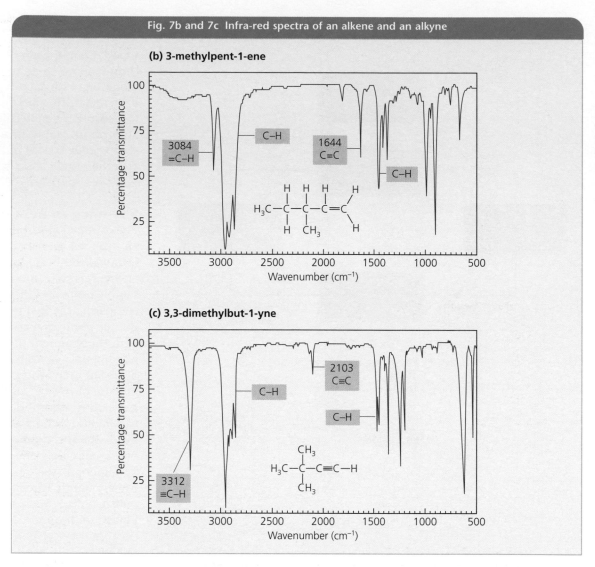

Fig. 7b and 7c Infra-red spectra of an alkene and an alkyne

The spectrum of 3-methylpent-1-ene (an alkene; Fig. 7b) contains two more main absorptions, when compared with the spectrum of the alkane. These two absorptions are due to stretching vibrations of the C–H bonds attached to the C=C double bond (the absorption at 3084 cm⁻¹) and stretching vibrations of the C=C bond (the absorption at 1644 cm⁻¹).

When looking at the spectrum of 3,3-dimethylbut-1-yne (an alkyne; Fig. 7c), the C–H stretching and bending absorptions are still present. The extra absorptions are those arising from stretching vibrations of the C–H attached to the C≡C bond (3312 cm⁻¹) and stretching vibrations of the C≡C bond (2103 cm⁻¹).

A comparison of the wavenumbers for the C=C and C≡C bonds shows that the shorter, stronger triple bond absorbs at a higher frequency (higher wavenumber).

By comparing the spectra of a large number of molecules of known structure, correlation tables of bond type versus infra-red absorption regions can be compiled. Table 1 is such a table, although far more extensive tables are available. However, Table 1 does include all the groups you are likely to meet during your advanced level studies. Such correlation tables allow particular bonds and groups of atoms ('functional groups') in a molecule to be recognised by matching experimental absorptions with data from the table.

9　Compare the absorptions caused by the stretching vibrations of the C–H bonds attached to C–C, C=C and C≡C groups. What pattern can you see?

10　Compare the spectra in Fig. 8 and, as far as possible, associate the absorptions labelled 1–11 with C–H, C–O, O–H or C=O bonds. Explain your deductions.

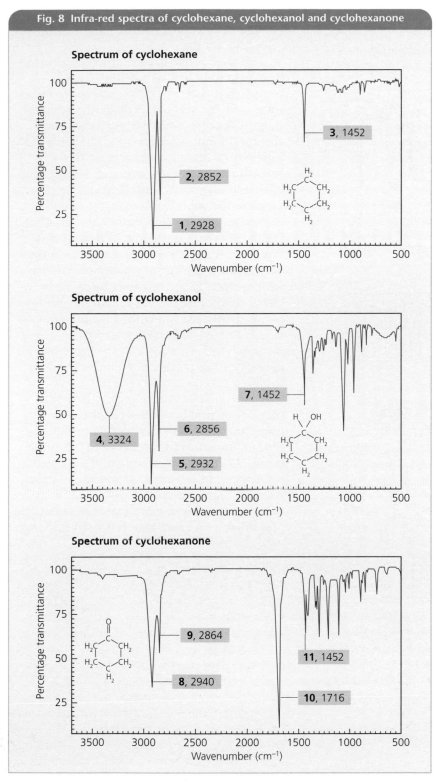

Fig. 8 Infra-red spectra of cyclohexane, cyclohexanol and cyclohexanone

Spectrum of cyclohexane

Spectrum of cyclohexanol

Spectrum of cyclohexanone

Table 1 An infra-red correlation table		
Bond	**Functional group**	**Wavenumber /cm^{-1}**
C–H	alkyne	3250–3310
C–H	alkyne	3250–3310
C–H	alkene	3000–3100
C–H	alkane	2850–2975
C–H	arene	3030–3080
C–H	aldehyde	2650–2880
N–H	amine (non H-bonded)	3320–3560
N–H	amide (non H-bonded)	3320–3560
N–H	amine (H-bonded)	3100–3400
O–H	alcohol (non H-bonded)	3580–3670
O–H	alcohol (H-bonded)	3230–3550
O–H	phenol	3100–3380
O–H	acid (H-bonded)	2500–3000
C≡N	nitrile	2210–2260
C≡C	alkyne	2100–2260
C=C	alkene	1620–1690
C=C	arene	1450–1600
C=O	acid chloride	1790–1815
C=O	ester	1730–1750
C=O	acid (alkyl)	1700–1725
C=O	acid (aryl)	1680–1700
C=O	aldehyde	1685–1740
C=O	ketone	1680–1725
C=O	amide	1630–1700
C–F	fluoro	1000–1400
C–Cl	chloro	600–800
C–Br	bromo	500–600
C–I	iodo	About 500
C–C	alkane	720–1175
C–O	ester	1180–1310
C–O	phenol	1120–1220
C–O	alcohol	1050–1150
C–N	amine	1030–1230
C–N	amide	1590–1650
C–H bend	arene	700–880
C–H bend	alkane	1365–1485

The values in Table 1 are for bond stretching vibrations, except for the bond bending vibration of the C–H bond. Notice that stretching vibrations require more energy than bending vibrations, so bond stretches occur at higher wavenumbers. Vibrations involving either stronger bonds (e.g. double

and triple bonds) or bonds to lighter atoms (e.g. hydrogen) occur at higher wave numbers. Thus single bonds to hydrogen (C–H, O–H, N–H) absorb between 4000 and 2500 cm^{-1}, double bonds absorb between 1500 and 2000 cm^{-1} and triple bonds absorb between 2000 and 2500 cm^{-1}.

Bonds involved in hydrogen bonding tend to give broad absorptions rather than sharp discrete absorptions. As a result, it is easy to recognise molecules containing –OH and –NH

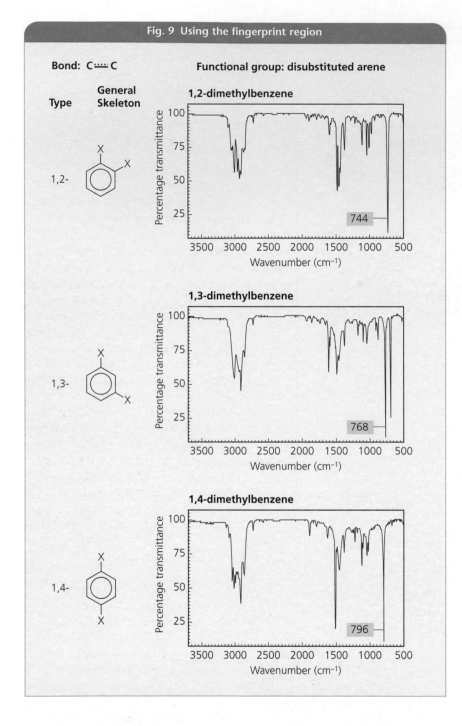

Fig. 9 Using the fingerprint region

Bond: C≡≡≡C   Functional group: disubstituted arene

Type | General Skeleton

1,2-

1,3-

1,4-

groups, i.e. alcohols, phenols (where OH is bonded directly to a benzene ring), carboxylic acids and amines.

The O–H bonds in water molecules cause hydrogen bonding. As a result, organic compounds that are 'wet' (contain traces of water) can be easily identified. Any residual water will cause a very strong, very broad absorption around 3500 cm⁻¹. The ratio of the intensity of the water absorption to the intensity of one of the stronger absorptions caused by the pure compound can be used to measure the water content.

The remainder of an infra-red spectrum (1500–500 cm⁻¹) is called the '**fingerprint' region** of the spectrum. Absorptions in the fingerprint region form a complex pattern, which depends on the molecule as a whole, rather than on particular functional groups or bonds. Using the fingerprint region allows closely related carbon skeletons to be distinguished. This is shown in Fig. 9 where, because of the presence of similar bonds, the infra-red spectra can be seen to be very similar from 3500 to 1500 cm⁻¹. However, they are distinctly different in their fingerprint regions (1500–500 cm⁻¹) because of the differences in arrangements of those bonds. Computer matching of the fingerprint regions of sample infra-red spectra and pre-recorded infra-red spectra of known compounds allow a particular molecule to be identified if an exact match is produced.

KEY FACTS

■ Using correlation tables, infra-red absorptions above 1500 cm⁻¹ can be associated with particular functional groups.

■ Using correlation tables, infra-red absorptions below 1500 cm⁻¹ (the fingerprint region) can be associated with particular general molecular skeletons.

■ Direct matching of the fingerprint region can identify particular molecules.

Using infra-red spectroscopy

Infra-red spectroscopy has many applications. By focusing on particular bonds with known infra-red absorptions, molecules that cause sweetness in ripened fruits can be recognised and measured without having to peel the fruit. Atmospheric pollutants, poisons and illegal drugs can be monitored in a similar fashion by focusing on a particular absorption. The method allows very closely related drugs to be distinguished, identified and measured quickly and accurately.

Analysing car exhaust gases using IR; for cars fitted with catalytic converters the maximum allowed level of CO is 0.3% and of hydrocarbons (HC) is 200 ppm (parts per million), for cars without catalysts the levels are 3.5% for CO and 1200 ppm for HC.

Because infra-red spectroscopy is such a fast process, it can be used to measure the pollution created by individual cars as they pass by an infra-red spectrophotometer positioned at the roadside. Such a device uses infra-red frequencies corresponding to the carbon–oxygen triple bond of carbon monoxide, the nitrogen–oxygen bonds of the various oxides of nitrogen and the carbon–hydrogen bonds of unburnt hydrocarbons, all of which are common pollutants. This type of system is also used to test the exhaust gases of cars during their annual MOT test.

It is also possible to identify the whole molecule by matching the whole infra-red spectrum to known spectra. A computer, using an appropriate database, usually does the storage and matching of spectra. However, it can be done manually using an overlay system where spectra of known compounds are printed on transparent sheets, which can be superimposed on unknown spectra in search of a match. Either way, the spectrum acts as a 'fingerprint' for the molecule. Identifying compounds by this direct matching technique is an everyday routine in forensic science and various chemical industries. Pure compounds, e.g. a drug such as aspirin, are identified directly, while the components of complex mixtures, e.g. cigarette smoke, can each be identified following separation by chromatography.

Using infra-red spectra

One important factor to remember when analysing infra-red spectra is to avoid trying to account for every absorption. The minor absorptions might be caused by impurities, background electronic noise from the spectrophotometer or non-tabulated vibrations. In fact, impurities can be identified and measured by the technique. For example, newly manufactured impure aspirin will contain some unreacted salicylic acid (2-hydroxybenzenecarboxylic acid) which is the main starting material (Fig. 10). This impurity will be detected by an absorption caused by the phenolic O–H group at about 3250 cm^{-1}. This absorption will disappear as the aspirin undergoes recrystallisation to produce a sample pure enough for safe use as a drug.

Fig. 10 Aspirin and salicylic acid

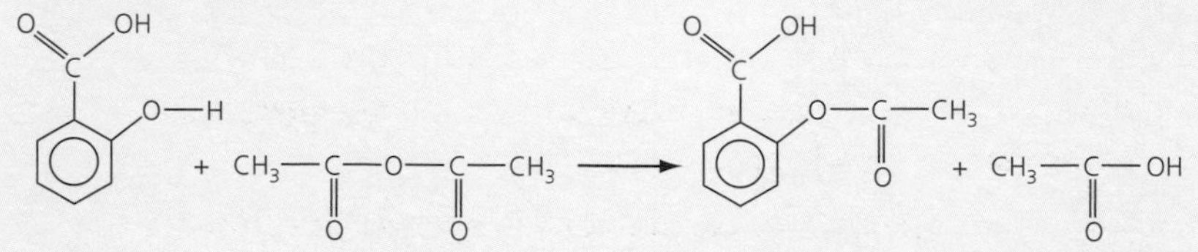

All that remains is to try applying the technique for yourself!

- Concentrate on the major absorptions.

- Be aware that one absorption may overshadow or merge with another.

- The *absence* of a particular absorption can tell you a lot. For example, if an infra-red spectrum does not show a sharp, strong absorption in the range 1630–1815 cm⁻¹, it does not contain a carbonyl group so it cannot be an aldehyde, ketone, acid or acid derivative.

- Unless a direct comparison with an authentic sample can be made, do not expect to make an absolutely definite identification using infra-red spectroscopy.

This usually requires additional information from other analytical techniques such as mass spectroscopy (see section 11.2) and nuclear magnetic resonance spectroscopy (see section 11.3).

Worked examples should help. Compare the infra-red spectra and structures of benzyl ethanoate, prop-2-en-1-ol and 2-methylbutanoic acid (Fig. 11). Absorptions corresponding to C–H bonds (from alkanes and alkenes), C=O bonds (two different types), C=C bonds, C–O bonds and O–H bonds can be distinguished. It can be seen that the infra-red spectra provide a lot of useful information about the structures of these molecules, but in none of the examples is the information sufficient for absolute identification.

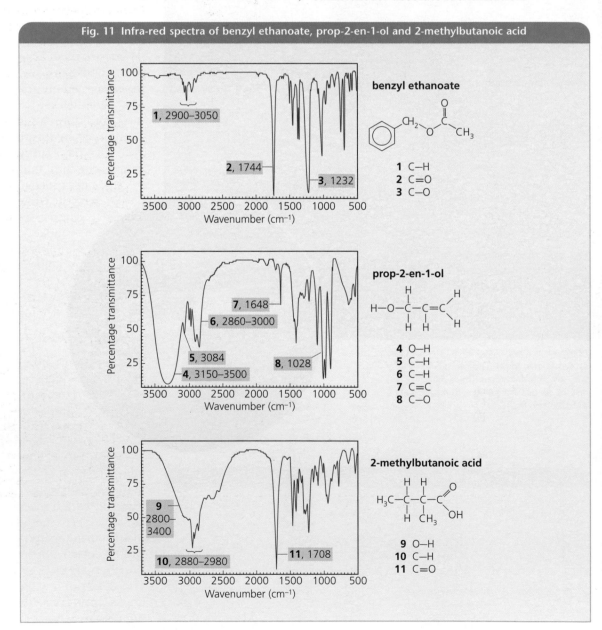

Fig. 11 Infra-red spectra of benzyl ethanoate, prop-2-en-1-ol and 2-methylbutanoic acid

benzyl ethanoate

1, 2900–3050
2, 1744
3, 1232

1 C–H
2 C=O
3 C–O

prop-2-en-1-ol

7, 1648
6, 2860–3000
5, 3084
8, 1028
4, 3150–3500

4 O–H
5 C–H
6 C–H
7 C=C
8 C–O

2-methylbutanoic acid

9 2800–3400
10, 2880–2980
11, 1708

9 O–H
10 C–H
11 C=O

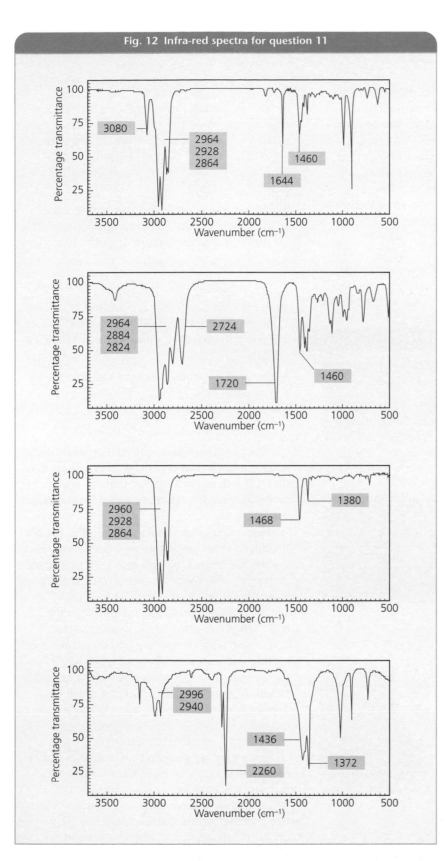

Fig. 12 Infra-red spectra for question 11

11 The infra-red spectra in Fig. 12 are of hex-1-ene, ethanenitrile, hexane and 3-methylbutanal, but not necessarily in that order.

a Draw a structure for each of these four molecules.

b For each structure, list the major bonds and groups of atoms present in each molecule and use Table 1 to list the wave-number(s) of the absorption(s) associated with each of these groups.

c By comparing these expected absorptions with the main actual absorptions in the spectra, decide which spectrum belongs to which compound.

Infra-red spectroscopy has applications in many different areas of science and technology, including pure research, forensic science and environmental science. It is used to help determine the structure of new molecules and to identify drugs, poisons and many other molecules.

Infra-red monitoring and measuring of atmospheric pollutants such as carbon dioxide, oxides of nitrogen, methane and chlorofluorocarbons are particularly important nowadays because of their involvement in the greenhouse effect and the consequent global warming.

Recordings made at the Mauna Loa Observatory in Hawaii of the levels of carbon dioxide in the atmosphere; in 1960 the average level was 316 ppm, by 1990 this had increased to 354 ppm. Carbon dioxide absorbs infra-red radiation at 1700 cm^{-1}.

11.2 Mass spectrometry

Mass spectrometry is an important tool in chemical investigations; it has many applications, both in research and general analysis and its use extends to forensic science, dating of geological and archaeological samples, investigations of foreign materials in food and drink, separation and measurement of isotopes and deducing reaction mechanisms.

A mass spectrometer is a device for detecting the presence of chemical particles, measuring their mass and recording their relative abundance. The particles may be atoms, molecules or molecular fragments. In brief, a mass spectrometer converts the chemical particles into ions (with charge z) and then records the m/z ratio for the particles (where m is the mass of the particle).

The separation and measurement of a sample in a mass spectrometer occurs in six main steps (Fig. 13).

1 The sample is introduced into the mass spectrometer. This may be an element, a compound or a mixture of both.

2 The sample is vaporised.

3 The sample is converted to positive ions.

4 The positive ions are accelerated by a fixed, known electric field. The positive ions are also 'collimated' (focused to form a narrow beam).

5 The ions are deflected by a variable radial magnetic field at 90° to the beam.

6 The ions are detected and the signal is amplified, measured and recorded.

The pressure inside the mass spectrometer is very low so that ions from the sample can reach the detector without colliding with 'air' molecules (and to remove other un-ionised material).

High energy electrons (step 3, Fig. 13) are produced when a current is passed through a coil of wire (C). The wire gets hot, giving off electrons. The electrons are attracted towards a positively charged electrode (A). This accelerates the electrons to high velocity, i.e. high kinetic energy. These electrons pass across the path of the vaporised sample and collide with some of the vaporised atoms or molecules. If an electron has sufficient energy, it will knock an electron off the atom or molecule. This loss of an electron produces a radical ion, $M^{+\bullet}(g)$, which is a single positive ion with an unpaired electron. A second electron may be knocked off, but this is a rare event. Ionisation of the sample is essential because subsequent stages (acceleration, deflection and detection) would not occur if the particles remained neutral.

Acceleration of the positive ions in an electric field follows ionisation (step 4, Fig. 13). A negative electrode causes this. This needs to occur because the next stage is deflection, which depends on the mass to charge ratio (m/z) and the ion's velocity.

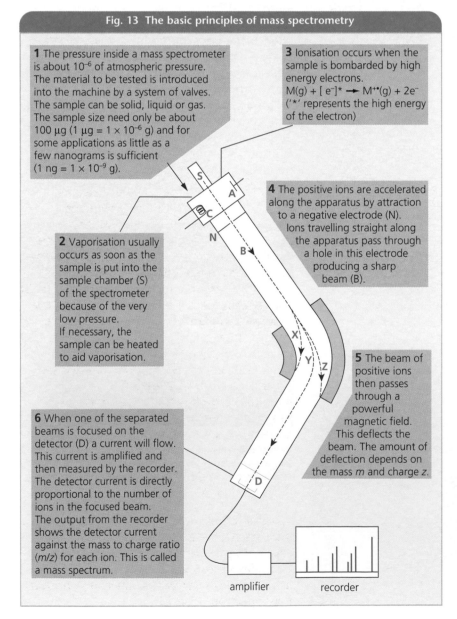

Fig. 13 The basic principles of mass spectrometry

1 The pressure inside a mass spectrometer is about 10^{-6} of atmospheric pressure. The material to be tested is introduced into the machine by a system of valves. The sample can be solid, liquid or gas. The sample size need only be about 100 µg (1 µg = 1×10^{-6} g) and for some applications as little as a few nanograms is sufficient (1 ng = 1×10^{-9} g).

2 Vaporisation usually occurs as soon as the sample is put into the sample chamber (S) of the spectrometer because of the very low pressure. If necessary, the sample can be heated to aid vaporisation.

3 Ionisation occurs when the sample is bombarded by high energy electrons.
$M(g) + [e^-]^* \rightarrow M^{+\bullet}(g) + 2e^-$
('*' represents the high energy of the electron)

4 The positive ions are accelerated along the apparatus by attraction to a negative electrode (N). Ions travelling straight along the apparatus pass through a hole in this electrode producing a sharp beam (B).

5 The beam of positive ions then passes through a powerful magnetic field. This deflects the beam. The amount of deflection depends on the mass m and charge z.

6 When one of the separated beams is focused on the detector (D) a current will flow. This current is amplified and then measured by the recorder. The detector current is directly proportional to the number of ions in the focused beam. The output from the recorder shows the detector current against the mass to charge ratio (m/z) for each ion. This is called a mass spectrum.

amplifier   recorder

During the deflection stage, for a particular magnetic field strength, heavier ions are deflected less than lighter ions. Also, ions with a double positive charge are deflected more than ions with a single positive charge (step 5, Fig. 13). Because the probability of a second electron being knocked off an ion is very small, mass spectrometry usually deals with ions with a single positive charge, i.e. $z = 1$. As a result, m/z for an ion is directly related to the mass of the ion.

Different ions experience different magnetic forces, depending on their m/z ratio, causing different degrees of deflection. The original beam is separated into a number of beams (e.g. X, Y and Z), each corresponding to a different m/z ion (step 6, Fig. 13). The strength of the magnetic field is varied until one of the separated beams, e.g.

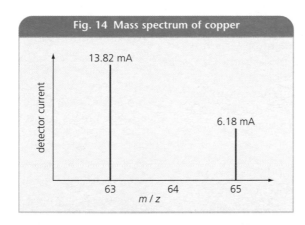

Fig. 14 Mass spectrum of copper

Fig. 15 Calculating the relative atomic mass of copper

1. Record the size of each peak. In this case the amount of each isotope present is proportional to the output of the detector (measured in mA).

   ^{63}Cu   13.82 mA
   ^{65}Cu   6.18 mA

2. Calculate the percentage abundance of each isotope.

   $$\% \ ^{63}\text{Cu} = \frac{13.82}{13.82 + 6.18} \times 100$$

   $$= 69.10\%$$

   $$\% \ ^{65}\text{Cu} = \frac{6.18}{13.82 + 6.18} \times 100$$

   $$= 30.90\%$$

3. Use the percentage abundances to calculate the average atomic mass of copper.

   $$A_r(\text{Cu}) = \frac{69.10}{100} \times 63 + \frac{30.90}{100} \times 65$$

   $$= 63.62 \text{ (to 4 s.f.)}$$

beam Y, Fig. 13 is focused on the detector (D) – a negative plate. The ions take electrons from the detector, resulting in a minute electric current, which can be amplified to aid measurement. The current is proportional to the number of electrons taken and hence to the number of ions with that m/z, giving a measure of the abundance of each ion. The magnetic field is varied so that each ion beam is successively focused and measured to give the relative abundance of each m/z value in the original sample after ionisation.

12 Positive ions are accelerated by attraction to a negative electrode (step 4 in Fig. 13). Some ions pass through a hole in the electrode. What do you suppose happens to the ions that do not pass through the electrode?

13 The separation of three ions is shown in Fig. 13. For strontium, which beam (X, Y, Z) could be:

a ^{86}Sr+•   b ^{88}Sr+•   c ^{87}Sr+•?

Mass spectrometry to distinguish between isotopes

Many of the techniques that use mass spectrometry involve the separation, measurement and comparison of compounds that differ only in the relative amount of different isotopes of the same element contained within the compound. For example, the only difference between cane sugar ethanol and grape sugar ethanol is the proportions of ^{12}C and ^{13}C in the ethanol molecule. Carbon compounds usually contain about 1.1% ^{13}C, but because the metabolism of these two plants (sugar cane and grape vine) is different, the sugars they produce contain slightly different amounts of ^{13}C. These differences are still present when the sugar is converted to ethanol by fermentation. Mass spectrometry is sufficiently sensitive to measure these differences and to allow the source of ethanol to be decided.

14 State the difference between the isotopes ^{12}C and ^{13}C.

When a sample of copper is introduced into a mass spectrometer, a mass spectrum containing two peaks is produced (Fig. 14). The **relative atomic mass** (A_r) of an element, in this case copper, can be calculated from its mass spectrum (Fig. 15).

Table 2 Isotopic abundance for zinc	
Isotope	**Abundance (%)**
^{64}Zn	48.89
^{66}Zn	27.81
^{67}Zn	4.11
^{68}Zn	18.56
^{70}Zn	0.62

15 Mass spectrometers can be calibrated to give percentage isotopic abundances directly. Use the isotopic abundances shown in Table 2 to calculate the relative atomic mass of zinc.

Mass spectrometry of molecular elements and compounds

The examples considered so far have been elements, which exist as giant atomic structures made from individual atoms, e.g. metals such as copper and zinc. Many non-metallic elements consist of diatomic molecules. Oxygen, O_2, chlorine, Cl_2, and nitrogen, N_2, should be familiar examples. What happens to these molecular elements when they are subjected to mass spectrometry?

When chlorine (Cl_2) is analysed by mass spectrometry, the spectrum (Fig. 16) contains peaks produced by the molecular ion $Cl_2^{+\bullet}$ (the superscript $\bullet$ indicates an unpaired electron on a radical ion) and also contains

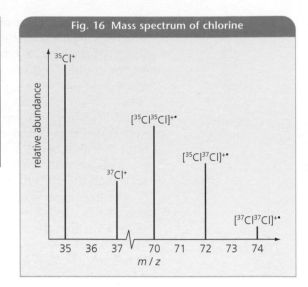

Fig. 16 Mass spectrum of chlorine

peaks produced by the $Cl^{+\bullet}$ ion. These peaks arise because some of the chlorine molecules have 'broken' in the spectrometer. This breaking up of molecules is called **fragmentation** and occurs for all types of molecules, both elements and compounds. Bombardment by high energy electrons in the mass spectrometer ionises the chlorine molecules and, since the ions have high energy, breaks some of the covalent bonds.

16 The ratio of ^{35}Cl to ^{37}Cl is 3:1. Calculate the ratio of peak heights for: $[^{35}Cl^{35}Cl]^{+\bullet}$, $[^{35}Cl^{37}Cl]^{+\bullet}$ and $[^{37}Cl^{37}Cl]^{+\bullet}$.

17 What feature would you expect to see in the mass spectrum of chloroethane?

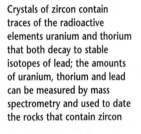

Crystals of zircon contain traces of the radioactive elements uranium and thorium that both decay to stable isotopes of lead; the amounts of uranium, thorium and lead can be measured by mass spectrometry and used to date the rocks that contain zircon

Molecular compounds, e.g. pentane, $CH_3CH_2CH_2CH_2CH_3$, may be ionised by electron bombardment to form the molecular radical ion $[C_5H_{12}]^{+\bullet}$.

$$CH_3CH_2CH_2CH_2CH_3 + [e^-]^* \rightarrow [CH_3CH_2CH_2CH_2CH_3]^{+\bullet} + 2e^-$$

but this is not the end of the story. The molecular ion is unstable. This results in some of its weaker (e.g. C–C more likely than C–H) covalent bonds breaking, just as some of the Cl–Cl bonds broke for chlorine. This produces a range of fragments, some of which are also positively charged; only the ions are subsequently detected with the radicals being evacuated. For example, a molecular radical ion $M^{+\bullet}$ will fragment to produce the ion X^+ and the radical $Y^\bullet$.

$$M^{+\bullet} \quad \rightarrow Y^\bullet + X^+$$

X^+ is detected and measured but the radical $Y^\bullet$ is not.

Although it is strictly correct to write molecular ions as $M^{+\bullet}$, they are normally written as M^+, and this is the convention that will be followed from now on.

As molecules get larger, the number of different bonds that can break increases, so a greater number of different fragments may result (Fig. 17). The fragment ions have different m/z ratios and, like isotopes, are separated by the mass spectrometer. This produces a mass spectrum (Fig. 18), with each signal corresponding to a different fragment ion from the original molecule. Usually, the signal occurring at the highest value on the m/z ratio axis is caused by the unfragmented ion (the **molecular ion** or **parent ion**), and this gives the **relative molecular mass** of the compound. For pentane, this occurs at mass to charge ratio of 72, which corresponds to the formula of C_5H_{12}.

Using their m/z values, many of the main fragments can be identified. They can then be pieced together, partly or completely, to deduce some or all of the original structure. The molecular ion peak in the mass spectrum only gives information about the molecular formula of this ion. For example, three isomers exist for the formula C_5H_{12}, which cannot be completely distinguished by considering only the molecular ion. However, pentane and methylbutane can both produce the fragments shown in Fig. 17 for pentane. The two lower fragments arise when the first C–C bond breaks, whereas cleavage of the second C–C bond produces the two upper fragments.

18 Which m/z ratio values would not be observed in the mass spectra of the other isomer of pentane, dimethylpropane?

19 What causes the weak signals (not to scale) at m/z ratios $(M + 1)$ for each major signal at m/z ratio M?

The relative heights of the peaks can also yield useful information because a more intense peak is associated with a more stable ion. Hence, tertiary carbocations will produce higher peaks than secondary carbocations which, in turn, produce higher peaks than primary carbocations. Hence, a very high peak at $m/z = 57$ is an indication of the tertiary $(CH_3)_3C^+$ carbocation rather than the secondary $(CH_3)_2C^+H$ carbocation or the primary $CH_3CH_2CH_2CH_2^+$ carbocation. Similarly, stable acylium ions (RCO^+) will give high peaks and provide a good indication that the R–CO– group is present in the sample molecule. For example, a relatively weak signal at $m/z = 43$ is more

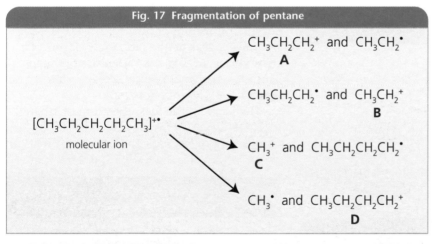

Fig. 17 Fragmentation of pentane

$[CH_3CH_2CH_2CH_2CH_3]^{+\bullet}$
molecular ion

$CH_3CH_2CH_2^+$ and $CH_3CH_2^\bullet$
A

$CH_3CH_2CH_2^\bullet$ and $CH_3CH_2^+$
B

CH_3^+ and $CH_3CH_2CH_2CH_2^\bullet$
C

$CH_3^\bullet$ and $CH_3CH_2CH_2CH_2^+$
D

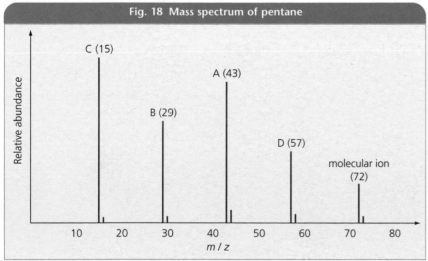

Fig. 18 Mass spectrum of pentane

likely to suggest the presence of a $CH_3CH_2CH_2$ group but a strong signal might suggest a CH_3CO group. Another particularly stable carbocation to look out for is the delocalised carbocation ($C_6H_5^+$) at $m/z = 77$, which would suggest the compound is a mono-substituted benzene.

Identifying compounds using mass spectrometry

The overall mass spectrum of a molecule can be used for identification of it. Thousands of mass spectra can be stored in an appropriate computer database. Rapid and accurate comparisons can be made, leading to identification of a sample.

When analysing a mass spectrum, there are certain m/z ratios to look out for. Their presence in a spectrum is a good indicator that the molecule contains that fragment. Some of the more common fragments to look for are summarised in Table 3.

Also, remember that the *absence* of a particular fragment can give a lot of information. For example, if the spectrum of a compound with molecular formula C_4H_{10} does not have a peak at m/z ratio of 29, it must be methylpropane (Fig. 19) rather than butane. This is because methylpropane does not contain a C_2H_5 group (m/z ratio of 29).

The peaks in the mass spectrum of ethanol (C_2H_5OH) (Fig. 20) can be accounted for by considering the various fragmentations that can occur. ($A_r[C] = 12$, $A_r[O] = 16$ and $A_r[H] = 1$).

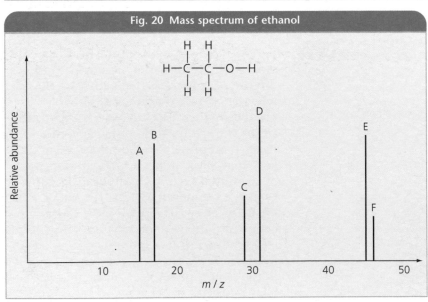

Fig. 19 C_4H_{10} – the two isomers

butane

methylpropane

Fig. 20 Mass spectrum of ethanol

Table 3 Common fragments found in mass spectra or lost during fragmentation	
Mass to charge value	**Possible fragments**
15	CH_3^+
17	HO^+
19	F^+
26	NC^+
29	$C_2H_5^+$, CHO^+
31	CH_3O^+, CH_2OH^+
35/37 (in a 3:1 ratio)	Cl^+
43	$C_3H_7^+$, CH_3CO^+
44	$CONH_2^+$
45	$COOH^+$
77	$C_6H_5^+$
79/81 (in a 1:1 ratio)	Br^+

Note: mass to charge values often differ by 14. This is usually caused by a difference of CH_2 in the structure.

- Signal A ($m/z = 15$) corresponds to a CH_3^+ fragment produced by breaking a C–C bond.

- Signal B ($m/z = 17$) corresponds to a HO^+ fragment produced by breaking a C–O bond.

- Signal C ($m/z = 29$) corresponds to a $C_2H_5^+$ fragment produced by breaking a C–O bond.

- Signal D ($m/z = 31$) corresponds to a CH_2OH^+ fragment produced by breaking a C–C bond.

- Signal E ($m/z = 45$) corresponds to a $C_2H_5O^+$ fragment produced by breaking an O–H bond.

- Signal F ($m/z = 46$) corresponds to the unfragmented molecular ion. It is simply the original ethanol molecule minus one electron. This signal (with the highest m/z ratio) gives the relative molecular mass of the compound.

 20 How would the mass spectrum of methoxymethane (CH_3–O–CH_3), an isomer of ethanol, differ from the mass spectrum of ethanol?

Another powerful approach is to examine the changes in m/z values rather than the actual m/z values. This gives very useful information about undetected free radicals that may be lost during fragmentation. For example, the presence of a methyl group is implied by detection of a fragment [F^+] at (M–15), even though a fragment may not be detected at $m/z = 15$. This is caused by the loss of a methyl free radical from the molecular ion [M^+].

$$[M^+] \rightarrow [F^+] + CH_3$$
$$m/z \quad\quad M \quad\quad\quad M–15$$

Similarly, an ethyl group or aldehyde group (CHO) is implied by a peak at (M – 29), a C–Cl group by (M – 35) and (M – 37) or a phenyl group at (M – 77). These sort of patterns can be looked for using the data from Table 3.

So far we have interpreted mass spectra of compounds where we have known the molecular structure. The next example involves an unknown molecule X, and shows how unknown structures can be deduced. Sample X is known to contain only carbon, hydrogen and oxygen and it produces the mass spectrum shown in Fig. 21.

The steps in deducing the structure of X are as follows:

1   The signal at 134 corresponds to the molecular ion. The molecular formula of X could well be $C_9H_{10}O$.

21 What other 'sensible' molecular formulae are possible for a compound containing C, H and O, with a molecular mass of 134?

2   The signal at 77 is very important as it suggests the presence of a phenyl (C_6H_5–) group. Alternatively, the signal at 57 also suggests this group because it represents the (M – 77) fragment.

3   If X contains a C_6H_5– group, the remainder of the molecule is formed from C_3H_5O. If evidence from infra-red spectroscopy (see Section 11.1) shows that X contains a C=O group but no C=C bond, this means that X could be any one of three isomers (Fig. 22).

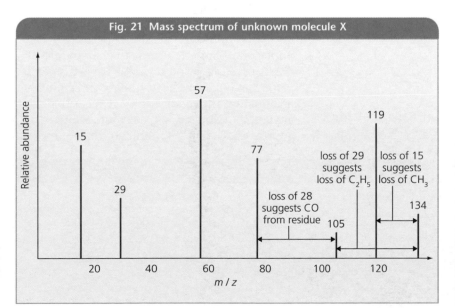

Fig. 21 Mass spectrum of unknown molecule X

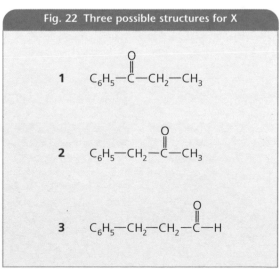

Fig. 22 Three possible structures for X

4 Isomers involving disubstituted benzenes are excluded because they do not contain an isolated phenyl group (Fig. 23).

5 C–OH (alcohol), C–O–C or C–O–C–C (ether) groups are excluded by the absence of fragments at M – 17 = 117, M – 31 = 103 and M – 45 = 89, respectively.

6 Isomer 3 is excluded as a possibility because of the absence of signals at m/z 43 and 91.

7 Isomer 2 is excluded as a possibility because of the absence of a signal at m/z 43, which would be expected to be strong for the particularly stable acylium ion, CH_3C^+O.

8 Isomer 1 is a particular possibility because of the presence of a very strong signal at m/z 57, which would be expected for the particularly stable acylium ion, $CH_3CH_2C^+O$.

9 If you consider the three possible isomers in turn and predict their possible fragmentation patterns (Table 4), it is possible to identify X as 1-phenylpropan-1-one, isomer 1.

10 It is reasonably certain that X is isomer 1 rather than one of the other two possibilities. In practice, other techniques such as infra-red spectroscopy (section 11.1), and particularly nuclear magnetic resonance spectroscopy, (section 11.3) are used to confirm the identification or sort out any ambiguities.

Fig. 23 Three impossible structures for X

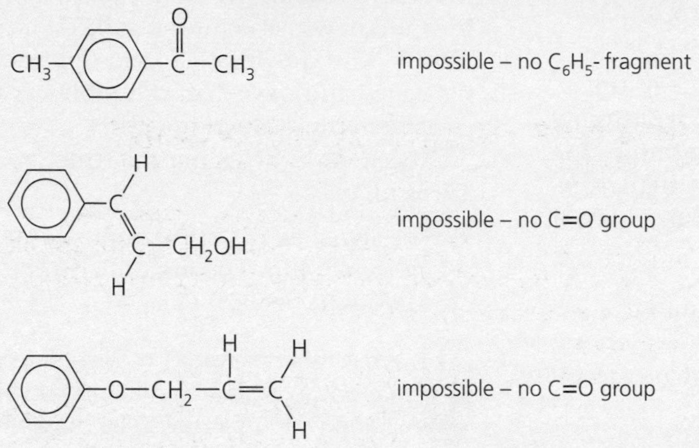

CH_3-⟨benzene⟩$-C(=O)-CH_3$ impossible – no C_6H_5- fragment

⟨benzene⟩$-CH=CH-CH_2OH$ impossible – no C=O group

⟨benzene⟩$-O-CH_2-C(=CH_2)-H$ impossible – no C=O group

High resolution mass spectrometry

Mass spectrometry can be used to deduce even more information. With an appropriately accurate and sensitive spectrometer, the mass of the molecular ion can be measured to seven significant figures. This is referred to as **'high resolution' mass spectrometry**. By using very accurate relative atomic masses (Table 5), it is then possible to distinguish molecules with very similar molecular mass values. For example, ethanoic acid (CH_3COOH), urea (also known as carbamide or aminomethanamide, NH_2CONH_2) and propanol (C_3H_7OH) would all produce a molecular ion with a mass to charge ratio of 60 using a low resolution mass spectrometer. Suppose a high resolution mass spectrometer is used and the relative mass is measured as 60.03235. Using equally accurate relative atomic mass data (see Table 5), the compound

Table 4 Fragmentation patterns for X (/= and/or)

Isomer	Structure	Peaks expected from breaking each bond		
		A	B	C
1	$C_6H_5-\overset{A}{C}(=O)-\overset{B}{C}H_2-\overset{C}{C}H_3$	77/57	105/29	119/15
2	$C_6H_5-\overset{A}{C}H_2-\overset{B}{C}(=O)-\overset{C}{C}H_3$	77/57	91/43	119/15
3	$C_6H_5-\overset{A}{C}H_2-\overset{B}{C}H_2-\overset{C}{C}(=O)-H$	77/57	91/43	105/29

 22 How would the fragmentation patterns of propan-1-ol and propan-2-ol differ?

Table 5 Acurate relative atomic masses

Element	Accurate A_r
carbon	12.00000
hydrogen	1.007825
oxygen	15.99491
nitrogen	14.00307

Table 6 Calculating molecular mass

For CH_3COOH	For NH_2CONH_2	For C_3H_7OH
2 x 12.00000 = 24.00000	1 × 12.00000 = 12.00000	3 × 12.00000 = 36.00000
4 x 1.007825 = 4.03130	4 × 1.007825 = 4.03130	8 × 1.007825 = 8.06260
2 x 15.99491 = 31.98982	1 × 15.99491 = 15.99491	1 × 15.99491 = 15.99491
	2 × 14.00307 = 28.00614	
Total = 60.02112	Total = 60.03235	Total = 60.05751

can be shown to be urea rather than ethanoic acid or propanol (Table 6).

Mass spectrometry is a powerful analytical method. Chemists have used it to help archaeologists date ancient artefacts very accurately and identify materials used by humans in past ages. It can be used to help distinguish between closely related molecules of importance in many areas of science and technology; for example, octane from methylheptane in petrochemistry, morphine from heroin in forensic science and aspirin from paracetamol in pharmaceutical chemistry. The research chemist will often use it to help determine the structure of new molecules. Who knows what other diverse applications for mass spectrometry will arise in the future?

11.3 Nuclear magnetic resonance (n.m.r.) spectroscopy

MRI scanners are powerful diagnostic tools used by doctors. MRI stands for magnetic resonance imaging. It is a technique that was first reported in 1976 and has developed very rapidly since then. The first commercial machines were available in 1983 and are now to be found in most major hospitals world-wide.

A false colour MRI scan section through a human head; how many structures can you identify?

However, MRI has its origins in a technique that chemists have been using regularly since the 1950s. This technique is called **nuclear magnetic resonance (n.m.r.) spectroscopy** and, along with information from other spectroscopic methods, it allows molecular structures to be deduced.

Using nuclear magnetic resonance

During n.m.r. spectroscopy, molecules in solution are placed in a strong magnetic field and then irradiated with radio-frequency electromagnetic radiation. At certain frequencies, the nuclei of some of the atoms in the molecules absorb the radio wave. These absorptions, called resonances, are used to identify the atoms, count them and work out their positions in the molecule relative to other atoms.

How does n.m.r. work?

The nuclei of most atoms have a property known as **nuclear spin**. It is only those atoms with both even atomic number and even mass number (e.g. $^{12}_{6}C$) that do *not* have 'spin'; such atoms are *not* n.m.r. active. Nuclei that do possess spin have their own **magnetic moment** and **magnetic field** and can be considered to behave as if they are small bar magnets (Fig. 24). It is this behaviour that leads to the production of n.m.r. spectra.

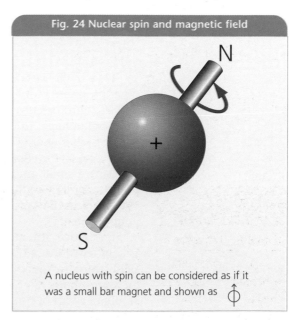

Fig. 24 Nuclear spin and magnetic field

N

+

S

A nucleus with spin can be considered as if it was a small bar magnet and shown as ⏀

Table 7 Some n.m.r. active nuclei
^1H
^{13}C
^{14}N
^{15}N
^{19}F
^{31}P

When nuclei of n.m.r.-active atoms (Table 7) are placed in a strong external magnetic field (Fig 25), the nuclei will align themselves with the direction of the applied magnetic field (Fig. 26). This is just like the behaviour of compass needles in a magnetic field. The tendency of nuclei to align with a magnetic field is not very useful by itself. However, the spinning nuclei have a second option: they can align themselves in the opposite direction to the applied magnetic field – this second spin state is at a slightly higher energy (Fig. 27).

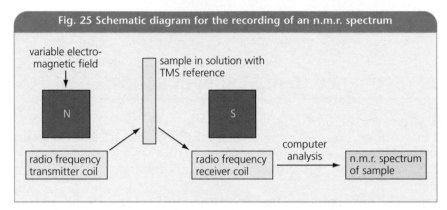

Fig. 25 Schematic diagram for the recording of an n.m.r. spectrum

variable electro-magnetic field

sample in solution with TMS reference

N

S

radio frequency transmitter coil

radio frequency receiver coil

computer analysis

n.m.r. spectrum of sample

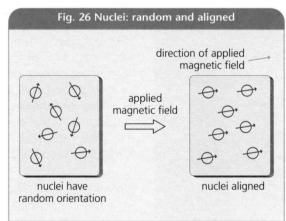

Fig. 26 Nuclei: random and aligned

direction of applied magnetic field

applied magnetic field

nuclei have random orientation

nuclei aligned

If the magnetised nuclei are subjected to an appropriate radio frequency radiation via a coil surrounding the sample, then **resonance** can occur. Nuclei in the α state are promoted to the higher energy β state. The excited nuclei can then relax back to the α state, releasing a small amount of energy as electromagnetic radiation. A second coil surrounding the sample can detect this energy (Fig. 25).

In practice, the radio frequency radiation is kept constant and the external magnetic field strength is gradually increased until, at an appropriate value of the field strength, resonance will occur. The nuclei in the sample will have been detected.

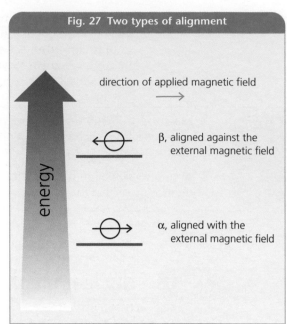

Fig. 27 Two types of alignment

direction of applied magnetic field

β, aligned against the external magnetic field

α, aligned with the external magnetic field

energy

Which atoms are investigated?

Here we concentrate only on the ^1H atom, which is particularly useful for investigation by n.m.r. because of the large number of hydrogen atoms in almost all organic compounds. This is often referred to as proton n.m.r. because the nucleus of this atom is a single proton.

Producing and interpreting a proton n.m.r. spectrum

Not all the ^1H nuclei in an organic sample will resonate at the same magnetic field strength for a particular radio frequency. ^1H nuclei which have different neighbouring atoms (said to have different **chemical environments**) absorb at slightly different external field strengths. The different environments are said to cause a '**chemical shift**' of the absorption. For example, the H of an O–H group, the H of a CH_2 group, the H of a CH_3 group and the H of an N–H group, all have different chemical shifts.

Chemical shifts are related to the electron density *near* the resonating nucleus. This provides a 'shield' between the resonating nucleus and the applied magnetic field. If the electron density around the resonating nucleus is low as a result of attachment to an electronegative group of atoms or a delocalised system, the nucleus is said to be 'deshielded' and will resonate at a lower applied magnetic field. For example, compare the ^1H atoms in C–H and O–H groups. The electronegative oxygen atom in an O–H group attracts electrons and reduces the electron density around the ^1H atom; it is a lot lower than around the ^1H atom of a C–H group. The ^1H nucleus in the O–H group is said to be deshielded and it reaches resonance at a lower external field strength than the ^1H nucleus of a C–H group.

All organic molecules – with the exception of 'symmetrical' compounds such as methane – contain ^1H atoms in different chemical environments. As the external magnetic field is varied for a fixed radio frequency, ^1H nuclei in different environments will resonate in turn. If the value of the external field is recorded as the different resonances occur, a spectrum, known as an n.m.r. spectrum, can be produced. The ^1H atoms of the molecule with any particular chemical environment have been detected separately from each other.

Samples are investigated in dilute solution. This separates the sample molecules from each other, preventing them from interacting with each other and causing very complex absorptions.

The choice of solvent is important. When investigating ^1H atoms, tetrachloromethane (CCl_4) or deuterated trichloromethane ($CDCl_3$) are commonly used because they are very powerful solvents for organic compounds and they do not contain ^1H atoms. This means they do not resonate and so do not interfere with the ^1H n.m.r. spectrum of the sample.

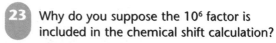

Fig. 28 Tetramethylsilane

During the production of an n.m.r. spectrum, **tetramethylsilane** (TMS, Fig. 28) is mixed with the sample. This is added to provide a reference point to which the n.m.r. spectrometer is tuned. The magnetic field is adjusted until the ^1H nuclei of TMS resonate: this is given a chemical shift value of 0. TMS is used for the following reasons:

- the hydrogen nuclei in TMS are highly shielded because silicon has a very low electronegativity; as a result, they resonate at a field strength well above that of any ^1H nuclei in common organic molecules;

- it gives one strong, sharp and easily detected absorption because it is caused by the combined effects of 12 equivalent ^1H atoms (equivalent atoms are those in identical chemical environments, usually due to symmetry within the molecule);

- it is non-toxic;

- it is cheap;

- it does not react with the sample;

- it is easily separated from the sample molecule because it has a low boiling point.

The chemical shifts of ^1H atoms in a sample molecule are measured and tabulated relative to the TMS reference absorption on the 'δ-scale' using:

$$\delta = \frac{B_{TMS} - B_{sample}}{B_{TMS}} \times 10^6$$

where B_{TMS} and B_{sample} represent the applied external magnetic field strengths at resonance for TMS or the sample respectively.

The value of the external magnetic field that causes a particular proton to resonate depends on the radio frequency that the n.m.r. spectrometer uses. This means that chemical shifts would vary from machine to machine, making comparisons difficult. The use of a reference compound (TMS) and a scale for chemical shift without units (the δ scale) avoids this problem. The use of the δ scale for measuring chemical shifts avoids the need to quote both the magnetic field strength and the radio frequency. The δ value for a particular resonance is the same for all n.m.r. spectrometers, whether they operate at radio frequencies of 60 MHz, 100 MHz or 250 MHz.

23 Why do you suppose the 10^6 factor is included in the chemical shift calculation?

In general, ^1H atoms bonded to electronegative atoms, e.g. O–H in alcohols and acids or N–H in amines are said to be deshielded and so absorb at lower field strengths, i.e. have larger chemical shifts, while ^1H atoms in alkanes and alkenes are shielded and so absorb at higher field strengths, i.e. have smaller chemical shifts (Fig. 29). A map of some of the more common ^1H absorptions is shown in Fig. 30.

Fig. 29 Explaining chemical shifts

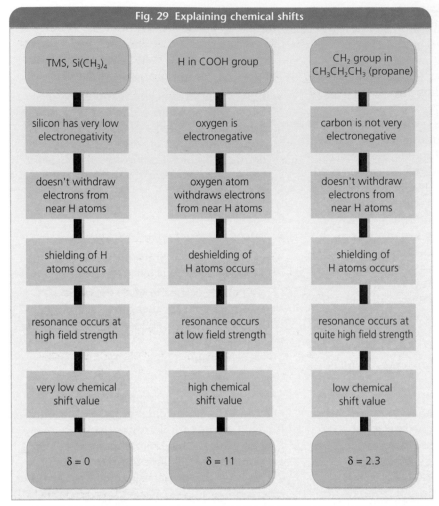

| TMS, Si(CH₃)₄ | H in COOH group | CH₂ group in CH₃CH₂CH₃ (propane) |

silicon has very low electronegativity | oxygen is electronegative | carbon is not very electronegative

doesn't withdraw electrons from near H atoms | oxygen atom withdraws electrons from near H atoms | doesn't withdraw electrons from near H atoms

shielding of H atoms occurs | deshielding of H atoms occurs | shielding of H atoms occurs

resonance occurs at high field strength | resonance occurs at low field strength | resonance occurs at quite high field strength

very low chemical shift value | high chemical shift value | low chemical shift value

δ = 0 | δ = 11 | δ = 2.3

Fig. 30 Chemical shifts of some common functional groups

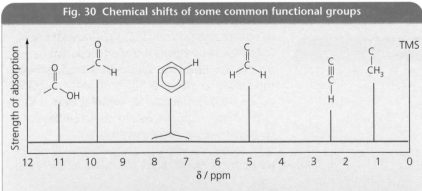

Fig. 31 A low resolution n.m.r. spectrum of ethanol

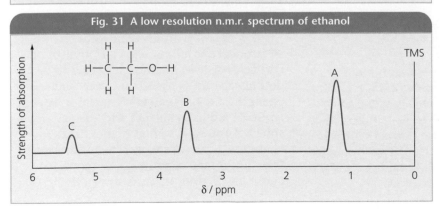

24 Why does the hydrogen atom of an aldehyde group (CHO) absorb at a much higher chemical shift than the hydrogen atom of an alkene group (C=C–H)?

Low resolution proton n.m.r. spectra

A low resolution n.m.r. spectrum of ethanol (Fig. 31) shows three absorptions (A, B and C) because the ethanol molecule contains three sets of ¹H atoms. Signal A is produced by the three equivalent ¹H atoms in the CH₃ group, signal B by the two equivalent ¹H atoms in the CH₂ group and signal C by the single ¹H atom in the OH group. Notice that signal C has the largest chemical shift because the ¹H atom is directly bonded to an electronegative oxygen atom. Signal B has a larger chemical shift than A because the CH₂ group is nearer to the electronegative oxygen atom than the CH₃ group is.

25 Explain why the three hydrogen atoms of the CH₃ group in ethanol are equivalent.

26 Deduce the number of absorptions in the low resolution NMR spectra of each of the following:
i benzene; **ii** propane; **iii** propanal; **iv** propanone; **v** ethanoic acid?

The strengths of the absorptions are proportional to the number of equivalent ¹H atoms causing the absorption. The strengths of the absorptions are measured by the area under each absorption peak. Hence, areas of absorptions A, B and C in spectrum of ethanol are in the ratio of 3:2:1. In modern spectrometers these are measured electronically and superimposed digitally or graphically on the main spectrum. This is called the **integrated spectrum**.

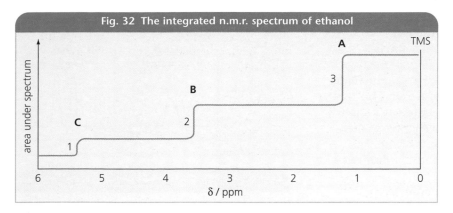

Fig. 32 The integrated n.m.r. spectrum of ethanol

The integrated spectrum of ethanol is shown in Fig. 32; the distances between the plateaus in the integrated spectrum represent the areas under the absorption peaks. The ratio of these distances gives the ratio of the number of ^1H atoms in each equivalent group.

27 Deduce the ratios of the areas under the absorptions for the following compounds

i ethanoic acid; **ii** propane;
iii methylbenzene.

High resolution proton n.m.r. spectra

The previous n.m.r. spectrum of ethanol (see Fig. 31) was highly simplified. Nevertheless, it did allow the sets of equivalent ^1H atoms to be identified and it did allow the number of protons in each set to be counted. When examined in more detail with a more sensitive spectrometer, some of the basic signals are split into groups of signals (Fig. 33). In the case of CH_3CH_2OH, the CH_3 signal is split into three, producing what is called a **triplet**, while the CH_2 signal is split to form a

quartet. Strangely, the OH signal is not split – it appears as a **singlet**. This splitting of the absorptions is caused by the influence of ^1H atoms bonded to neighbouring atoms and the influence is called a **coupling effect**.

You should remember that each spinning ^1H nucleus generates a slight magnetic field. When a particular ^1H nucleus resonates in an applied magnetic field, the actual magnetic field that acts on it is the sum of the applied field and the fields from its neighbours. Thus the ^1H atoms are linked through space

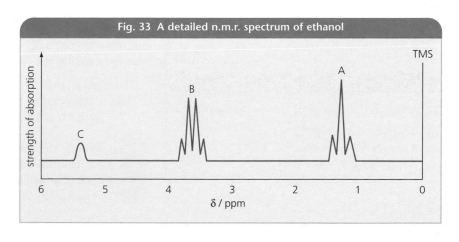

Fig. 33 A detailed n.m.r. spectrum of ethanol

(**coupled**) by the interactions of their own magnetic fields (Figs. 34 and 35).

The magnetic field from a single neighbouring ^1H nucleus (Fig. 34) may act with or against the applied magnetic field. There is a 50:50 chance of either. This means the nucleus will resonate either at a slightly lower applied field or at a slightly higher applied field. The end result is that the original resonance is split into a **doublet** in a ratio of 1:1.

Suppose there are two equivalent neighbouring ^1H atoms (Fig. 35) instead of just one. Their magnetic fields could be both with the applied field, both against it or one with while the other is against. This means a ^1H nucleus is coupled to two neighbours in three different ways. The result is a **triplet**. If the magnetic fields from the two neighbours are both with the applied field the group will absorb at slightly lower field. Conversely, if the magnetic fields from the neighbours are both against the applied field the group will absorb at slightly higher field.

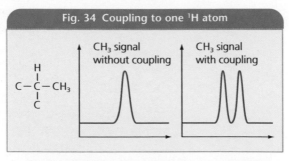

Fig. 34 Coupling to one ^1H atom

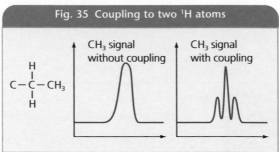

Fig. 35 Coupling to two ^1H atoms

Finally, if the magnetic field from one of the neighbours is with the applied field and the field from the other is against, they will cancel each other out and the group will absorb at its own characteristic field strength. The intensity of the triplet will be 1:2:1 (Fig. 35). These are the relative probabilities of the neighbour's fields adding with, cancelling out or adding against the applied field. This is the situation for the CH_3 group in ethanol. The absorption associated with these three ^1H atoms is split into a triplet because of the coupling effect of the two ^1H atoms bonded to the neighbouring carbon atom.

The number of atoms in the group itself has no influence on splitting because all such atoms absorb at the same field strength; they do not couple with each other. Furthermore, because the magnetic fields involved are so weak, coupling with 'next-door-but-one' neighbours is negligible.

Table 8 summarises the pattern of significant couplings, all of which can be summarised by the '**n + 1**' rule which states:

The n.m.r. absorption of a proton which has n equivalent neighbouring protons will be split into n + 1 peaks.

28 Calculate the splitting effect experienced by a ^1H atom adjacent to a CH_3 group? Check your answer by looking at part of the spectrum of ethanol (Fig. 33).

Fig. 36 How triplet signals are formed

Direction of applied field	Direction of fields from 2 neighbours	Proportion	Effect on resonance position
→	← ←	1	higher field
→	← → or → ←	2	no effect
→	→ →	1	lower field

Table 8 Coupling patterns

Number of equivalent neighbouring (coupled) ^1H atoms	Splitting effect	Resulting signal	Ratio
1	into 2 signals	doublet	1:1
2	into 3 signals	triplet	1:2:1
3	into 4 signals	quartet	1:3:3:1
n	into n + 1 signals	n + 1 peaks	

To further our understanding of n.m.r. it is worthwhile to consider the following questions.

- Why is the OH signal from ethanol *not* split into three by coupling with the 1H atoms of the neighbouring CH_2 group?

- What happens when coupling occurs with neighbours on two sides?

The H of the OH group absorbs at low field because of the high electronegativity of the oxygen atom. The signal is not split for a related reason. The polarity of the O–H bond causes the H atom to be very readily exchanged with other H atoms from neighbouring hydrogen-bonded molecules. The exchanged H atom has a 50:50 chance of having the same spin direction as the one displaced and, since this exchange occurs more rapidly than the actual resonance effect, the coupling with neighbours averages to zero and no splitting can be detected. This rapid exchange can be shown by adding deuterated water (heavy water, D_2O) to the sample. The O–H absorption gets weaker as the H is exchanged for D (D is not sensitive to n.m.r. under the conditions that are used to record ^1H spectra).

The H atoms of C–H bonds do not exchange because carbon and hydrogen have very similar electronegativities. Consequently, the CH_2 signal in ethanol is split into a quartet by the neighbouring CH_3 group but is not split further by its neighbour on the other side as this is the exchanging H of the OH group.

Coupling with neighbours on two sides can happen. With propane, for example, the CH_2 signal is seen as a heptet (Fig. 37) because it has six neighbouring H atoms, three on either side in two equivalent CH_3 groups. This is associated with the symmetry of the propane molecule. The CH_3 groups produce identical triplets by coupling with the CH_2 group.

However, for propanal (Fig. 38), three sets of signals are seen. The ratios (3:2:1) from the integrated spectrum show these are caused by the ^1H atoms of the CH_3 group, the CH_2 group and the CHO group. The CH_2 signal is split into a quartet by coupling with the CH_3 group and the H of the CHO group on the other side then splits each part of the quartet into a doublet. This produces a 'quartet of doublets'. The CHO and CH_3 signals are both triplets because of coupling with the central CH_2 group. Thus the spectrum gets much more complicated for non-symmetrical molecules.

This double splitting of the CH_2 signal can also be seen in the ethanol spectrum if conditions are changed to prevent exchange occurring for the OH group. This can be done by cooling the sample to about –70 °C. At such temperatures the exchange process occurs more slowly than the resonance effect, resulting in the coupling between the

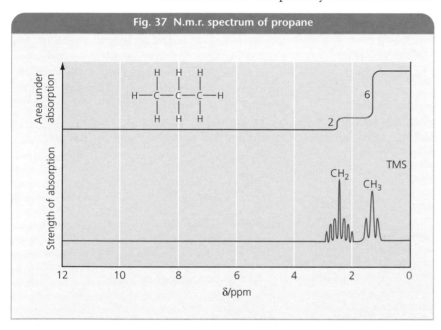

Fig. 37 N.m.r. spectrum of propane

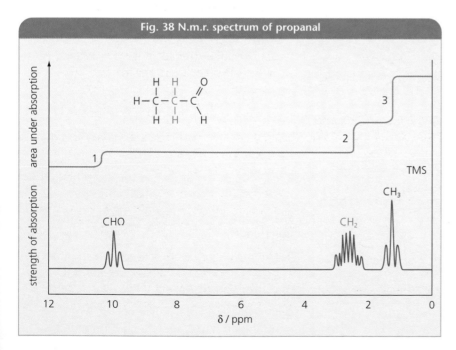

Fig. 38 N.m.r. spectrum of propanal

OH and CH_2 groups becoming detectable. The OH signal is split by the CH_2 group to form a triplet and the CH_2 signal becomes a 'quartet of doublets'.

28 What sort of splitting will be seen for the CH_2 group of ethanol when its n.m.r. spectrum is measured at −70 °C?

KEY FACTS

- N.m.r. signal intensities are proportional to the number of hydrogen atoms in the group responsible for the signal. These are measured by the integrated n.m.r. spectrum.

- Each n.m.r. signal may be split as a result of coupling with neighbouring atoms.

- n equivalent neighbouring ^1H atoms cause an n.m.r. signal to be split into $n + 1$ peaks.

Conclusions

N.m.r. has been used by chemists for several decades, but they are still developing the technique and applying it to more complex and sophisticated examples.

Three-dimensional structures of large molecules, such as proteins and other natural products, can be determined, because atoms can be identified that are close to each other as a result of the way the molecule is folded.

However, even though n.m.r. gives a lot of information about molecular structures, it does not give all the answers. It needs to be used in conjunction with information acquired by other chemical and spectroscopic methods.

APPLICATION A

Full structural determinations

Substance X contains 45.9% carbon, 8.9% hydrogen and 45.2% chlorine.

1 Calculate the empirical formula of substance X.

When warmed with water and then treated with silver nitrate solution, X produces a white precipitate, which is easily soluble in dilute ammonia.

2 What is the significance of this reaction and the associated observations?

X's mass spectrum shows peaks at m/z ratios 15, 35, 37, 43, 78 and 80. The peaks at 35 and 37 are in the ratio 3:1, as are those at 78 and 80.

3 What relative molecular mass is suggested by this spectrum? Explain your answer.

4 Hence deduce the molecular formula of X.

5 Use this molecular formula to draw and name all the possible structures for X.

6 What major absorptions would you expect to observe in the infra-red spectrum for X?

7 Why is it not possible to identify X using a limited number of peaks?

8 How might the infra-red spectrum be used to identify X?

The diagram shows the low resolution n.m.r. spectrum for X.

9 Use this spectrum to deduce which of the possible isomers given in question 5 corresponds to X.

10 Describe the appearance of the high resolution n.m.r. spectra for compound X. Explain your answer.

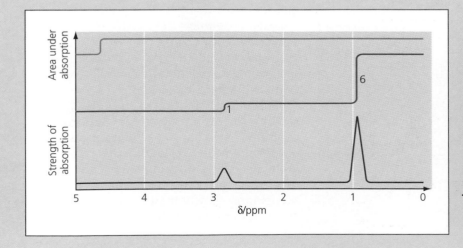

Substance Y contains 69.8% C, 11.6% H and 18.6% O.

1 Calculate the empirical formula of compound Y.

Substance Y has significant infra-red absorptions at 2960 and 1690 cm⁻¹ and does not form a red precipitate when reacted with Fehling's solution.

2 What can be inferred from these pieces of information?

Substance Y's mass spectrum shows significant peaks at *m/z* ratios 15, 43, 71 and 86.

3 What is the likely value for the relative molecular mass of substance Y?

4 Hence, deduce the molecular formula of substance Y.

5 Which groups are suggested by these mass spectrum data?

6 Draw and name the structure which is consistent with all of the above data and observations.

EXAMINATION QUESTIONS

1 Give the reagent and conditions needed to produce hex-1-ene from hexan-1-ol. The infra-red spectrum of one of these two compounds is shown below. Use the spectrum and the table of infra-red absorption data in the Data section to identify this compound. State two regions, other than the fingerprint region, where the infra-red spectrum of the other compound would be different. (5)

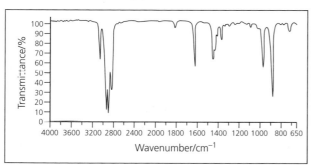

NEAB CH03 February 1997 Q7(c)

2

a The infra-red spectrum shown below is that of compound **X**, which has the molecular formula $C_4H_8O_2$.

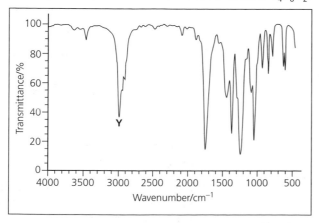

i) Use the table of infra-red absorption data in the Data section to help you identify the bond responsible for the absorption marked **Y**. (1)

ii) Draw the structures of the two carboxylic acids having the molecular formula $C_4H_8O_2$ and explain why **X** cannot be either of these. (4)

b Compound **X** has three peaks with ratio of areas 3:2:3 in its low-resolution proton n.m.r. spectrum. Draw two possible structures for compound **X**. (2)

AQA CH03 June 1999 Q6(part)

3 Chlorine exists as a mixture of the isotopes ³⁵Cl and ³⁷Cl, which are present in the ratio of 3:1 respectively. Deduce the number of molecular ion peaks in the mass spectrum of 1,1,1-trichloroethane and calculate the *m/z* value of the molecular ion peak with the lowest abundance. (2)

AQA CH03 June 1999 Q5(a)(iii)

4 Given that bromine exists naturally as a 1:1 mixture of isotopes ⁷⁹Br and ⁸¹Br, draw on the axes below the molecular ion peaks in the mass spectrum of $CH_3CH_2CHBrCH_3$. (3)

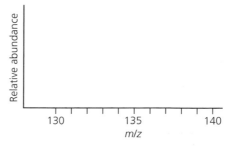

NEAB CH03 Feb 1997 Q3(d)

5 An ester, **B**, was hydrolysed under acidic conditions and the products were separated. The mass spectra of the products are shown below.
Use the spectra to determine the relative molecular mass, M_r, of the alcohol and of the carboxylic acid formed and hence draw their structures. (6)

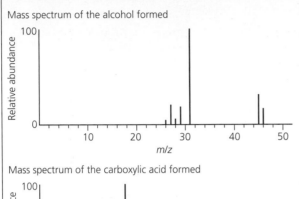

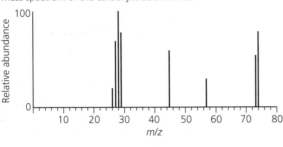

NEAB CH03 June 1997 Q2(c)

6 Although the relative molecular mass of bromoethane is 109, there is no peak at $m/z = 109$ in its mass spectrum. Equally sized peaks at $m/z = 108$ and $m/z = 110$ do, however, occur. Explain these observations. (3)
AQA CH03 March 1999 Q2(c)

7
a The structure of 2,4-dimethylpentane, an isomer of heptane, is shown below.

$$CH_3-CH-CH_2-CH-CH_3$$
$$\quad\;\; |\qquad\qquad |$$
$$\quad\;\; CH_3\qquad\; CH_3$$

Explain why the low-resolution n.m.r. spectrum of 2,4-dimethylpentane show three peaks with relative areas 6:1:1. (3)
b Another isomer of heptane has the structure shown below.

$$\qquad\;\; CH_3\quad CH_3$$
$$\qquad\;\; |\qquad\; |$$
$$CH_3-C-\!\!-\!\!-C-CH_3$$
$$\qquad\;\; |\qquad\; |$$
$$\qquad\;\; CH_3\quad H$$

How many peaks are present in its low-resolution n.m.r. spectrum? (1)
NEAB CH03 Feb 1997 Q6(d,e)

8 The proton n.m.r. spectrum of ethyl ethanoate has peaks at δ 4.1, 2.0 and 1.2
Given that the peak at δ 2.0 is a singlet, deduce the splitting patterns of the other two peaks. You may find the data provided in the table helpful. (2)

Proton chemical shift data

Type of proton	δ/ppm
RCH_3	0.7–1.2
R_2CH_2	1.2–1.4
R_3CH	1.4–1.6
$RCOCH_3$	2.1–2.6
$ROCH_3$	3.1–3.9
$RCOOCH_3$	3.7–4.1

NEAB CH06 June 1998 Q3(d)

9 The proton n.m.r. spectrum of a chloroalkyl ketone, **A**, C_5H_9ClO, is shown below.

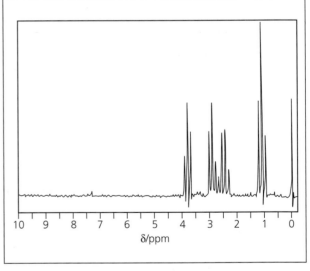

The measured integration trace gives the ratio 1.2 to 1.2 to 1.2 to 1.8 for the peaks at δ 3.8, 2.8, 2.4 and 1.1, respectively.
Refer to the spectrum above and to n.m.r. information given in data books, in order to answer the following questions.
a How many different types of proton are present in compound **A**? (1)
b What is the actual ratio of the numbers of each type of proton? (1)
c The peaks at δ 2.4 and δ 1.1 arise from the presence of an alkyl group. Identify the group and explain the splitting pattern. (3)
d What can be deduced from the splitting of the peaks at δ 3.8 and δ 2.8? (1)
e Deduce the structure of compound **A**. (1)
NEAB CH06 March 1998 Q6

10

a The proton n.m.r. spectrum of an organic compound provides information about chemical structure. Four important aspects are the *number of absorptions*, the *positions of the absorptions*, the *relative intensities of the absorptions (areas)* and the *splitting of absorptions* into several peaks.
What information can be deduced from each of these features of a proton n.m.r. spectrum? (7)

b Explain why tetramethylsilane (TMS) is suitable for use as an internal standard for proton n.m.r spectra. (3)

c Compound **A**, $C_5H_{10}O$, reacts with $NaBH_4$ to give **B**, $C_5H_{12}O$. Treatment of **B** with concentrated sulphuric acid yields compound **C**, C_5H_{10}. Acid-catalysed hydration of **C** gives a mixture of isomers, **B** and **D**. Fragmentation of the molecular ion of **A**, $[C_5H_{10}O]^+$, leads to a mass spectrum with a major peak at *m/z* 57. The infra-red spectrum of compound A has a strong band at 1715 cm^{-1} (Spectrum 1) and the infra-red spectrum of compound **B** has a broad absorption at 3350 cm^{-1} (Spectrum 2). The proton n.m.r. spectrum of **A** has two signals at δ 1.06 (triplet) and 2.42 (quartet), respectively (Spectrum 3).

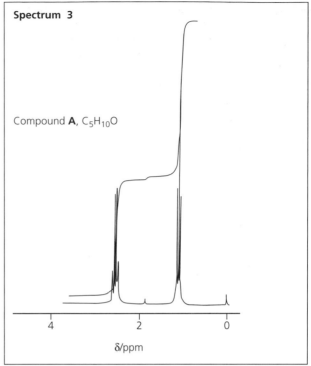

Spectrum 3

Compound **A**, $C_5H_{10}O$

δ/ppm

Use the analytical and chemical information provided to deduce structures for compounds **A**, **B**, **C** and **D**, respectively. Refer to question 1 for infra-red absorption data.
Include in your answer an equation for the fragmentation of the molecular ion of **A** and account for the appearance of the proton n.m.r. spectrum of **A**. Explain why isomers **B** and **D** are formed from compound **C**. (20)
NEAB CH06 Feb 1996 Q7

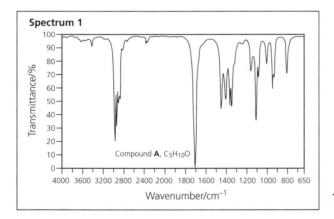

Spectrum 1

Compound A, $C_5H_{10}O$

11 Explain how you could distinguish between samples of 1-bromopropane and 2-bromopropane using:

a infra-red spectroscopy, (2)

b low-resolution proton n.m.r. spectroscopy. (2)
NEAB CH03 June 1998 Q5(b)

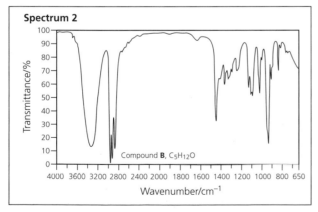

Spectrum 2

Compound B, $C_5H_{12}O$

12 Thermodynamics

Many chemical reactions are accompanied by a measurable heat energy change called an enthalpy change – symbol ΔH (enthalpy comes from the Greek word for warm). Reactions that release energy are called exothermic: ΔH is negative; reactions that absorb energy are called endothermic: ΔH is positive. Enthalpy changes occur because of energy changes when bonds break and new bonds form during the chemical reaction.

A good example of an exothermic reaction is combustion. Heat is produced! Energy can also be released when some substances are mixed with water, such as concentrated sulphuric acid.

When methane combines with oxygen, a lot of energy is released.

When concentrated sulphuric acid is added to water, so much heat is released that the water nearly boils. Always add the acid to water and not the water to acid, and stir.

If you have ever mixed up plaster of Paris, you may have noticed that the mixture becomes warm, but not nearly as hot as the sulphuric acid.

Some clays mixed with water also become warm, which makes them useful for 'hot mud' facial packs in applications such as deep-cleansing of the pores of the skin. When your skin is warm, the pores open. This allows you to sweat, a process that is important for regulating body temperature. A deep cleansing product that generates a little heat, stimulates blood flow and opens up the pores. By mixing cleansing oils with the heating material, grease and dirt deep inside the pores can be removed. Of course, the mixture needs to be safe to the skin and easily rinsed off.

12.1 Enthalpy changes

Chapter 5 of *AS Chemistry* covered enthalpy changes for reactions involving covalent compounds. In this chapter we will consider reactions involving electrovalent (ionic) compounds.

The enthalpy change, ΔH, for a system is the heat taken in or given out during a chemical or physical change that takes place at constant pressure.

The overall enthalpy change in a reaction is often the sum of several endothermic and exothermic processes.

A reaction, such as making sodium chloride from its elements, can be used to show the endothermic and exothermic steps. In the formation of sodium chloride from its elements, several changes are involved:

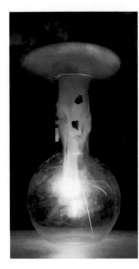
The reaction of sodium and chlorine is highly exothermic.

- sodium atoms in the solid state separate to form gaseous atoms;
- gaseous atoms form gaseous positive sodium ions;
- gaseous chlorine molecules split into gaseous chlorine atoms;
- gaseous chlorine atoms form gaseous negative chloride ions;
- gaseous positive and negative ions attract each other and form a solid crystal lattice.

When the element sodium reacts with the element chlorine, a lot of energy is released. The heat from this reaction can be measured, and it is the sum of a number of different enthalpy changes. The overall reaction is exothermic:

$$\Delta H^{\ominus}_{f,298} = -411 \text{ kJ mol}^{-1}$$

$\Delta H^{\ominus}_{f,298}$ is called the **standard enthalpy of formation** (Fig. 1). It is the enthalpy change when one mole of sodium chloride is formed from its elements in their standard states under standard conditions:

$$Na(s) + \tfrac{1}{2} Cl_2(g) \rightarrow NaCl(s) \quad \Delta H^{\ominus}_{f,298} = -411 \text{ kJ mol}^{-1}$$

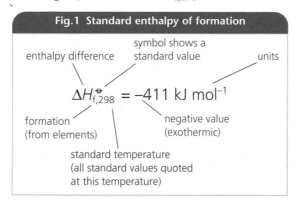

Fig.1 Standard enthalpy of formation

enthalpy difference | symbol shows a standard value | units

$$\Delta H^{\ominus}_{f,298} = -411 \text{ kJ mol}^{-1}$$

formation (from elements) | negative value (exothermic)

standard temperature (all standard values quoted at this temperature)

Standard enthalpy changes, $\Delta H^{\ominus}$

The size of any enthalpy change depends on the amount of substance used and on the conditions of measurement. Chemists have agreed **standard amounts** and **standard conditions** so that comparisons can be made between different measurements. The symbols for standard enthalpy changes are always followed by the sign $^{\ominus}$.

Amount of substance

The standard amount used by chemists is the **mole** (see *AS Chemistry*, page 46), and this will be quoted in the units as per mole using the symbol **mol^{-1}**.

Standard states

The size of an enthalpy change for a given reaction alters if either the pressure or the temperature at which it is measured is altered. In order to make comparisons between different sets of data, it is convenient to quote enthalpy changes measured under agreed **standard conditions**.

- **Standard pressure** The standard pressure chosen is 100 kPa (1 bar).

- **Standard temperature** The most common reference temperature used is 298 K. Once this is specified, the enthalpy change is the **standard enthalpy change**, $\Delta H^{\ominus}$(298 K) (see *AS Chemistry*, page 76).

Standard quantities at temperatures other than 298 K can be quoted if this is more convenient for a particular reaction. So $\Delta H^{\ominus}$ (1000 K) refers to a standard molar enthalpy change at a temperature of 1000 K. If no temperature is stated, it is assumed that the reference temperature is 298 K. Thus $\Delta H^{\ominus}$ on its own is the same as $\Delta H^{\ominus}$(298 K).

Forming sodium chloride

The whole process of forming sodium chloride from its elements involves a number of enthalpy changes (some of which may occur simultaneously). These changes are summarised below and in Fig. 6 (page 159).

Enthalpy of atomisation

Energy is supplied to solid sodium metal to form gaseous sodium atoms. The energy required to vaporise one mole of solid atoms is called the enthalpy of atomisation. For a metallic (or atomic) lattice, this is simply the enthalpy of sublimation. It is an endothermic process (ΔH is positive), because the atoms are attracted to each other and need to be separated.

$$Na(s) \rightarrow Na(g) \quad \Delta H^{\ominus}_{sub} = +107 \text{ kJ mol}^{-1}$$

The enthalpy of atomisation is related to the forces of attraction between atoms. Magnesium has a greater enthalpy of atomisation than sodium (Table 1), because the greater ionic charge attracts the two delocalised electrons in the metallic lattice of magnesium more strongly. More energy must be supplied to overcome these forces.

Table 1 Enthalpies of atomisation	
Element	$\Delta H^{\ominus}_{at}$ /kJ mol^{-1}
lithium	+159
sodium	+107
magnesium	+149
aluminium	+326
potassium	+89

Aluminium has an even greater ionic charge than magnesium, so the enthalpy of atomisation is even greater. Potassium is a larger atom than sodium and has more inner electron shells to screen the outer delocalised electrons from the nucleus. Also, because it is a larger atom than sodium, the electrons are further away from the nucleus, so the force of attraction is weaker and the enthalpy of atomisation is less than that for sodium.

The enthalpy of atomisation is the standard enthalpy change which accompanies the formation of one mole of gaseous atoms.

Ionisation

The next stage involves removing the outer electron from each gaseous sodium atom to form a gaseous sodium ion. This change is the first ionisation enthalpy for sodium. It too is an endothermic process (ΔH_i is positive).

$$Na(g) \rightarrow Na^+(g) + e^- \quad \Delta H_i^\ominus = +496 \text{ kJ mol}^{-1}$$

 Explain why removing an electron from an atom is an endothermic process.

Many metals lose more than one electron from their outer shell when they form ions, so further ionisation enthalpies may need to be considered, e.g. the ionisation enthalpy required to form the Mg^{2+} ion is the sum of the first and second ionisation enthalpies:

$$Mg(g) \rightarrow Mg^+(g) + e^- \quad \Delta H_i^\ominus = +738 \text{ kJ mol}^{-1}$$
$$Mg^+(g) \rightarrow Mg^{2+}(g) + e^- \quad \Delta H_i^\ominus = +1451 \text{ kJ mol}^{-1}$$

Total enthalpy required for the overall process:

$$Mg(g) \rightarrow Mg^{2+}(g) + 2e^- = +2189 \text{ kJ mol}^{-1}.$$

It is theoretically possible to remove all the electrons from atoms. These successive ionisation enthalpy values can be used to indicate which group of the Periodic Table the element is in (Table 2 and Fig. 2).

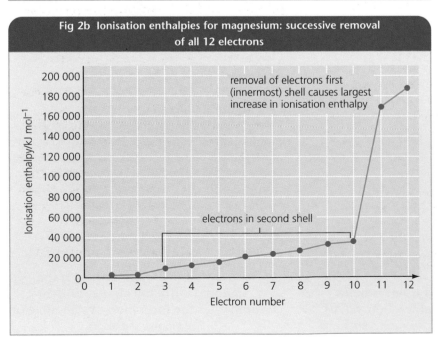

Fig 2a Ionisation enthalpies for magnesium: successive removal of the first 10 electrons

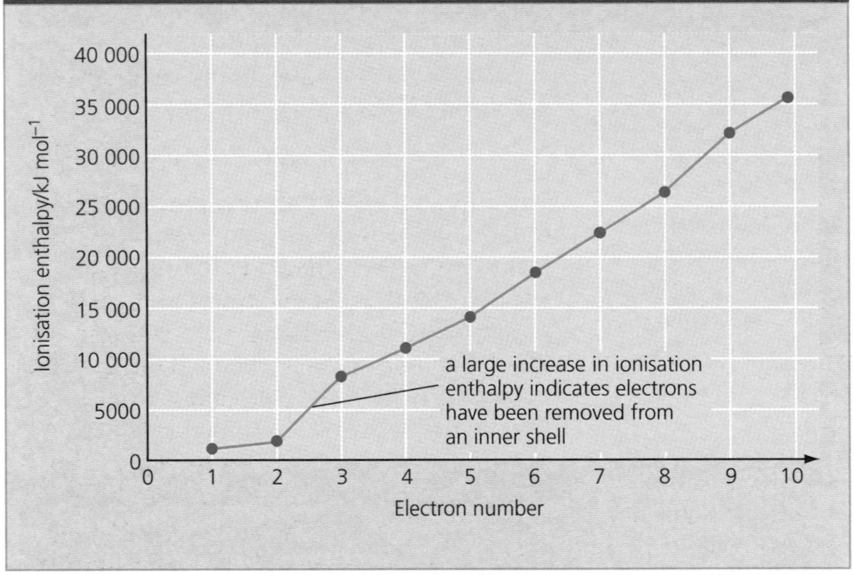

a large increase in ionisation enthalpy indicates electrons have been removed from an inner shell

Fig 2b Ionisation enthalpies for magnesium: successive removal of all 12 electrons

removal of electrons first (innermost) shell causes largest increase in ionisation enthalpy

electrons in second shell

Table 2 Successive molar ionisation enthalpies for magnesium	
No. of ionisation	$\Delta H_i^\ominus$/kJ mol^{-1}
1	+738
2	+1451
3	+7733
4	+10541
5	+13629
6	+17995
7	+21704
8	+25657
9	+31644
10	+35463
11	+169996
12	+189371

The energy needed to remove one electron from each atom in one mole of magnesium atoms is 738 kJ mol^{-1}. When an electron is removed from an Mg atom to form an Mg$^+$ ion, the repulsion between the second and third shells is decreased, so the outer shell moves nearer to the nucleus.

Because the remaining outer-shell electron is now closer to the nucleus, the force of attraction between it and the nucleus is greater than in the neutral atom. Therefore, the second ionisation enthalpy (+1451 kJ mol^{-1}) is greater than the first ionisation enthalpy (+738 kJ mol^{-1}).

The third electron must be removed from an inner orbital that is much nearer to the nucleus. Electrons in this level are attracted far more strongly to the nucleus so far more energy is required to remove them. There is therefore a large increase in ionisation energy between the second and the third values. This indicates 2 electrons in the outer shell.

The ionisation enthalpy is the standard enthalpy change for the removal of an electron from a species in the gas phase to form a positive ion and an electron, both also in the gas phase.

2 The successive ionisation enthalpies in kJ mol⁻¹ for three different elements are:

a 801 2427 3660 25026 32828

b 1086 2353 4621 6223 37832 47278

c 1314 3388 5301 7469 10989 13327 71337 84080

Determine which group of the Periodic Table each of the elements is in, stating reasons.

Atomisation of chlorine

Chloride ions are formed in two stages. The first of these involves splitting gaseous chlorine molecules into gaseous atoms. Atomisation of chlorine is another endothermic process.

$$\tfrac{1}{2} Cl_2(g) \rightarrow Cl(g) \qquad \Delta H_{at}^{\ominus} = +121 \text{ kJ mol}^{-1}$$

Here, the molar enthalpy of atomisation is defined as the enthalpy change needed to produce one mole of atoms in their standard states from the element in its standard state. This value is half of the value for the Cl–Cl bond enthalpy.

Sometimes the enthalpy of dissociation is used to describe a change from gaseous molecules to gaseous atoms. For chlorine, it is the enthalpy required to dissociate one mole of $Cl_2(g)$ molecules in their standard state into their constituent atoms.

$$Cl_2(g) \rightarrow 2Cl(g) \qquad \Delta H_{diss}^{\ominus} = +242 \text{ kJ mol}^{-1}$$

The bond dissociation enthalpy is the standard enthalpy change that accompanies the breaking of a covalent bond in a gaseous molecule to form two free radicals also in the gaseous phase

 3 Why is atomisation an endothermic process?

Table 3 shows a comparison of bond dissociation enthalpy values for different elements. The values for the different halogens are quite similar because they all have single covalent bonds that are being broken. The values decrease as the halogen atoms get larger, since the forces holding the atoms together are weaker. The values for O_2 and N_2 are larger than for any of the halogens because atomising oxygen involves breaking two bonds (σ and π) and for nitrogen three bonds (σ and two π). See *AS Chemistry*, page 189.

Table 3 Bond dissociation enthalpy values for different elements					
Molecule	Cl_2	Br_2	I_2	O_2	N_2
$\Delta H_{at}^{\ominus}$ **/kJ mol⁻¹**	121	112	107	249	472

In each of the changes described so far, the particles have gained energy. Other changes can release energy.

Electron affinity

A gaseous chlorine atom can form a negative gaseous chloride ion.

$$Cl(g) + e^- \rightarrow Cl^-(g) \qquad \Delta H_{ea}^{\ominus} = -349 \text{ kJ mol}^{-1}$$

The enthalpy change for this process is its **electron affinity**. Electron affinities can be exothermic for the atoms to the right-hand side of the Periodic Table (the non-metallic elements). These are atoms with a strong attractive force produced by a relatively high nuclear charge (for that period).

Do not confuse electron affinity with ionisation energy. All ionisation enthalpies are endothermic because there must be an energy input for the electron to move away from the attractive force of the nucleus. If one electron is added to a highly electronegative atom, energy may be released (exothermic). However, if more than one electron is required to complete the outer shell, the second or third electron affinities are endothermic. This is because energy is needed to overcome the increasing amount of repulsion against a further electron going into the shell. Since an ion has a lower nuclear charge than the corresponding atom with the same number of electrons, attraction by an 'ionic' nucleus is less than that for an 'atomic' nucleus. Since the positive nuclear charge in the ionic nucleus cannot fully offset the increased repulsive forces of putting yet another electron into occupied orbitals, the process is endothermic (see Fig. 3).

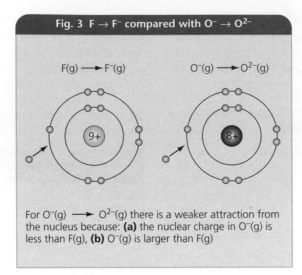

Fig. 3 F → F⁻ compared with O⁻ → O²⁻

F(g) ⟶ F⁻(g)   O⁻(g) ⟶ O²⁻(g)

For O⁻(g) ⟶ O²⁻(g) there is a weaker attraction from the nucleus because: **(a)** the nuclear charge in O⁻(g) is less than F(g), **(b)** O⁻(g) is larger than F(g)

For the fluorine atom and the singly charged oxygen ion in Fig. 3, the change in electronic structure is the same for each process. Fluorine has 9 protons and oxygen has 8. At the same time, O⁻ is larger than F, so the force of attraction between the nucleus and outer electrons is less in the O⁻ ion.

Electron affinity and electronegativity

Electron affinity occurs only in atoms with relatively high **electronegativity**. The higher the electronegativity, the stronger the force of attraction for electrons. Because a chlorine atom in the gas phase has a strong affinity for an electron, the formation of a chloride ion gives out heat to the surroundings. The electron affinity for this atom is exothermic.

However, electron affinities can be endothermic. Forming a gaseous O^{2-} ion from a gaseous O^- ion is endothermic. Electron affinities of some negative ions are shown in Table 4.

Electron affinity is the standard enthalpy change when an electron is added to an isolated atom in the gas phase.

Lattice enthalpy

The strongly attractive electrostatic forces between chloride ions and sodium ions in the gaseous state pull the ions together to form a close-packed, solid ionic lattice (Fig. 4).

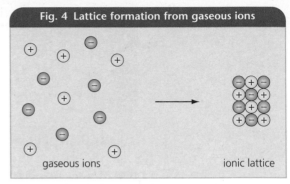

Fig. 4 Lattice formation from gaseous ions

gaseous ions   ionic lattice

$$Na^+(g) + Cl^-(g) \rightarrow NaCl(s) \quad \Delta H_L^\ominus = -786 \text{ kJ mol}^{-1}$$

A considerable amount of energy is released in this process, and the particles move to a much lower energy level. This energy is called the lattice formation enthalpy.

The strength of the attraction can be related to several features of the ions, which means that the amount of energy released is related to: the charges on the ions, their size and the type of lattice formed. A larger exothermic lattice enthalpy is favoured by a greater charge on the ions, smaller ions and a closer alignment in the lattice.

Because the sodium ion is much smaller than the original sodium atom, the two newly formed Na⁺ and Cl⁻ ions can align much more closely. This produces much stronger forces of attraction, so we talk about positive and negative ions attracting each other.

Note that, in the above example, the lattice enthalpy is stated for the *formation* of solid NaCl from its gaseous ions. The value stated is negative, showing that it is an exothermic process. Lattice enthalpy values can be quoted for the lattice *dissociation* process:

$$NaCl(s) \rightarrow Na^+(g) + Cl^-(g) \quad \Delta H_L^\ominus = +786 \text{ kJ mol}^{-1}$$

Here, the lattice enthalpy is positive; energy must be supplied to break up the solid lattice and form gaseous ions.

The enthalpy of lattice dissociation is the standard enthalpy change which accompanies the separation of one mole of a solid ionic lattice into its gaseous ions.

Table 4 Electron affinities for some non-metal atoms			
	$\Delta H_{ea}^\ominus$ /kJ mol⁻¹ (1st electron affinity)		$\Delta H_{ea}^\ominus$ /kJ mol⁻¹ (2nd electron affinity)
H(g) → H⁻(g)	−72		
F(g) → F⁻(g)	−328		
Cl(g) → Cl⁻(g)	−348		
Br(g) → Br⁻(g)	−324		
I(g) → I⁻(g)	−295		
N(g) → N⁻(g)	0		
O(g) → O⁻(g)	−141	O⁻(g) → O²⁻(g)	+798
S(g) → S⁻(g)	−200	S⁻(g) → S²⁻(g)	+640
P(g) → P⁻(g)	−72		

Enthalpy of formation

The energy released during the formation of 1 mole of sodium chloride from its constituent elements in their standard states under standard conditions is called the standard molar enthalpy of formation (Fig. 5). The first law of thermodynamics tells us that energy cannot be created or destroyed, but it can be stored or transferred. When energy is transferred, it produces an effect. All the energy changes relating to the stages in the formation of NaCl, or in any other chemical reaction, can be linked by a **Born–Haber cycle** (Fig. 6). The sum of all the energy changes is the enthalpy of formation.

According to **Hess's law** (see *AS Chemistry*, page 81), the overall enthalpy change in a complete cycle must be zero, i.e. the sum of the steps $\Delta H^{\ominus}_{sum} = 0$. Using Hess's law, the overall energy change is independent of the number of steps (Fig. 7).

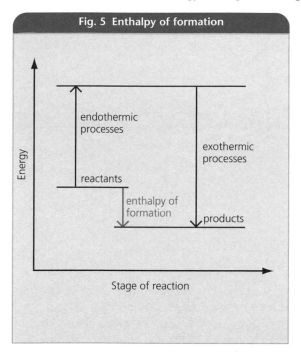

Fig. 5 Enthalpy of formation

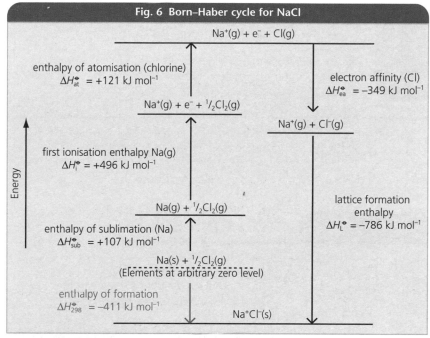

Fig. 6 Born–Haber cycle for NaCl

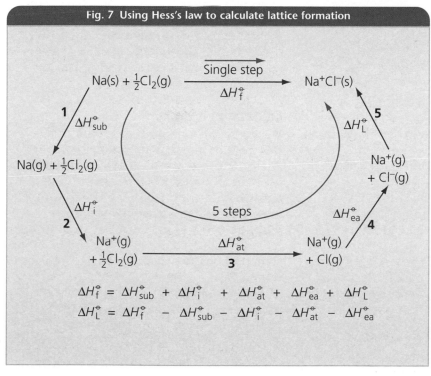

Fig. 7 Using Hess's law to calculate lattice formation

$$\Delta H^{\ominus}_f = \Delta H^{\ominus}_{sub} + \Delta H^{\ominus}_i + \Delta H^{\ominus}_{at} + \Delta H^{\ominus}_{ea} + \Delta H^{\ominus}_L$$

$$\Delta H^{\ominus}_L = \Delta H^{\ominus}_f - \Delta H^{\ominus}_{sub} - \Delta H^{\ominus}_i - \Delta H^{\ominus}_{at} - \Delta H^{\ominus}_{ea}$$

$\Delta H^{\ominus}$ (single step) $= \Delta H^{\ominus}$ (5 steps)

$\Delta H^{\ominus}_f = $ steps $1 \rightarrow 5$

$\Delta H^{\ominus}_f = \Delta H^{\ominus}_f - $ steps $1 \rightarrow 4$

$\quad = -411 - 107 - 496 - 121 - (-349)$

$\quad = -786$ kJ mol^{-1}

Because the lattice enthalpy value is so high, the enthalpy of formation is strongly exothermic and so the compound formed is very stable compared with its elements. Energy cycles are a useful way of calculating the overall energy change in a reaction. They can also be used to help us to understand why some compounds exist and others do not.

If we construct a Born–Haber cycle for the hypothetical compound MgCl(s), we can estimate the enthalpy of formation.

$$\Delta H^{\ominus}_f = Mg(s) \rightarrow \quad Mg(g) \rightarrow Mg^+(g)$$
$$\qquad\qquad +148 \qquad +740$$

$$Cl_2(g) \rightarrow Cl(g) \rightarrow Cl^-(g)$$
$$\qquad +121 \qquad -349$$

$Mg^+(g) + Cl^-(g) \rightarrow MgCl(s)$
                      $-815(est)$

$\Delta H_f^{\ominus} \, MgCl(s) = 148 + 740 + 121 - 349 - 815$

                      $= -155 \text{ kJ mol}^{-1}$

The enthalpy of formation for $MgCl_2(s)$ is in fact -641 kJ mol^{-1}. This is a much larger negative value than for the hypothetical $MgCl(s)$, so more energy can be released if $MgCl_2(s)$ is formed, making $MgCl_2(s)$ more stable than $MgCl(s)$. This means that $MgCl_2(s)$ is the chloride formed when magnesium reacts with chlorine.

The standard enthalpy of formation is the enthalpy change involved in the production of one mole of a compound from its elements under standard conditions, reactants and products being in their standard states.

12.2 Heat from solution

The process of dissolving

Water molecules are polar because of electronegativity differences within the molecule (Fig. 8).

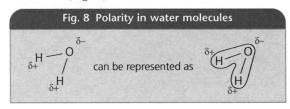

Fig. 8 Polarity in water molecules

can be represented as

The polar ends of the water molecules are attracted to the ions in the ionic lattice of a solid. This attraction distorts the charge cloud of the positive and negative ions in the lattice and reduces the forces holding them together (Fig. 9).

The ions move from the lattice into solution and become surrounded by water molecules – the ions are then said to be hydrated. Positive cations are attracted to the negative end of the dipole (the oxygen atom in the water molecule). Similarly, negative anions are attracted to the positive end of the dipole (the hydrogen atoms in the water molecule). The force of attraction between water molecules and the cations and anions means that energy is released.

The arrangement of water molecules around the cations and anions leads to further layers of water molecules being attracted, releasing yet more energy. The hydrated ions tend to have a particular number of water molecules associated with them. When the ions move through the solution, these water molecules move with them and this is called the hydration sheath. The number of water molecules attached to the ion is called its **hydration number** (see Table 5).

Enthalpy of hydration

When some compounds are put in water, there can be a measurable enthalpy change. There are two processes that can take place between ions and water: hydration of gaseous ions which is the **enthalpy of hydration**, and dissolving a solid compound to form a solution, which is the **enthalpy of solution**.

Both positive and negative ions of an ionic compound will be hydrated in the presence of water.

Some compounds, such as concentrated sulphuric acid, produce a lot of heat when mixed with water. Other compounds, such as anhydrous copper sulphate, evolve less heat when added to water.

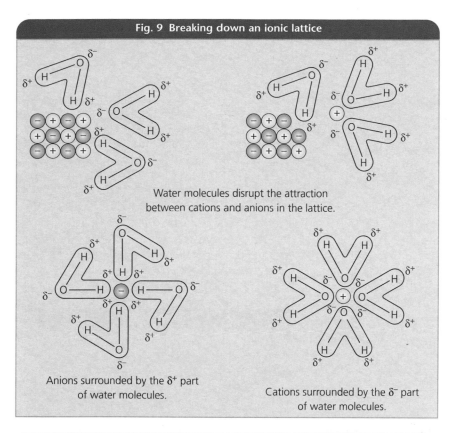

Fig. 9 Breaking down an ionic lattice

Water molecules disrupt the attraction between cations and anions in the lattice.

Anions surrounded by the δ+ part of water molecules.

Cations surrounded by the δ− part of water molecules.

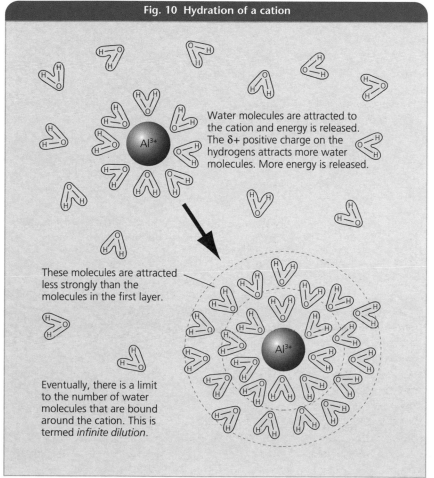

Fig. 10 Hydration of a cation

Water molecules are attracted to the cation and energy is released. The δ+ positive charge on the hydrogens attracts more water molecules. More energy is released.

These molecules are attracted less strongly than the molecules in the first layer.

Eventually, there is a limit to the number of water molecules that are bound around the cation. This is termed *infinite dilution*.

For ions to dissolve in water, they need to be separated from one another. The ions in an ionic lattice are attracted by strong forces, so separating them is an endothermic process. For any compound to dissolve in water, a large amount of energy is needed to compensate for this endothermic change. Part of the energy is released when the ions become surrounded with water molecules. Ions and water molecules are attracted to each other, so this process releases energy. It is called the enthalpy of hydration ΔH_{hyd}. (For any solvent, the enthalpy released when a solution forms is referred to as the solvation enthalpy.)

The enthalpy of hydration is the standard enthalpy change for the process:

$$X^{+/-}(g) \xrightarrow{\text{water}} X^{+/-}(aq) \quad \Delta H^{\ominus} = \Delta H^{\ominus}_{hyd}$$

The enthalpy of hydration, ΔH_{hyd}, is the enthalpy change when one mole of gaseous ions completely dissolves in water, meaning that one mole of gaseous ions is surrounded by water molecules. Ions with strong forces of attraction will release large amounts of energy while others with weaker forces will not become as warm. Where ionic compounds have hydration energies higher than their lattice energies, dissolving is an exothermic process.

The amount of water used in hydration will affect the amount of energy released. For all the energy to be released, the ion must have all the possible layers of water molecules attached. Water molecules close to the central ion are strongly attracted and a lot of energy is released. The molecules that are further away from the central ion exert weaker forces and lesser amounts of energy are released. The force of attraction is weaker at each layer until the water molecules are far enough away to not be attracted to the ion. This is said to be a state of **infinite dilution** (Fig. 10). Once infinite dilution is reached, no further energy will be released due to hydration.

When anhydrous copper sulphate mixes with water, six water molecules form covalent bonds with the lone pairs on the oxygens, then this complex ion is hydrated. The hydration enthalpy released is greater than the energy needed to break down the lattice.

The energy output is greater than the energy input, so the mixture feels warmer than the original components (Fig. 11). Fig. 12 shows an energy profile for hydration.

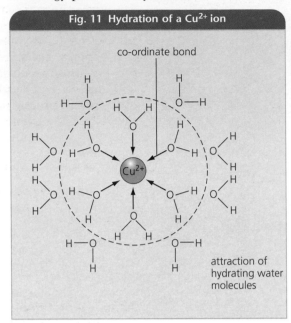

Fig. 11 Hydration of a Cu^{2+} ion

co-ordinate bond

Cu^{2+}

attraction of hydrating water molecules

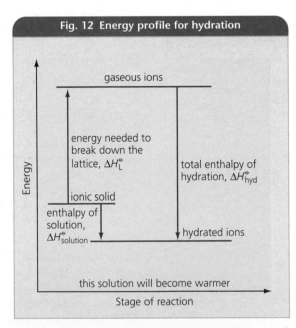

Fig. 12 Energy profile for hydration

gaseous ions

energy needed to break down the lattice, $\Delta H_L^{\ominus}$

total enthalpy of hydration, $\Delta H_{hyd}^{\ominus}$

ionic solid

enthalpy of solution, $\Delta H_{solution}^{\ominus}$

hydrated ions

this solution will become warmer

Stage of reaction

Energy

Factors affecting enthalpy of hydration

The force of attraction between ions and water molecules depends upon the charge and the size of the ion. Consider the ions Na^+, Mg^{2+} and Al^{3+}. The cations all have the same electronic structure, but they have different nuclear charges and sizes (Fig. 13).

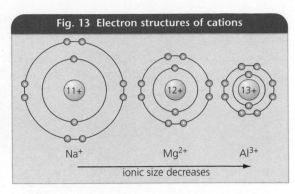

Fig. 13 Electron structures of cations

11+   12+   13+

Na^+ Mg^{2+} Al^{3+}

ionic size decreases

The positive nuclear charge increases from Na^+ to Al^{3+}. Therefore, the force of attraction for the negative pole of the water dipole is greater for Al^{3+} than for Na^+; Al^{3+} attracts water molecules very strongly and has a greater enthalpy of hydration than Na^+ (Fig. 14).

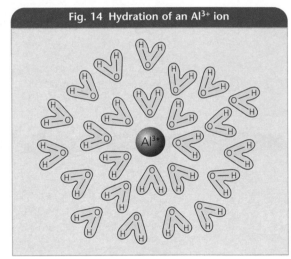

Fig. 14 Hydration of an Al^{3+} ion

Al^{3+}

For ions in the same period, the ionic charge increases as you move across from left to right, whilst the ionic size decreases. Therefore, the enthalpy of hydration increases considerably. For ions in the same group – where ionic charge is constant and ionic size increases down the group – there is a less dramatic change and the enthalpy of hydration decreases down the group (compare the values for Na^+, K^+ and Rb^+ in Table 5).

Table 5 Enthalpies of hydration and hydration numbers for different cations					
	Na^+	Mg^{2+}	Al^{3+}	K^+	Rb^+
Enthalpy of hydration/kJ mol^{-1}	−390	1891	−4613	−305	−281
Average hydration number	5	15	26	4	4

Table 6 Hydration enthalpies for some negative ions	
Ion	$\Delta H^{\ominus}_{hyd}$/kJ mol^{-1}
F$^-$	−457
Cl$^-$	−381
Br$^-$	−351
I$^-$	−307
OH$^-$	−460

Table 5 shows that 2+ and 3+ ions have much larger hydration enthalpies than the 1+ ions. Ionic charge is a more significant factor than ionic size for producing larger hydration enthalpies.

Negative ions are also hydrated, but these will attract the positive end of the water dipole (Fig. 9). The attraction of oppositely charged species will release energy and contribute to the overall value for the enthalpy of solution.

The following are calculations for enthalpies of hydration for sodium chloride and magnesium chloride, illustrated in Fig. 15.

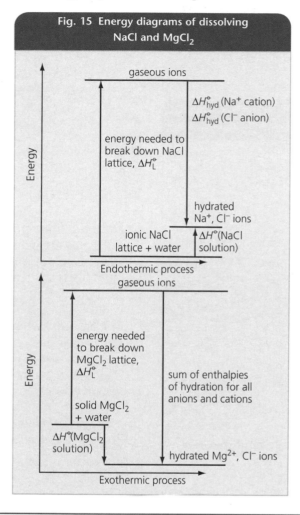

Fig. 15 Energy diagrams of dissolving NaCl and MgCl$_2$

NaCl

$$\begin{aligned}\Delta H^{\ominus}_{solution} &= \Delta H^{\ominus}_{hyd(cation)} + \Delta H^{\ominus}_{hyd(anion)} - \Delta H^{\ominus}_{L} \\ &= \quad -390 \quad + \quad (-381) - (-780) \\ &= +9 \text{ kJ mol}^{-1}\end{aligned}$$

MgCl$_2$

$$\begin{aligned}\Delta H^{\ominus}_{solution} &= \Delta H^{\ominus}_{hyd(cation)} + \Delta H^{\ominus}_{hyd(anion)} - \Delta H^{\ominus}_{L} \\ &= \quad -1891 \quad + \quad 2(-381) - (-2526) \\ &= -127 \text{ kJ mol}^{-1}\end{aligned}$$

Enthalpy of solution

The enthalpy of solution, $\Delta H^{\ominus}_{solution}$, is the enthalpy change when one mole of solute is dissolved to form a solution of infinite dilution.

The enthalpy of solution is the standard enthalpy change for the process in which one mole of an ionic solid dissolves in an amount of water large enough to ensure that the dissolved ions are well separated and do not interact with one another.

KEY FACTS

- Water molecules are attracted to cations and anions.

- The hydration enthalpy of an ion is greater for an ion with a higher charge to size ratio.

- The enthalpy of hydration, $\Delta H^{\ominus}_{hyd}$, is the enthalpy change when one mole of gaseous ions is completely surrounded by water molecules.

- If $\Delta H^{\ominus}_{hyd}$ is greater than $\Delta H^{\ominus}_{L}$, energy will be released when ions are separated to form a solution.

12.3 Calculating enthalpy changes

For some reactions it is difficult to measure the enthalpy change experimentally. Again, we can calculate them from other known data using Hess's law (*AS Chemistry*, page 81).

An example is the reaction:

$$NH_3(g) + HCl(g) \xrightarrow{\Delta H} NH_4Cl(s)$$

At one end, cotton wool is soaked in concentrated ammonia solution and at the other hydrogen chloride solution. The gases diffuse along the tube to form white ammonium chloride.

We can use $\Delta H_f^{\ominus}$ values to calculate the enthalpy change in this reaction (Table 7).

Table 7 Enthalpy values for ammonia reacting with hydrogen chloride			
Compound	$NH_3(g)$	$HCl(g)$	$NH_4Cl(s)$
$\Delta H_f^{\ominus}$ /kJ mol^{-1}	–46.1	–92.3	–314.4

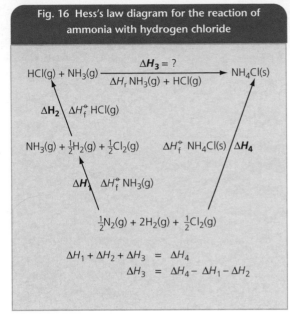

Fig. 16 Hess's law diagram for the reaction of ammonia with hydrogen chloride

$$\Delta H_1 + \Delta H_2 + \Delta H_3 = \Delta H_4$$
$$\Delta H_3 = \Delta H_4 - \Delta H_1 - \Delta H_2$$

The steps are as follows:

1 Look at Table 4, page 158, and work out two ways of producing NH_4Cl from the elements:

(a) $N_2(g)/H_2(g)/Cl_2(g) \rightarrow NH_4Cl(s)$
(1 stage directly)
(b) $N_2(g)/H_2(g)/Cl_2(g) \rightarrow HCl(g) + NH_3(g) \rightarrow NH_4Cl(s)$ (2 stages indirectly)

2 Draw a Hess's law diagram like the one in Fig. 16.
You should have arrows going clockwise and anticlockwise.
Check you have used matching quantities for the process.
Label each step with: ΔH_1 ΔH_2 ΔH_3 ΔH_4

3 Clockwise enthalpy = anticlockwise enthalpy
changes   changes
indirect route   direct route
$$\Delta H_1 + \Delta H_2 + \Delta H_3 = \Delta H_4$$

4 Substitute the values and calculate ΔH_3:
$$\Delta H_3 = \Delta H_4 - \Delta H_1 - \Delta H_2$$
$$= -314.4 - (-46.1) - (-92.3)$$
$$= -176.0 \text{ kJ mol}^{-1}$$

4 Calculate the energy change for the reaction:

$$2NaHCO_3(s) \rightarrow Na_2CO_3(s) + CO_2(g) + H_2O(l)$$
$\Delta H_f CO_2(g) = -393.5$ kJ mol^{-1},
$H_2O(l) = -285.8$, $Na_2CO_3(s) = -1130.7$,
$NaHCO_3(s) = -950.8$

Mean bond enthalpies

A method of calculating enthalpy changes in reactions is to use **mean bond enthalpy** values. This is the enthalpy change when one mole of bonds of a specified type is broken. They are covered in Section 5.6 of Chapter 5 in *AS Chemistry*, where the discussion describes how mean bond enthalpies are normally quoted in data books.

If a molecule of ammonia is dissociated, the bonds are broken successively (Fig. 17). When this happens, one electron from the covalent pair is retained by each atom. These species are free radicals (see *AS Chemistry*, page 184).

The standard molar bond dissociation enthalpy $\Delta H_{diss}^{\ominus}$ is the enthalpy change when one mole of bonds of the same type in gaseous molecules under standard conditions is broken, producing free radicals.

The enthalpy change for each stage in Fig. 17 will be different because each stage is in a different environment each time a hydrogen is removed. The mean bond enthalpy is simply one-third of the total dissociation energy change. This is an approximation but provides a useful guide:
$$1/2 N_2(g) + 3/2 H_2(g) \rightarrow NH_3(g)$$

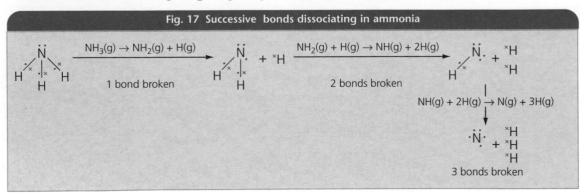

Fig. 17 Successive bonds dissociating in ammonia

$NH_3(g) \rightarrow NH_2(g) + H(g)$ 1 bond broken

$NH_2(g) + H(g) \rightarrow NH(g) + 2H(g)$ 2 bonds broken

$NH(g) + 2H(g) \rightarrow N(g) + 3H(g)$

3 bonds broken

Table 8 Mean bond enthalpies					
Bond	N≡N	H–H	N–H	Cl–H	Cl–Cl
Bond enthalpy /kJ mol^{-1}	945.4	435.9	391.0	432.0	243.4

We can use the data in Table 8 to check if this approximation is acceptable.

The enthalpy change for the formation of ammonia is:

a Breaking the bonds in $N_2(g)$ and $H_2(g)$ to form $N(g) + 3H(g)$
$1/2 N_2(g) = 1/2 \times 945.4 = +472.7$
$3/2 H_2(g) = 3/2 \times 435.9 = +653.9$
total = $+1126.6$ kJ mol^{-1}

b Forming the bonds in $NH_3(g)$:
$3N–H = 3 \times -391 = -1173$
total enthalpy change for bond breaking and bond forming:
= $+1126.6 + (-1173)$
= -46.4 kJ mol^{-1}
Experimental value for $\Delta H_f = -46.1$ kJ mol^{-1}

The value is quite close to the experimental value, so mean bond enthalpy values are satisfactory here.

5 Use Table 8 to calculate the enthalpy of formation for HCl(g). The value quoted for HCl(g) is -92.3 kJ mol^{-1}. Comment on the two values.

Using mean bond enthalpy values can help us to understand the bonding in certain compounds. The bond enthalpy values are considered separately and it is assumed that there is no interaction between adjacent bonds or atoms. When this is the case, the values agree with experimental data. If the values do not agree, we need to consider how the bonding is affecting the values. These discrepancies can occur with both covalent and ionic compounds.

One famous example is Kekulé's description of the bonding in benzene. He said that the molecule consisted of alternate single and double bonds. Addition of hydrogen to a double bond gives an enthalpy change of -120 kJ mol^{-1}, so hydrogenation of three separate double bonds would give -360 kJ mol^{-1}. The enthalpy change for the hydrogenation of benzene is only -208 kJ mol^{-1}. The molecule is more stable than predicted by 152 kJ mol^{-1}. The extra stability is now explained by delocalisation of the electrons in benzene (Fig. 18).

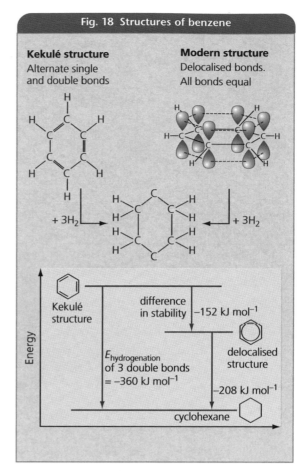

Fig. 18 Structures of benzene

Kekulé structure
Alternate single and double bonds

Modern structure
Delocalised bonds. All bonds equal

Another example is the ethanoate ion CH_3COO^-, as shown in Fig. 19.

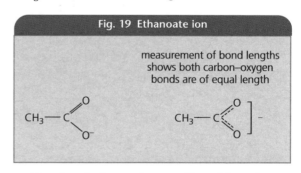

Fig. 19 Ethanoate ion

measurement of bond lengths shows both carbon–oxygen bonds are of equal length

Bond enthalpy values and bond lengths indicate two identical C–O bonds, not a single bond and a double bond.

For ionic compounds, it is assumed that the ions are completely separate particles. However, many ionic compounds demonstrate some covalent character in the ionic bond. It is possible to calculate a theoretical lattice energy from the attractive and repulsive forces. When the positive and negative ions have a larger difference in electronegativity, there is good agreement between the theoretical and experimental values from Born–Haber cycles (Table 9).

Table 9 A selection of lattice enthalpy values			
Lattice	Enthalpy values/kJ mol⁻¹		
	Experimental value	Theoretical value	Difference
NaCl	780	770	−10
MgCl₂	2526	2326	−200
CaCl₂	2258	2223	−35
SrCl₂	2156	2127	−29

If there is some covalent character in the ionic crystal, there are greater forces of attraction and a greater energy value. Look at Table 9. NaCl gives good agreement (−10), but MgCl₂ does not (−200); Mg^{2+} is a much more polarising ion and has a greater electronegativity than sodium. The larger Group II ions such as calcium and strontium have less polarising power than magnesium, so the values are closer to the theoretical values.

12.4 Entropy

Spontaneous changes

Why do things happen the way they do?

Why does ice melt? Why does your tea go cold? If a metal bar is hot at one end and you leave it, the heat will be conducted throughout the bar until all the bar is at the same temperature. These are all examples of spontaneous changes. If you have a metal bar at a uniform temperature, you wouldn't expect it to become hotter at one end and colder at the other.

Spontaneous changes seem to happen without us having to do anything: we don't have to put energy into them to make them happen. There are other spontaneous changes – a gas expands from high pressure to low pressure and not the reverse, e.g. a balloon will let all the air out unless you tie a knot in it, but the balloon will not blow itself up. Also, if a smelly gas is in an open container in one corner of a room, the gas will spread out and fill the room. It probably will not go back into the container spontaneously. Although it might be possible if all the particles just happened to move in that direction at the same time, it's just not very likely.

Some of these changes can be reversed if there is an energy input, e.g. you can blow up a balloon if you expend some energy and use the energy to make the particles move closer together.

Evaporation and condensation are reversible but there is an enthalpy change if either process occurs. Iron, if left, will rust spontaneously, but iron oxide can be changed back into iron if it is heated to a high temperature with carbon. A rechargeable battery can be changed back to its original state if electricity is passed through it in the reverse direction.

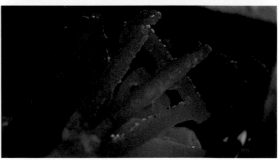

The reaction that occurs when iron rusts is reversed when rust is heated in the absence of oxygen.

To understand why these things happen, we need to consider something about probability because all physical and chemical changes are governed by the laws of probability.

If you buy a lottery ticket, your six numbers have just as much chance of coming up as anyone else's, but it is much more likely that someone else will win rather than you. That is because about 20 million sets of numbers are chosen every week. Your combination is one possibility, but there are another 19 999 999 possibilities, so it is much more likely not to be you.

If you look at the bromine gas in the photo, when you remove the card the gas diffuses through both jars until the jars appear identical. This is simplified in Fig. 20.

After about an hour, bromine in the lower gas jar has just reached the upper jar. It takes several hours for the gases to reach uniform distribution.

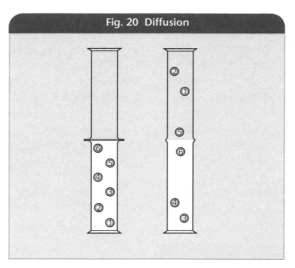

Fig. 20 Diffusion

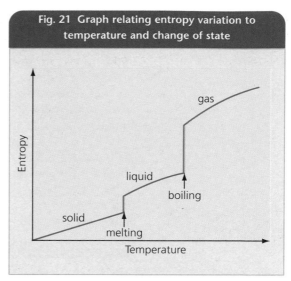

Fig. 21 Graph relating entropy variation to temperature and change of state

There is only one arrangement with all the particles in the lower jar and only one arrangement for all the particles in the upper jar, but there are numerous other combinations.

The total number of possibilities for 6 molecules in 2 jars is:

$$2 \times 2 \times 2 \times 2 \times 2 \times 2 = 2^6 = 64$$

Therefore, you are very unlikely to see all the particles in the top or bottom jar. In the photo there are something like 10^{22} molecules, so the number of ways of arranging 10^{22} molecules in 2 jars would be $(2^{10})^{22}$, a vast number. The chances of all the molecules being in one jar are so remote that all you will ever see is a random arrangement of the molecules between the two jars.

An ordered state, such as all the particles in one jar or the other, has fewer possible arrangements than a disordered state, where they are arranged randomly between both jars, so is less likely to happen. If bricks are tipped out of a lorry, then an ordered state would be a nice, neat brick wall and a disordered state would be a pile of rubble. There are few arrangements for a neat brick wall, but millions of arrangements for a pile of rubble, so that's what you will get.

In physical and chemical changes, the processes will always go spontaneously from an ordered state to a disordered state. However, when we are considering physical and chemical processes, we must consider the total order for both the systems: the change taking place and its surroundings.

The degree of disorder in a system is called its **entropy** (from the Greek word for change). As the disorder increases, so does the entropy value, and an increase in entropy relates to how many ways the particles can be arranged and how many ways the energy can be distributed between the particles. As the temperature increases, the number of energy levels available in the particles increases, so the entropy increases. There will be a sudden change of entropy if there is a change of state (Fig. 21).

We can examine some of the examples mentioned above and apply the idea of entropy.

If a crystal melts, there is a dramatic change from an ordered state in the crystal to a disordered state in the liquid. If the liquid then changes to a gas, there will be an even larger increase in disorder.

A beaker of hot water has a higher entropy value than a cold one but if it cools down it warms up the surrounding particles in the room. There are then many more ways of distributing the energy, so if we consider the system and its surroundings, the total entropy will increase and the process will be spontaneous.

If the iron bar is hot at one end, statistics shows that there are fewer arrangements for the distribution of energy than if all the particles in the bar were used, so the energy will distribute itself uniformly along the bar.

When a crystal dissolves in water, it goes from a highly ordered state to a much more disordered state, so the entropy will increase.

Absolute entropies

Since entropy is linked to disorder and decreases with temperature decrease, at 0 K (absolute zero) for a perfectly ordered pure crystal, the entropy value should be zero. From this it is possible to calculate absolute standard entropy values $S^\ominus$ and hence calculate standard entropy changes $\Delta S^\ominus$.

You cannot calculate absolute enthalpy values, only relative changes in enthalpy values from one system to another. The units for standard entropy values are J K^{-1} mol^{-1}. Note that J is used and not kJ as in enthalpy changes. The entropy values are quite small when compared to values for enthalpy changes.

Calculating entropy changes $\Delta S^{\ominus}_{298}$
These are the differences between the total entropies of the products and reactants and they can be calculated from standard entropy values like the ones in Table 10.

Table 10 Standard entropy values			
$S^{\ominus}_{298}$ /J K^{-1} mol^{-1}			
H_2(g)	131	diamond	2.4
O_2(g)	205	graphite	5.7
N_2(g)	192	HCl(g)	187
Cl_2(g)	223	HNO_3(l)	156
H_2O(g)	189	NO_2(g)	240
H_2O(l)	70	NaCl(s)	72

$$H_2(g) + Cl_2(g) \rightarrow 2HCl(g)$$
$$\Delta S^{\ominus} = \Sigma S^{\ominus}_{products} - \Sigma S^{\ominus}_{reactants} \quad (\Sigma = \text{sum of})$$
$$\Sigma S^{\ominus}_{products} = 2 \times 187 = 374$$
$$\Sigma S^{\ominus}_{reactants} = 131 + 223 = 354$$
$$\Delta S^{\ominus} = +20 \text{ J K}^{-1} \text{ mol}^{-1}$$

In the above reactions all the molecules are gases and individually have quite high entropy values, but there is only a small increase in entropy when they react.

$$4HNO_3(l) \rightarrow 4NO_2(g) + O_2(g) + 2H_2O(l)$$
$$\Delta S^{\ominus} = \Sigma S^{\ominus}_{products} - \Sigma S^{\ominus}_{reactants}$$
$$\Sigma S^{\ominus}_{products} = (4 \times 240) + 205 + (2 \times 70) = 1305$$
$$\Sigma S^{\ominus}_{reactants} = 4 \times 156 = 624$$
$$\Delta S^{\ominus} = +681 \text{ J K}^{-1}$$

For 1 mole of HNO_3(l) reacting, the entropy change will be +170 J K^{-1} mol^{-1}.

There is a considerable increase in disorder in this reaction and therefore a large entropy change. Nitric acid is a liquid, so has a lower entropy value and when the reaction occurs gas molecules with high entropy values are formed in the products.

KEY FACTS

■ Entropy is a measure of the amount of disorder in a system.

■ At 0K, in a perfectly ordered crystal, the entropy value is zero.

■ Entropy increases with increase in temperature.

■ Entropy increases from solid to liquid to gas.

■ Standard entropy are measured in J K^{-1} mol^{-1}.

12.5 Free energy change

Look at the values for H_2O in Table 10, and you will see that the entropy for water vapour is 189 J K^{-1} mol^{-1} and the value for liquid water is 70 J K^{-1} mol^{-1}. If entropy must always increase, can water vapour form liquid water? Yes, it can if we consider the surroundings as well. Steam will condense on a cold surface and when this happens energy transfers from the vapour to the cold surface. This increases the entropy of the surface, so the total

entropy of the system *and* the surroundings increases. The water has decreased in entropy by 119 J K^{-1} mol^{-1} but the entropy increase in the surroundings is 148 J K^{-1} mol^{-1} (calculated from the enthalpy transferred, see Fig. 22.)

$$\Delta S^{\ominus}_{total} = \Delta S^{\ominus}_{system} + \Delta S^{\ominus}_{surroundings}$$
$$= -119 + 148$$
$$= +29 \text{ J K}^{-1} \text{ mol}^{-1}$$

So the overall entropy increases.

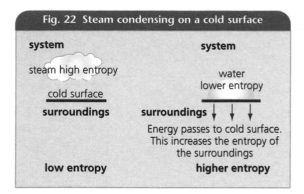

Fig. 22 Steam condensing on a cold surface

The *total* entropy increase gives rise to the expression $-T\Delta S^{\ominus}_{total}$ and to the standard free energy change $\Delta G^{\ominus}$, also called the **Gibbs free energy** change whose units are kJ mol^{-1}. We can use the changes in enthalpy and entropy values of the system to calculate the total entropy change for the system plus its surroundings, to determine whether the process will be spontaneous.

For a process to be spontaneous, the free energy change must be zero or have a negative value.

Reaction feasibility

We often use the term feasibility to mean whether a reaction or process can take place or not spontaneously. Even if a process has a negative $\Delta G^{\ominus}$ value, the calculation does not give any indication about the rate at which it takes place, e.g. diamond is thermodynamically less stable than graphite, so diamond should spontaneously revert to graphite, but the reaction is infinitely slow so it doesn't happen.

A chemical or physical change is said to be feasible if the value for $\Delta G^{\ominus}$ is negative or zero and this feasibility depends upon the relative magnitude of the enthalpy and entropy terms. The standard enthalpy change, $\Delta H^{\ominus}$, and the standard entropy change, $\Delta S^{\ominus}$, are linked to $\Delta G^{\ominus}$ by the expression:

$$\Delta G^{\ominus} = \Delta H^{\ominus} - T\Delta S^{\ominus}$$

The enthalpy change will affect the entropy of the surroundings. If energy is released in a reaction it will be absorbed by the surroundings, increasing the entropy of the surroundings.

The magnitude of entropy values is much smaller than that for enthalpy values. Enthalpy and free energy are quoted using kJ and entropy using J, so it is important in calculations to use compatible units (usually the entropy term is divided by 1000 to use kJ throughout).

If the combined enthalpy and entropy terms give a zero or negative value for $\Delta G^{\ominus}$, the process is feasible. By looking at the equation, you can see that, if $\Delta H^{\ominus}$ is large and positive, it is not possible under usual conditions to achieve a negative value for $\Delta G^{\ominus}$, because entropy values are smaller, so these endothermic reactions will not be feasible.

Many reactions release heat. These have negative $\Delta H^{\ominus}$ values, so $\Delta G^{\ominus}$ will probably be negative, and the reaction feasible. Reactions will go spontaneously from a high-energy state to a low-energy state and when they do this they release energy. You might not expect a reaction to go spontaneously in the opposite direction, i.e. from a low-energy state to high-energy. It's a bit like asking water to flow uphill. However, some changes are spontaneous *and* endothermic; the most common of these involve reactions that produce gases, e.g. hydrogencarbonate with acid, dissolving certain compounds in water, e.g. ammonium nitrate, or changing state, e.g. melting ice. These changes involve a large increase in the disorder of the particles in the system. This increased disorder can enable endothermic reactions to occur.

If $\Delta H^{\ominus}$ is small and positive, the combined $-T\Delta S^{\ominus}$ term can be large enough to compensate, giving a negative $\Delta G^{\ominus}$ value and the process may take place spontaneously. Increasing the temperature will increase the magnitude of this combined term, so often these processes are temperature dependent. However, remember that temperature also affects rate and this is not the same as feasibility.

Example: Calculate the feasibility of the reaction at the temperature 298 K (25 °C)

$H_2(g) + O_2(g) \rightarrow H_2O(g)$

$\Delta H^{\ominus} = -242 \text{ kJ mol}^{-1}$

$\Delta S^{\ominus} = -147 \text{ J K}^{-1} \text{ mol}^{-1}$

$T = 298 \text{ K}$

$\Delta G^{\ominus} = \Delta H^{\ominus} - T\Delta S^{\ominus}$

$\quad = -242 - 298\left(\dfrac{-147}{1000}\right)$

$= -242 + 44$

$\quad = -198 \text{ kJ mol}^{-1}$

There is a decrease in entropy in changing from 1.5 moles of gas molecules to 1.0 moles, but this is quite a small term compared to the negative enthalpy term $\Delta H^{\ominus}$ so, overall, $\Delta G^{\ominus}$ is negative. Even though entropy is decreasing, the reaction is still feasible due to the large negative enthalpy change.

If gases are evolved from solids during a reaction, there will be a considerable increase in entropy, e.g. neutralising reactions with carbonates or hydrogencarbonates that evolve carbon dioxide.

$H^+(aq) + HCO_3^-(aq) \rightarrow H_2O(l) + CO_2(g).$

Enthalpy change:

$\Delta H^{\ominus} = \Delta H^{\ominus}_{\text{products}} - \Delta H^{\ominus}_{\text{reactants}}$

$\Delta H^{\ominus}_{\text{products}} = -679$

$\Delta H^{\ominus}_{\text{reactants}} = -692$

$\Delta H^{\ominus} = +13 \text{ kJ mol}^{-1}$

This is a positive enthalpy change, so the reaction is endothermic. The system is moving from lower to higher energy, it takes in heat from the surroundings, which will therefore decrease in entropy.

The entropy change for the reaction is:

$\Delta S^{\ominus} = S^{\ominus}_{\text{products}} - S^{\ominus}_{\text{reactants}}$

$S^{\ominus}_{\text{products}} = 283 \text{ J K}^{-1} \text{ mol}^{-1}$

$S^{\ominus}_{\text{reactants}} = 91 \text{ J K}^{-1} \text{ mol}^{-1}$

$\Delta S^{\ominus} = +192 \text{ J K}^{-1} \text{ mol}^{-1}$

The entropy of the system (the reaction) is increasing. It is a large entropy increase because a gas is being produced during the reaction.

The feasibility of the reaction is determined by combining these two terms, i.e. finding the Gibbs free energy.

Note: the entropy term is divided by 1000 to give units in kJ not J.

The free energy change can be calculated from these values.

$\Delta H^{\ominus} = +13 \text{ kJ mol}^{-1}$

$\Delta S^{\ominus} = +192 \text{ J K}^{-1} \text{ mol}^{-1}$

$T = 298 \text{ K}$

$\Delta G^{\ominus} = \Delta H^{\ominus} - T\Delta S^{\ominus}$

$\quad = +13 - 298\left(\dfrac{192}{1000}\right)$

$\quad = +13 - 57$

$\quad = -44 \text{ kJ mol}^{-1}$

This calculation gives a negative value for $\Delta G^{\ominus}$.

At room temperature the combined $T\Delta S^{\ominus}$ term of 57 kJ K^{-1} mol^{-1} is sufficient to enable the endothermic reaction of neutralising hydrogencarbonates to take place.

Whether a compound dissolves or not depends largely upon the difference between the lattice enthalpy and the sum of the hydration enthalpies. Some possible changes are shown in Fig. 23 (this page and page 171).

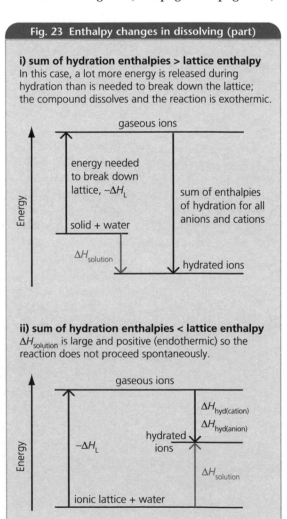

Fig. 23 Enthalpy changes in dissolving (part)

i) sum of hydration enthalpies > lattice enthalpy
In this case, a lot more energy is released during hydration than is needed to break down the lattice; the compound dissolves and the reaction is exothermic.

gaseous ions

Energy

energy needed to break down lattice, $-\Delta H_L$

solid + water

$\Delta H_{\text{solution}}$

sum of enthalpies of hydration for all anions and cations

hydrated ions

ii) sum of hydration enthalpies < lattice enthalpy
$\Delta H_{\text{solution}}$ is large and positive (endothermic) so the reaction does not proceed spontaneously.

gaseous ions

Energy

$-\Delta H_L$

$\Delta H_{\text{hyd(cation)}}$

$\Delta H_{\text{hyd(anion)}}$

hydrated ions

$\Delta H_{\text{solution}}$

ionic lattice + water

Fig. 23 Enthalpy changes in dissolving (continued)

iii) small endothermic enthalpy of solution
However, when $\Delta H_{solution}$ is only slightly positive it is possible for the salt to dissolve even though the process is endothermic.

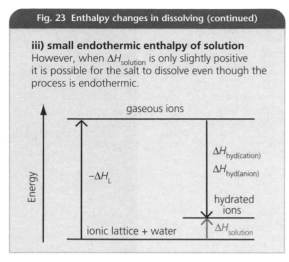

When a crystal dissolves, it changes from an ordered state and forms a solution of randomly arranged ions (i.e. less ordered arrangement), so the overall entropy increases (Fig. 24). Increasing ΔS will make the $-T\Delta S$ term more significant and it will be sufficient to produce a negative ΔG value.

Fig. 24 Dissolving a salt

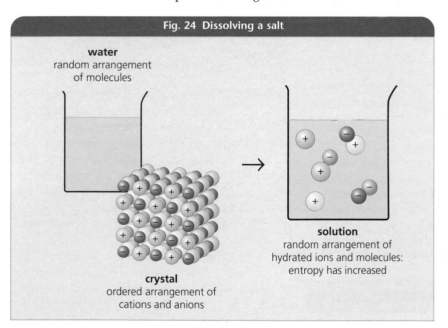

water
random arrangement of molecules

crystal
ordered arrangement of cations and anions

solution
random arrangement of hydrated ions and molecules: entropy has increased

Ammonium chloride is a salt with a positive enthalpy of solution (i.e. endothermic), but when it dissolves there is an increase in entropy. The ΔH value for dissolving ammonium chloride is quite low (+14.4 kJ mol^{-1}), so the increase in disorder produced by the regular crystal dissolving to form freely moving hydrated ions enables the change to happen spontaneously – a negative value for ΔG (Table 11).

Table 11 Entropy values for dissolving ammonium chloride

Species	$NH_4Cl(s)$	$NH_4^+(aq)$	$Cl^-(aq)$
S/J K^{-1} mol^{-1}	94.6	113.4	56.5

Example: Calculating entropy change on dissolving

$NH_4Cl(s) \rightarrow NH_4^+(aq) + Cl^-(aq)$
$94.6 \rightarrow \quad\quad 113.4 \quad + \quad 56.5$

Entropy increase = 113.4 + 56.5 − 94.6 = +75.3 J
At 298 K the combined entropy term $T\Delta S^\ominus$ = 298 × 75.3 = 22.4 kJ

$$\begin{aligned} \Delta G^\ominus &= \Delta H^\ominus - T\Delta S^\ominus \\ &= 14.4 - 22.4 \\ &= -8.0 \text{ kJ mol}^{-1} \end{aligned}$$

$\Delta G^\ominus$ is negative, so at 298 K ammonium chloride will dissolve spontaneously in water.

Temperature has a significant effect on whether the process is spontaneous. At a given temperature T, the $-T\Delta S$ term may not be large enough to give a negative value for ΔG. In some cases it is possible to raise T to the point where the reaction does become spontaneous. However, you should be aware that, even when a reaction is said to be feasible, it doesn't necessarily mean that a reaction will proceed; the activation energy may still be very high and therefore the rate very low.

Example: Calculating ΔH, ΔS and ΔG for the thermal decomposition of calcite

Study the data in Table 12, for the thermal decomposition of calcite:

$CaCO_3(s) \rightarrow CO_2(g) + CaO(s)$

Table 12 Entropy and enthalpy data for the thermal decomposition of calcite

	S/J K^{-1} mol^{-1}	$\Delta H_f^\ominus$/kJ mol^{-1}
$CaCO_3(s)$	92.9	−1206.9
$CO_2(g)$	213.6	−393.5
$CaO(s)$	39.7	−635.1

Calculate:
a the standard enthalpy change;
b the standard entropy change;
c the Gibb's free energy change at 298 K.
d Above what temperature might the reaction proceed spontaneously?

enthalpy change = −635.1 − 393.5 − (−1206.9)
                  = −1028.6 − (−1206.9)
                  = +178.3 kJ mol^{-1}

Remember the units for entropy include J (not kJ), i.e. J K^{-1} mol^{-1}

entropy change for $CaCO_3(s) \rightarrow CaO(s) + CO_2(g)$

$$92.9 \quad \rightarrow \quad 39.7 \; + \; 213.6$$

entropy increase $= 39.7 + 213.6 - 92.9$
$= 160.4$ J mol^{-1}

At 298 K, the combined entropy term $-T\Delta S^{\ominus} =$
$298 \times 160.4/1000$
$= 47.8$ kJ mol^{-1}

at 298 K, $\Delta G^{\ominus} = \Delta H^{\ominus} - T\Delta S^{\ominus}$
$= 178.3 - 47.8$
$= +130.5$ kJ mol^{-1}

$\Delta G^{\ominus} = \Delta H^{\ominus} - T\Delta S^{\ominus}$ is positive.

The reaction is not thermodynamically feasible at this temperature. To find the temperature at which feasibility occurs: When $\Delta G^{\ominus} = 0$, the reaction becomes just feasible.

$\Delta G^{\ominus} = \Delta H^{\ominus} - T\Delta S^{\ominus}$
$T\Delta S^{\ominus} = \Delta H^{\ominus}$ if $\Delta G^{\ominus} = 0$
$T = \Delta H^{\ominus} / \Delta S^{\ominus}$
$= 178.3/0.1604$
$= 1111$ K

This means that the calcite must be heated to at least 1111 K. Note this is only the temperature at which the reaction becomes feasible and it gives no indication of the rate.

6 Calculate the temperature at which the thermal decomposition of sodium hydrogencarbonate becomes feasible. The stoichiometry for the equation does not affect the feasibility temperature since halving the number of moles will affect both ΔH and ΔS equally (Table 13).

$$2NaHCO_3(s) \rightarrow Na_2CO_3(s) + H_2O(g) + CO_2(g)$$

Table 13 Entropy and enthalpy data for the thermal decomposition of sodium hydrogencarbonate		
	$S^{\ominus}$/J K^{-1} mol^{-1}	$\Delta H_f^{\ominus}$/kJ mol^{-1}
NaHCO$_3$(s)	102	−951
Na$_2$CO$_3$(s)	135	−1131
CO$_2$(g)	214	−394
H$_2$O(g)	189	−242

Entropy and changing state

Ice melts at 0 °C and water freezes at 0 °C, so it is an example of a system at equilibrium. Under these conditions, $\Delta H^{\ominus} = T\Delta S^{\ominus}$ and $\Delta G^{\ominus} = 0$. If the temperature remains constant and a small amount of heat is put into the

system, the energy of the system increases (and $\Delta G^{\ominus}$ would be increasingly positive). This can be compensated for by increasing the entropy and therefore the combined $-T\Delta S^{\ominus}$ term. Thus $\Delta G^{\ominus}$ is still 0. If energy goes into the system, the entropy must increase to keep $\Delta G^{\ominus} = 0$. The system can do this by changing from a solid to a liquid, i.e. if energy is put in the ice melts. Heat must be supplied to break down the order of the crystals, i.e. it is an endothermic process. The value for the enthalpy change for ice melting is $\Delta H = +6.01$ kJ mol^{-1} and the entropy increase when ice melts is 22 J K^{-1} mol^{-1}.

At 0 °C the enthalpy term is 6.01 kJ mol^{-1}. At 0 °C the entropy change on changing state is $273 \times 22.0 = 6.01$ kJ K^{-1} mol^{-1}.

$\Delta G^{\ominus} = \Delta H^{\ominus} - T\Delta S^{\ominus}$
$= 0$ kJ mol^{-1}

At 0 °C, the arrangement of water molecules is more random. This increases the value of the combined entropy term $T\Delta S$, so that it is equal to or greater than 6.01 kJ mol^{-1}. This gives a zero $\Delta G^{\ominus}$ value and the ice melts. The endothermic process of ice melting means that there is a cooling effect on the

APPLICATION   Limiting sports injury damage

Sports injuries such as muscle strains need to be cooled down to help prevent swelling. In most circumstances, a simple ice pack is enough to do the job, but storing ice in a 'first response box' – a large, portable first aid kit – would be impractical. What sports coaches and physiotherapists need is a pack that can be stored at room temperature and will cool down when required.

Central Scientific have been asked to design a portable, easily used pack that cools 'on demand', dissolving a stable compound seemed the most promising method.

When ionic compounds are formed, a series of endothermic and exothermic processes take place. These processes involved in making salts can be summarised in a Born–Haber cycle and can be used to calculate the relative stabilities of compounds.

1 From the values in the table below, all values in kJ mol^{-1}, construct a Born–Haber cycle for KCl(s) to calculate the standard molar enthalpy of formation. Define all the terms you use, and explain why each one is either exothermic or endothermic.

In kJ mol^{-1}:	Atomis'n of K	1st ionis'n of K	Atomis'n of Cl	Electron affinity of Cl	Lattice enthalpy
	89	+419	122	−348	−711

Construct a Born–Haber cycle for KCl$_2$(s) to show why this compound will not be found naturally.

In kJ mol^{-1}:	Atomis'n of K	1st ionis'n of K	2nd ionis'n of K	Atomis'n of Cl	Electron affinity of Cl	Lattice enthalpy (estimated)
	89	+419	3051	122	−348	−2350

Whether a compound dissolves or not depends largely upon the difference between the lattice enthalpy and the sum of the hydration enthalpies. Some possible changes are shown in Fig. 23.

2 Dissolving ammonium chloride in water will produce a cooling effect. Draw an enthalpy cycle (see Fig. 12) for the hydration of NH$_4$Cl.
Use the data below to calculate the enthalpy of solution, $\Delta H_{solution}$:

$\Delta H_L = -676$ kJ mol^{-1}
$\Delta H_{hyd(cation)} + \Delta H_{hyd(anion)} = -664$ kJ mol^{-1}

3 The above process is endothermic. Why can an endothermic process such as dissolving ammonium chloride happen spontaneously?

In an endothermic reaction, heat is drawn from the surroundings. If a cooling pack is applied to an injury, it will draw heat from that part of the body, so cooling it down. Another compound that would produce a cooling effect on mixing with water is ammonium nitrate. The researchers' data book gave the enthalpy of solution value for ammonium nitrate as +26.5 kJ mol^{-1}. This more endothermic reaction would produce a greater cooling effect than ammonium chloride.

The development team consulted with medical staff to find out what temperature the pack had to cool down to. They recommended a temperature of 1.5 °C to be maintained for at least 15 minutes. Central Scientific planned to make a pack of approximately 200 g, and this needed to cool down from 18 °C to 1.5 °C. The enthalpy change will be given by the expression:

$H = mc\Delta T$

where m = the mass of the pack, c = the specific heat capacity and ΔT = the temperature change.

4 For a 200 g pack, assuming a specific heat capacity of 4.2 J g^{-1} K^{-1}, calculate the enthalpy change taking place.

5 Calculate the number of moles of ammonium nitrate that will produce this amount of cooling.

surroundings. Each 18 g (1 mole) of ice at 0 °C takes in 6.01 kJ of energy when it melts.

If energy is removed from water at 0 °C, ice forms, the entropy term decreases (ice is more ordered than water) and $-T\Delta S^{\ominus}$ becomes less negative. However, since ΔH is now exothermic and therefore negative when ice melts, ΔG will be zero. The water is able to freeze to a more ordered structure, with a lower entropy.

Example: Calculating the entropy change when water boils

Calculate the entropy change when water boils. The molar enthalpy of vaporisation of water is +44.0 kJ mol^{-1}.

$$H_2O(l) \rightarrow H_2O(g)$$

At equilibrium:

$$\Delta H^{\ominus} = T\Delta S^{\ominus} \text{ and } \Delta G^{\ominus} = 0$$

So
$$\Delta S^{\ominus}_{vap} = \Delta H^{\ominus}_{vap}/T_{vap}$$
$$\Delta H^{\ominus}_{vap} = +44 \text{ kJ mol}^{-1}$$
$$T_{vap} = 373 \text{ K}$$
$$= 44 \times 10^3/373$$
$$= 118 \text{ J K}^{-1} \text{ mol}^{-1}$$

The entropy value for changing from liquid to vapour: 118 J K^{-1} mol^{-1} is much greater than that for changing a solid to a liquid: 22 J K^{-1} mol^{-1}. Both melting and boiling are endothermic, but the enthalpy change for vaporisation is also much greater.

KEY FACTS

■ Above or at 0 °C, $\Delta H^{\ominus} = +6.01$ kJ mol^{-1}, but $-T\Delta S^{\ominus}$ is greater than or equal to 6.01 kJ mol^{-1}, so $\Delta G^{\ominus}$ is zero/negative; ice → water

■ Below or at 0 °C, $\Delta H^{\ominus} = -6.01$ kJ mol^{-1}, but $-T\Delta S^{\ominus}$ decreases by less than or equal to +6.01 kJ mol^{-1}, so $\Delta G^{\ominus}$ is zero/negative; water → ice.

■ The feasibility of a process is determined by the standard Gibbs free energy, $\Delta G^{\ominus}$:
$\Delta G^{\ominus} = \Delta H^{\ominus} - T\Delta S^{\ominus}$

■ If $\Delta G^{\ominus}$ is negative at temperature T, the process will be spontaneous.

■ A system is at equilibrium when $\Delta G^{\ominus}$ is zero.

EXAMINATION QUESTIONS

1

a Construct a Born–Haber cycle for the formation of sodium oxide, Na$_2$O, and use the data given below to calculate the second electron affinity of oxygen.

Na(s) → Na(g)	$\Delta H^{\ominus} = +107$ kJ mol^{-1}
Na(g) → Na$^+$(g) + e$^-$	$\Delta H^{\ominus} = +496$ kJ mol^{-1}
O$_2$(g) → 2O(g)	$\Delta H^{\ominus} = +249$ kJ mol^{-1}
O(g) + e$^-$ → O$^-$(g)	$\Delta H^{\ominus} = -141$ kJ mol^{-1}
O$^-$(g) + e$^-$ → O^{2-}(g)	$\Delta H^{\ominus}$ to be calculated
2Na$^+$(g) + O^{2-}(g) → Na$_2$O(s)	$\Delta H^{\ominus} = -2478$ kJ mol^{-1}
2Na(s) + $\frac{1}{2}$O$_2$(g) → Na$_2$O(s)	$\Delta H^{\ominus} = -414$ kJ mol^{-1}

b Explain why the second electron affinity of oxygen has a large positive value. (2)

c Explain, by reference to steps from relevant Born–Haber cycles, why sodium forms a stable oxide consisting of Na$^+$ and O^{2-} ions but not oxides consisting of Na$^+$ and O$^-$ or Na^{2+} and O^{2-} ions. (4)
NEAB June 2000 CH04 Q6

2

a The reaction given below does not occur at room temperature.

$$CO_2(g) + C(s) \rightarrow 2CO(g)$$

Use the data given below to calculate the lowest temperature at which this reaction becomes feasible.

	C(s)	CO(g)	CO$_2$(g)
$\Delta H^{\ominus}_f$/kJ mol^{-1}	0	−110.5	−393.5
$S^{\ominus}$/J K^{-1} mol^{-1}	5.7	197.6	213.6 (8)

b The reaction shown below is very endothermic.

$$N_2(g) + O_2(g) \rightarrow 2NO(g) \quad \Delta H^{\ominus} = +180.4 \text{ kJ mol}^{-1}$$

i) Use the expression given below to explain why the reaction does not occur to a significant extent at room temperature.

$$\ln K = -\Delta H^{\ominus}/RT + \Delta S^{\ominus}/R$$

ii) Give one example of where this reaction occurs and explain why it causes environmental problems. (4)

c When an electrical heating coil was used to supply 3675 J of energy to a sample of water which was boiling at 373 K, 1.50 g water were vaporised. Use this information to calculate the entropy change for the process:

$$H_2O(l) \rightarrow H_2O(g)$$
(3)

NEAB June 2000 CH04 Q7

EXAMINATION QUESTIONS

3

a Some mean bond enthalpy values are shown below.

Bond	Mean bond enthalpy/kJ mol^{-1}
H–H	436
C–H	413
C–C	348
C=C	612

i) The hydrogenation of 1-ethylcyclohexene occurs according to the equation:

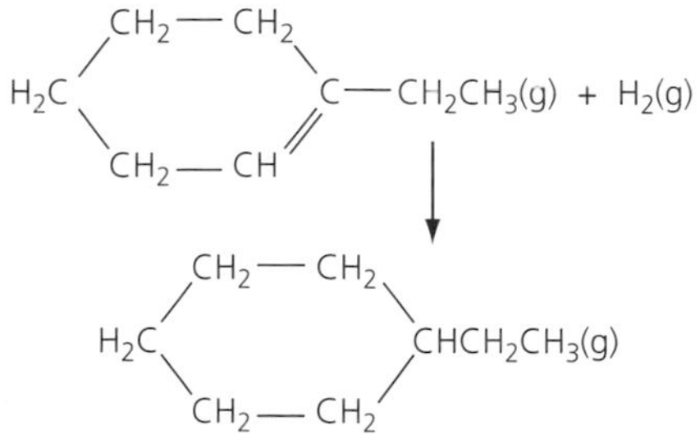

Use mean bond enthalpy values to determine the standard enthalpy of hydrogenation of this compound to form ethylcyclohexane.

ii) Deduce the enthalpy change when the hypothetical compound:

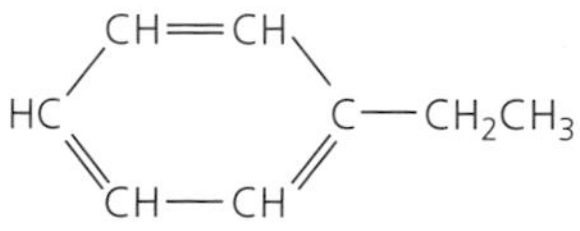

HC $\quad$ C—CH$_2$CH$_3$ (1-ethylcyclohexa-1,3,5-triene)

is fully hydrogenated to form ethylcyclohexane.

iii) The standard enthalpy change for the hydrogenation of ethylbenzene, C$_6$H$_5$CH$_2$CH$_3$, is −210 kJ mol^{-1}. Which of the two compounds, ethylbenzene or 1-ethylcyclohexa-1,3,5-triene, is more stable? Explain your answer. (6)

NEAB March 1999 CH04 Q3 (part)

4

a Define the term *standard enthalpy of formation*. (3)

b Explain what is meant by the *enthalpy of hydration of an ion*. (2)

c The standard enthalpy of formation of methane is −75 kJ mol^{-1} and the bond enthalpy of the H–H bond is +436 kJ mol^{-1}. The enthalpy change which accompanies the formation of gaseous carbon atoms from solid graphite is +717 kJ mol^{-1}.
Determine a value for the mean bond enthalpy of the C–H bond in methane. (4)

d Use the data in the table below to determine the standard enthalpy of hydration of the potassium ion. (3)

Cl$^-$(g) → Cl$^-$(aq)	$\Delta H^{\ominus} = -364$ kJ mol^{-1}
KCl(s) → K$^+$(g) + Cl$^-$(g)	$\Delta H^{\ominus} = +718$ kJ mol^{-1}
KCl(s) → K$^+$(aq) + Cl$^-$(aq)	$\Delta H^{\ominus} = +17$ kJ mol^{-1}

NEAB June 1997 CH04 Q1

5 Potassium chlorate(V) decomposes on heating according to the equation:

$$4KClO_3(s) \rightarrow 3KClO_4(s) + KCl(s) \quad \Delta H^{\ominus} = +16.8 \text{ kJ mol}^{-1}$$

The standard molar entropy, $S^{\ominus}$, for each species involved in the reaction is shown below.

	$S^{\ominus}$/J K^{-1} mol^{-1}
KClO$_3$(s)	112
KClO$_4$(s)	134
KCl(s)	83

a Calculate the standard entropy change for the above reaction. (2)

b Calculate the standard free energy change at 298 K for the above reaction. (3)

c i) In terms of free energy, state the necessary condition for spontaneous change.

ii) Assuming that ΔH and ΔS do not vary with temperature, determine the lowest temperature at which the above reaction will become feasible. (4)

NEAB June 1997 CH04 Q2

6

a In an excess of oxygen, graphite and hydrogen burn to form carbon dioxide and steam respectively. In a limited supply of oxygen, graphite burns to form carbon monoxide only.

i) Write an equation (including state symbols), to indicate which of these combustions is most likely to lead to a positive entropy change on going from reactants to products. Give a reason for your choice.

ii) Write an equation (including state symbols), to indicate which of these combustions is least likely to lead to a positive entropy change on going from reactants to products. Give a reason for your choice. (4)

b Write an equation relating entropy change, free energy change, and enthalpy change. (1)

c i) In terms of ΔG, what is meant by the term *feasible reaction*?

ii) Consider a reaction which, at a given temperature, has ΔG positive and ΔS negative. Will raising the temperature make ΔG less positive? Explain your answer. (4)

NEAB February 1997 CH04 Q2

13 Periodicity

Linus Pauling was a brilliant theoretical and an investigative chemist. No other modern-day chemist has worked on such a wide range of research topics. He combined his experimental work with a high level of theoretical interpretation, but his real talent was in his ability to translate these into simple concepts, which are of use to a wide range of scientists.

He studied crystals by X-ray diffraction, identified the alpha helix in many proteins, worked on the structure of metals, the theory of magnetism, van der Waals forces, and human illnesses. One of the lasting legacies of his research, however, was his *scale of electronegativity*, where he combined his practical studies of the angles and position of atoms in molecules with theoretical calculations. The order of elements in the *Pauling scale* is exactly reflected in the *Periodic Table*. In recognition of all his work, Pauling was awarded the Nobel Prize.

Linus Pauling gained the Nobel Prize for his research into the nature of the chemical bond and its application to elucidating the structure of complex substances

13.1 Electronegativity and periodicity

Linus Pauling defined **electronegativity** as the tendency of an atom involved in a bond to attract the bonding electrons to itself. So, when two different atoms are bonded together, the one with the higher electronegativity takes a larger share of the electron density. Pauling assigned an electronegativity value to every element, assigning the value 4 to fluorine, the most electronegative element of all. The match-up of electronegativity to the Periodic Table can be seen below.

Electronegativities of selected elements

1.0 Li	1.5 Be											2.0 B	2.5 C	3.0 N	3.5 O	4.0 F
1.0 Na	1.2 Mg											1.5 Al	1.8 Si	2.1 P	2.5 S	3.0 Cl
0.9 K	1.2 Ca	1.3 Sc	1.4 Ti	1.5 V	1.6 Cr	1.6 Mn	1.7 Fe	1.7 Co	1.8 Ni	1.8 Cu	1.6 Zn	1.7 Ga	1.9 Ge	2.1 As	2.4 Se	2.8 Br

The **larger the difference** between the electronegativities of the two bonded atoms, **the more polar the bond**. When one element from the left of the Table combines with one from the right, an **ionic bond** will be formed, whereas two elements from the right-hand side of the Table will form a **covalent bond**. So, in NaCl, an ionic compound, there is a large difference in electronegativity between the two elements in the compound, and therefore a very unequal share of the electrons within the bond. The pull on the electrons is towards the chlorine, which forms Cl⁻. The Na forms Na⁺ and they are held together by an ionic bond.

Pauling said that other compounds will have bonding types intermediate between these. A compound such as aluminium chloride will have some ionic character and some covalent character. It will have polar bonds. These are indicated by an arrow with the head pointing towards the more negative end of the molecule and a tail to the more positive end, e.g. Al→Cl. Sometimes polarity is written as δ+ and δ− (see Chapter 12, Fig. 8).

13.2 Properties and periodicity

Periodicity relates to how the properties of elements and compounds change in patterns across and down the Periodic Table and uses Pauling's ideas to help understand them. Section 4.3 in *AS Chemistry* looked at the change in chemical properties going down a Group; in this chapter we will look at how the properties change for the elements across Period 3 from sodium to chlorine.

For the properties of the elements in the Periodic Table, we discuss trends from the left-hand side (Group I) to the right-hand side (Group VII). There are only small changes at the atomic level (micro), but these can produce quite startling differences in the chemical and physical properties of compounds or elements (macro).

Table 1 Atomic changes in Period 3								
Group number	I	II	III	IV	V	VI	VII	VIII/0
Elements in Period 3	Na	Mg	Al	Si	P	S	Cl	Ar
Nuclear charge	11	12	13	14	15	16	17	18
[Ne] electron configuration	[Ne] $3s^1$	[Ne] $3s^2$	[Ne] $3s^23p^1$	[Ne] $3s^23p^2$	[Ne] $3s^23p^3$	[Ne] $3s^23p^4$	[Ne] $3s^23p^5$	[Ne] $3s^23p^6$
Atomic radius/nm	0.191	0.160	0.143	0.118	0.110	0.102	0.099	0.095
First ionisation energy/ kJ mol^{-1}	496	738	578	789	1012	1000	1251	1521
Electronegativity (Pauling)	1.0	1.2	1.5	1.8	2.1	2.5	3.0	–
Formula of oxides	Na_2O Na_2O_2	MgO	Al_2O_3	SiO_2	P_4O_6 P_4O_{10}	SO_2 SO_3	Cl_2O Cl_2O_7	
Formula of chloride	NaCl	$MgCl_2$	Al_2Cl_6	$SiCl_4$	PCl_3 PCl_5	SCl_2	Cl_2	–

The compounds printed in blue are not required by the A2 Specification.

Look at Table 1 and you will see definite patterns. It is very useful to be familiar with these trends in behaviour because they will allow you to make sound predictions about how particular elements and compounds will react. The reactions of elements depend upon their electronic structure, so if the electronic structure of their constituents is changing in a pattern, the properties will also change in a pattern. The reactivities and properties of elements depend upon a combination of things: nuclear charge, size of the atom, the number of outer electrons and the number of shielding electrons (for detail on electronic structure of atoms, see pages 16–18 in *AS Chemistry*), and these help us to explain and predict the properties of many compounds.

Linus Pauling used electronegativity to explain trends in the Periodic Table (for electronegativity see page 64 of *AS Chemistry*). You can see from Table 1 that electronegativity increases across the period from left to right, which means that elements to the right will attract the electrons in a bond more strongly.

In a compound, the ionic or covalent character depends upon the electronegativity difference between the two elements. If there is a large difference in electronegativity, the ions formed stay as separate ions. If the electronegativity difference is smaller, distortion of the charge cloud of the negative ion occurs, and some covalent character in the ionic bond results. This effect is called **polarisation** (Fig. 1) and is discussed on page 31 of *AS Chemistry*.

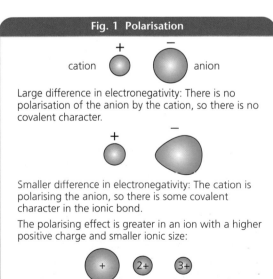

Fig. 1 Polarisation

cation   anion

Large difference in electronegativity: There is no polarisation of the anion by the cation, so there is no covalent character.

Smaller difference in electronegativity: The cation is polarising the anion, so there is some covalent character in the ionic bond.

The polarising effect is greater in an ion with a higher positive charge and smaller ionic size:

+   2+   3+

increasing polarising power

The ability of an atom to attract electrons will be the same in different compounds, but the overall polarisation of the covalent bond will differ, depending on the two elements forming the bond.

Na–Cl

Na would have a relatively weak attraction for electrons in a covalent bond and Cl would have a strong attraction, so the electron charge will be found almost completely on the Cl, and this is written $Na^+ Cl^-$. The bond is ionic.

Si–Cl

Si has a much stronger attraction for the electrons in a covalent bond than Na has, so the electrons in the Si–Cl bond are shared. However, the Cl has a higher electronegativity than Si, so there is some polarity in the bond, resulting in partial ionic character. We describe molecules like these as having polar bonds or possessing dipoles. This is written:

$$\overset{\delta+}{Si}-\overset{\delta-}{Cl} \text{ or } Si \leftrightarrow Cl$$

Cl–Cl

Two Cl atoms will have identical electronegativities, so there will be no polarity in the bond. Cl–Cl forms a completely non-polar (100% covalent) bond. Fig. 3 shows changes in oxidation state of oxides and chlorides across Period 3.

1 Deduce the oxidation state of Cl in PCl_5 and in Cl_2O_7.

2
a Why do atomic radii decrease across Period 3?
b Explain why the melting point of magnesium is greater than that of sodium.
c Explain why the first ionisation energy of sodium is the lowest in Period 3.

Sodium and magnesium lose all their outer electrons during ionic/electrovalent bonding and the oxidation number equals the number of electrons lost. Aluminium forms both ionic and covalent compounds in oxidation state +3. The elements silicon, phosphorus and sulphur form mainly covalent compounds, but they can still be considered as having positive oxidation states. Chlorine forms ionic and covalent compounds in oxidation states from −1 to +7.

The difference in the type of bonding which occurs can be explained by electronegativity differences. Ionic bonding occurs between elements with a large difference in electronegativity, and covalency between elements with a small difference in electronegativity. Between these two extremes, the compounds formed have both ionic and covalent characteristics. The formulas of compounds formed often relate to the number of electrons needed to empty or complete the outer electron shells. Other oxidation states can be related to electrons in s, p and d orbitals and follow more complex rules.

As has been shown in Table 1, the ionisation energies of the elements generally increase across Period 3 (Fig. 4).

Fig. 2 Factors affecting electronegativity

The electronegativity value depends on:
• nuclear charge
• size of atom
• shielding electrons

These factors influence the chemical properties of an element.

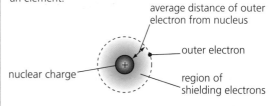

Fig. 3 Changes in oxidation states in oxides and chlorides

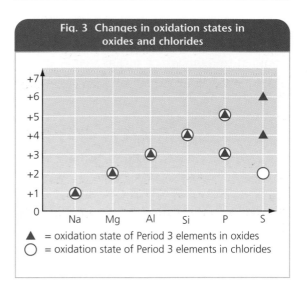

▲ = oxidation state of Period 3 elements in oxides
○ = oxidation state of Period 3 elements in chlorides

Fig. 4 First ionisation energy for Period 3 elements

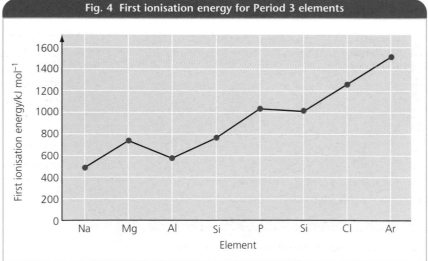

Sodium has the lowest ionisation energy of any of the elements in Period 3, so its attraction for its outer electrons is the weakest. The energy needed to move the electron to a higher level or even remove it completely is the lowest. In any period, the elements with the lowest ionisation energies are metals and show properties that reflect the weaker forces of attraction for electrons. In a metal, the outer electrons are delocalised forming a 'sea' of mobile electrons (Fig. 5).

attraction will depend upon the size of the ion, ionic charge and number of mobile electrons. A greater attraction will produce a stronger metallic bond, which results in properties such as a higher melting point and greater metal hardness.

The low first ionisation energy of the sodium atom means that it has weaker forces holding the outer electron in place than other atoms in the same period, so a relatively small amount of energy is needed to remove the electron completely. For sodium, the relatively lower nuclear charge means the electron shells are spaced out more and the atomic radius is much larger (0.191 nm) compared with 0.160 nm and 0.143 nm for magnesium and aluminium. This means that the sodium has relatively weak forces holding the atoms together in the metal:

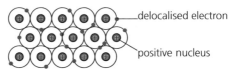

Fig. 5 A model of metallic bonding

delocalised electron

positive nucleus

The outer electrons are delocalised. The electrons are free to drift from one metal ion to another through the outer energy levels.

The greater the number of delocalised electrons, the stronger the lattice.

This structure (Fig. 5) gives metals their typical properties of high electrical and thermal conductivities. The mobility of the electrons means they can easily transfer energy throughout the structure. Also, the attraction between the electrons and the central ions holds the structure together, and the strength of this

- It is soft: the crystal structures in the metal are easily separated.
- It has a low melting point: the atoms can be easily moved from a structured arrangement in a solid to a random arrangement in a liquid.
- It has a low density: the large size of the atom gives a large volume to the structure.

KEY FACTS

- Electronegativity is the tendency of an atom involved in a covalent bond to attract the bonding electrons to itself.

- Electronegativity increases across Period 3.

- Electronegativity of an element depends upon nuclear charge, shielding electrons and size of atom.

13.3 Elements in water

Sodium reacts with water to form aqueous sodium hydroxide and hydrogen

When sodium is put into water, it floats (it has a low density), it reacts violently (it relatively easily loses one outer electron) and liberates hydrogen. The large amount of energy released during the reaction melts the sodium.

$$2Na(s) + 2H_2O(l) \rightarrow 2NaOH(aq) + H_2(g)$$

This reaction is written as an ionic equation:

$$2Na(s) + 2H_2O(l)$$
$$\rightarrow 2Na^+(aq) + 2OH^-(aq) + H_2(g)$$

Because the NaOH(aq) is completely dissociated and all the OH^-(aq) ions are released, the resulting solution will be strongly alkaline, with a pH of approximately 13.

Compared with sodium, magnesium has a smaller atomic radius, a higher ionic charge and two delocalised electrons. This results in stronger forces holding the magnesium lattice together. The two mobile electrons available give a greater electrical conductivity than sodium has. The greater forces of attraction mean that the two outer electrons are more difficult to remove, so it is much less reactive than

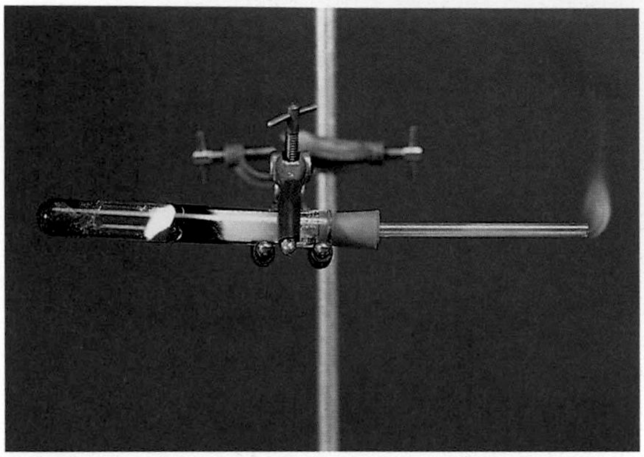

The reaction of magnesium with steam

sodium. The reaction is still very exothermic, but because of the high activation energy the magnesium must be heated strongly to start the reaction, and the water must be heated to form steam, which is passed over the hot magnesium.

The reaction happens when you pass steam over heated magnesium. The products are hydrogen and magnesium oxide:

$$Mg(s) + H_2O(g) \rightarrow MgO(s) + H_2(g)$$

Aluminium continues the trend. It is smaller than magnesium, has a higher electrical conductivity and is less reactive in water.

Table 2 The properties of oxides in Period 3							
Element	**Na**	**Mg**	**Al**	**Si**	**P**	**S**	
Formula of oxide	Na_2O	MgO	Al_2O_3	SiO_2	P_4O_{10}	SO_2	SO_3
State at 25 °C	solid	solid	solid	solid	solid	gas	liquid
Melting point/K	1548 (sublimes)	3125	2345	1883	853 > 1atm	200	290
Boiling point/K	–	3873	3253	2503 (sublimes)	573	263	318
Electrical conductivity when molten	good	good	good	none	none	none	none
Structure	giant ionic	giant ionic	giant ionic	giant molecule	simple molecule	simple molecule	simple molecule
Enthalpy of formation 298 K/kJ mol^{-1} per mole of O atoms	–416	–602	–559	–455	–298	–149	–132
Adding water	reacts and forms hydroxide ions in solution	slightly soluble, dissolved oxide forms a few hydroxide ions in solution	insoluble but amphoteric	insoluble but acidic	acidic reacts and gives H$^+$ ions in solution	acidic reacts and forms weak acid H_2SO_3 with a few H$^+$ ions in solution	acidic reacts and forms strong acid H_2SO_4 with H$^+$ ions in solution
Typical pH of aqueous solution of oxide	13	8	7, i.e. no reaction	7, i.e. no reaction	2	3	1

Formation of oxides

All the elements in Period 3 react with oxygen and usually form an oxide with the highest possible oxidation state (see Table 2).

The metals are all highly reactive when you heat them in pure oxygen, they all glow brightly during the reaction.

$$4Na(s) + O_2(g) \rightarrow 2Na_2O(s) \text{ (in limited air)}$$
$$2Mg(s) + O_2(g) \rightarrow 2MgO(s)$$
$$4Al(s) + 3O_2(g) \rightarrow 2Al_2O_3(s)$$

Sodium burning in oxygen

Sodium oxide is a white solid powder

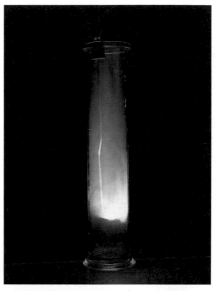

Aluminium burns in oxygen to form aluminium oxide, a white solid powder

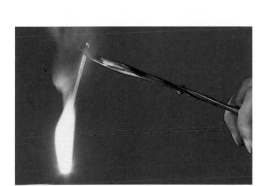

Magnesium ribbon burning in air

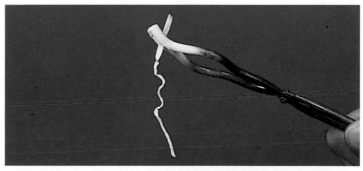

Magnesium oxide is a white solid

All the metals produce ionic oxides due to the large differences in electronegativity. Sodium and magnesium oxides are basic and aluminium oxide is amphoteric, illustrating the increasing electronegativity of the metal (Table 1).

Silicon, phosphorus and sulphur all react with oxygen. They are non-metallic elements, there is a small electronegativity difference, so they form covalent oxides (Fig. 6).

$$Si(s) + O_2(g) \rightarrow SiO_2(s)$$
$$4P(s) + 5O_2(g) \rightarrow P_4O_{10}(s)$$
$$S(s) + O_2(g) \rightarrow SO_2(g)$$

Phosphorus burns in air to give phosphorus(V) oxide, P_4O_{10}, a white solid powder

Sulphur burns in oxygen with a blue flame to give a colourless gas

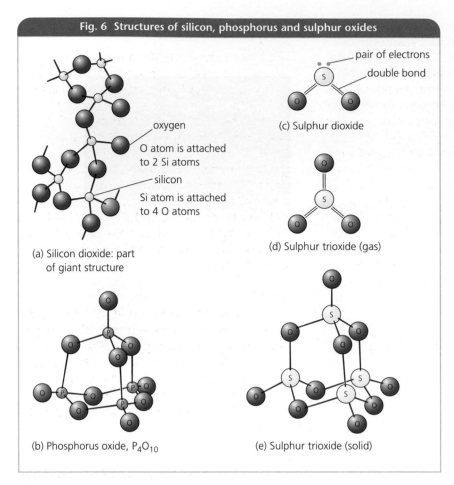

Fig. 6 Structures of silicon, phosphorus and sulphur oxides

oxygen

O atom is attached to 2 Si atoms

silicon

Si atom is attached to 4 O atoms

(a) Silicon dioxide: part of giant structure

(b) Phosphorus oxide, P_4O_{10}

pair of electrons

double bond

(c) Sulphur dioxide

(d) Sulphur trioxide (gas)

(e) Sulphur trioxide (solid)

In the absence of a catalyst, sulphur forms SO_2 and not SO_3 (see the Contact process on page 109 in *AS Chemistry*), so it does not use its highest possible oxidation state, as the other elements do.

All these oxides form acidic solutions in water. Fig. 7 summarises the properties of Period 3 elements.

Oxide properties related to bonding

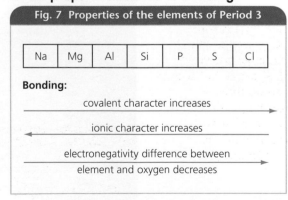

Fig. 7 Properties of the elements of Period 3

Na	Mg	Al	Si	P	S	Cl

Bonding:

covalent character increases →

← ionic character increases

electronegativity difference between element and oxygen decreases →

Refer to Table 2 for the melting points of the oxides. Melting points are an indication of

the forces of attraction between atoms, ions or molecules. The bonding in these oxides can be grouped into three types.

Ionic oxides

The forces between ions tend to be stronger than the forces between molecules, which are usually weak van der Waals forces. For ions with higher charges, there will be a greater force of attraction, which leads to a stronger bond. The greater charge on the positive ion, however, gives a greater polarising effect, increased covalency and therefore a lower melting point.

For the ions from Na^+ to Mg^{2+} to Al^{3+}, the increasing ionic charge and the decreasing size of the ions gives increasing electrostatic attractive forces and higher melting points from Na to Mg. However, the greater polarising effect for Al gives a lower melting point than for Mg.

Covalent oxides – macromolecular

If the molecule is a giant structure, the forces holding the major particles together are forces between atoms. These are covalent bonds and are extremely strong. This gives high melting points. The high melting point of $SiO_2(s)$ illustrates this.

Covalent oxides – simple molecules

Moving on across Period 3, after $SiO_2(s)$, the melting points decrease significantly because the compounds are simple covalent molecules and so there are weaker forces of attraction between the molecules. SO_2 consists of simple, discrete molecules with weak forces of attraction between them, so the melting and boiling points are low. It is a gas at room temperature. SO_3 can form groups of molecules (Fig. 6e) so it can be solidified by cooling to just below room temperature. P_4O_{10} is a larger molecule than SO_2, so there are stronger van der Waals forces of attraction and a higher melting point.

Electrical conductivity in the molten oxides or in compounds dissolved in water is an indication of ionic character. The molten oxides show good conductivity for sodium and magnesium and aluminium oxides but poor conductivity for the others. This indicates significant ionic bonding for sodium and magnesium and aluminium oxides but covalent bonding for oxides from silicon to sulphur.

The electrical conductivities in aqueous solutions are more complex. Sodium oxide dissolves in water and reacts to give sodium hydroxide – the ions produced conduct electricity. Magnesium oxide is only sparingly soluble and aluminium oxide is insoluble, so there are no mobile ions to provide any conductivity for these two compounds. $SiO_2(s)$ is covalent and insoluble so will not conduct, $P_4O_{10}(s)$ and $SO_2(g)$ are both covalent (so will not conduct in the molten state), but they both react with water to give acidic solutions so the hydrogen cations and the anions produced conduct electricity.

3 Suggest why sodium fluoride has a lower melting point (1266 K) than magnesium fluoride (1534 K).

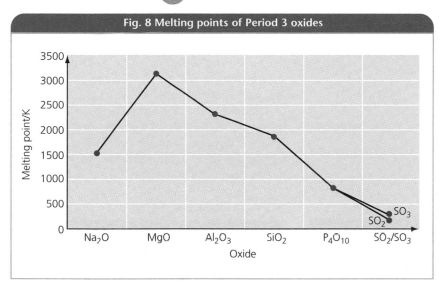

Fig. 8 Melting points of Period 3 oxides

4 Look at Fig. 8 above and explain how the bonding in the oxides of silicon and sulphur produces such a difference in their melting points.

Reactions of Period 3 oxides with water

Table 2 (page 180) shows the change in pH for the solutions of the oxides across Period 3. There is an evident chemical trend, but it is masked by the change in solubilities. A substance will only change the pH of water if it dissolves. The oxide ion is too highly charged to exist on its own in water. It attracts water molecules and hydrolyses to give $OH^-(aq)$ ions (Fig. 9).

$$O^{2-}(aq) + H_2O(l) \rightarrow 2OH^-(aq)$$

Again, the oxides can be considered in three groups.

Ionic oxides
Na_2O
The oxide is very soluble in water, so many $OH^-(aq)$ ions will be formed from its reaction with water as outlined above. This will give an alkaline solution with a very high pH. We can consider the system as one with interactions within the X–O–H system, where X is the Period 3 element. There is equilibrium between the various possible species. There is a large difference in electronegativity between Na and O, so this will be an ionic bond, and the O–H bond with a smaller difference in electronegativity will be covalent. We have Na^+ ^-O–H and an alkaline solution.

MgO
The oxide is only slightly soluble, which can be related in part to the higher lattice energy from the Mg^{2+} ion (compared with Na^+). If only a small amount of Mg^{2+} dissolves, only a few O^{2-} ions are dissociated, so only a few OH^- ions will be formed and the solution will only be slightly alkaline, having a pH of only 8.

Al_2O_3
This oxide is insoluble, so no O^{2-} ions will be dissociated, no OH^- ions will be formed and the pH will be 7. Although Al_2O_3 does not change the pH of water from 7, it will react with acids and with alkalis such as sodium hydroxide, hence it is amphoteric.

The electronegativity difference between Al and O is less than between either Na and O or Mg and O, so there is more covalent character in the Al–O bond than in Na–O or Mg–O. This allows Al_2O_3 to show some acidic characteristics, as well as it being a metal oxide and therefore showing basic characteristics.

Fig. 9 Hydrolysis of oxide ion

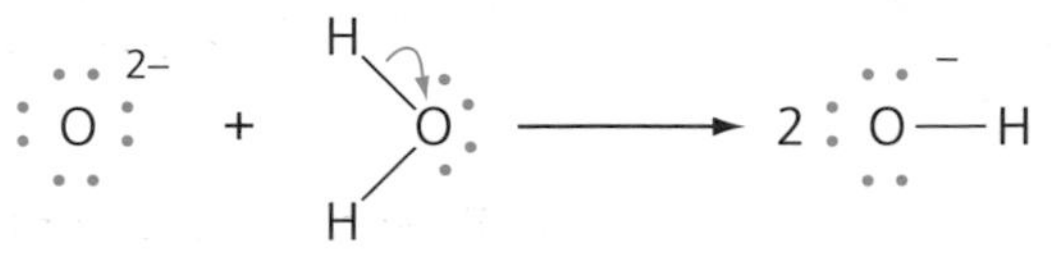

Aluminium oxide will react with compounds that are more basic, such as NaOH(aq), and it will react with compounds that are more acidic, such as H_2SO_4(aq).

In acid:

$$Al_2O_3(s) + 3H_2SO_4 \rightarrow Al_2(SO_4)_3 + 3H_2O(l)$$

Or as an ionic equation:

$$Al_2O_3(s) + 6H^+(aq) \rightarrow 2[Al(OH)_6]^{3+}(aq) + 3H_2O(l)$$

In alkali:

$$Al_2O_3(s) + 2NaOH(aq) + 3H_2O(l) \rightarrow 2NaAl(OH)_4(aq)$$

Or as an ionic equation:

$$Al_2O_3(s) + 2OH^-(aq) + 3H_2O(l) \rightarrow 2Al(OH)_4^-(aq)$$

with alkali:		with acid:
$[Al(OH)_4]^-$	$\leftarrow Al_2O_3 \rightarrow$	$[Al(H_2O)_6]^{3+}$
behaves as an acid		behaves as a base

Covalent oxide – macromolecular SiO_2

This is a macromolecule, a stable compound, and it is insoluble in water. Also, there is only a small difference in electronegativity between Si and O, so there is minimal attraction for the water molecule. H^+ and OH^- ions are not formed and the pH will remain neutral.

Covalent oxides – simple molecules

These acidic oxides hydrolyse in water and give the corresponding acids. The mechanism of how the oxides hydrolyse are all different and are laid out in Fig. 10.

$$P_4O_{10}(s) + 6H_2O(l) \rightarrow 4H_3PO_4(aq)$$

$$SO_2(g) + H_2O(l) \rightarrow H_2SO_3(aq)$$

$$SO_3(g) + H_2O(l) \rightarrow H_2SO_4(aq)$$

As the electronegativity difference decreases across Period 3, the X–O bond is strengthening and the likelihood of it ionising is decreasing, so the likelihood of the O–H bond ionising is increasing. When this happens, the O still retains the electrons and H^+(aq) is formed. This gives an increase in $[H^+]$ and a more acidic solution, so a decrease in pH (see Fig. 11).

Fig. 10 Hydrolysis of covalent oxides: phosphorus(V) and sulphur(VI) oxides

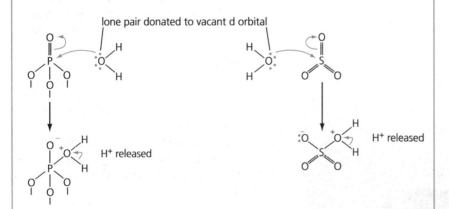

lone pair donated to vacant d orbital

H^+ released

H^+ released

Fig. 11 Summary of alkaline and acidic solutions

Low electronegativity of X gives alkaline solution.	High electronegativity of X gives acidic solution.
$X^+ \ ^-O{-}H$	$X{-}O^- \ H^+$

The degree of acidity is related to the electronegativity difference between the Period 3 element and oxygen.

KEY FACTS

- The decreasing electronegativity differences across Period 3 determine the change in chemical characteristics of the oxides.

- The ionic character of the oxides decreases from left to right.

- The basic character of the oxides in water decreases from left to right.

- Sodium and magnesium are basic oxides, aluminium oxide is amphoteric, silicon, phosphorus and sulphur oxides are acidic.

13.4 Formation of chlorides

Element	Na	Mg	Al	Si	P	
Formula of chloride	NaCl	$MgCl_2$	Al_2Cl_6	$SiCl_4$	PCl_3	PCl_5
State at 25 °C	solid	solid	solid	liquid	liquid	solid
Melting point/K	1074	987	463	203	161	435 decomposes
Electrical conductivity when molten	good	good	poor	none	none	none
Structure	giant ionic	giant ionic	simple molecule	simple molecule		simple molecule
Enthalpy of formation 298 K/kJ mol^{-1} per mole of Cl atoms	−411	−321	−235	−160	−107	
With water	soluble	soluble Very slight hydrolysis	hydrolyses Gives an acidic solution (HCl)	hydrolyses Gives an acidic solution (HCl)	hydrolyses Gives an acidic solution (HCl)	
Typical pH of aqueous solution of chloride	7	6.5	3–5	0–1	0–1	

Table 3 Properties of chlorides in Period 3

All the above chlorides of Period 3 can be made by heating the element in chlorine. Under these conditions, the chloride with the highest oxidation state is formed.

$$2Na(s) + Cl_2(g) \rightarrow 2NaCl(s)$$
$$Mg(s) + Cl_2(g) \rightarrow MgCl_2(s)$$
$$2Al(s) + 3Cl_2(g) \rightarrow Al_2Cl_6(s)$$
$$Si(s) + 2Cl_2(g) \rightarrow SiCl_4(l)$$
$$2P(s) + 5Cl_2(g) \rightarrow 2PCl_5(s)$$

Enthalpy of formation of chlorides

The enthalpy of formation of the compounds formed when Period 3 elements react with chlorine decreases from left to right as the electronegativity difference between the elements and chlorine decreases.

Table 3 shows that elements to the left of the Periodic Table have a more exothermic value for the enthalpy of formation for the chlorides. The value ultimately depends on the combination of factors such as ionic charge, ionic size and the lattice structure.

Sodium has the lowest nuclear charge in this period, so the endothermic processes of atomisation and first ionisation are the lowest. The endothermic processes for Na therefore need a smaller energy input compared with that for the other elements in Period 3.

Sodium burning in chlorine

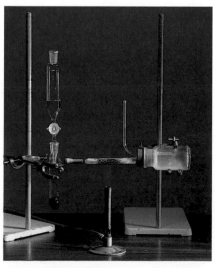

Chlorine reacting with aluminium to form white aluminium chloride

185

A Born–Haber cycle for NaCl(s) and $MgCl_2$(s) (see Chapter 12) shows the relative values of the various exothermic and endothermic processes (Fig. 12).

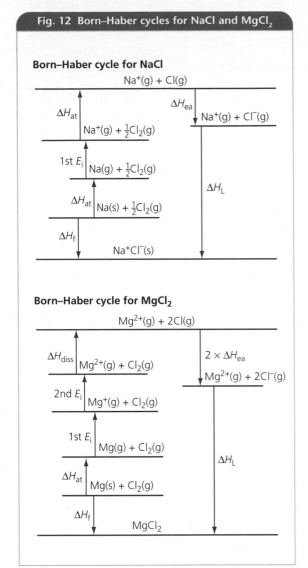

Fig. 12 Born–Haber cycles for NaCl and $MgCl_2$

Born–Haber cycle for NaCl

$Na^+(g) + Cl(g)$

ΔH_{at}

ΔH_{ea} $Na^+(g) + Cl^-(g)$

$Na^+(g) + \frac{1}{2}Cl_2(g)$

1st E_i $Na(g) + \frac{1}{2}Cl_2(g)$

ΔH_L

ΔH_{at} $Na(s) + \frac{1}{2}Cl_2(g)$

ΔH_f

$Na^+Cl^-(s)$

Born–Haber cycle for $MgCl_2$

$Mg^{2+}(g) + 2Cl(g)$

ΔH_{diss} $Mg^{2+}(g) + Cl_2(g)$

$2 \times \Delta H_{ea}$

$Mg^{2+}(g) + 2Cl^-(g)$

2nd E_i $Mg^+(g) + Cl_2(g)$

1st E_i $Mg(g) + Cl_2(g)$

ΔH_L

ΔH_{at} $Mg(s) + Cl_2(g)$

ΔH_f

$MgCl_2$

Structure and bonding related to properties

The properties of the chlorides formed by Period 3 elements depend on the type of bonding present in the compounds and this, in turn, is related to the electronegativity differences between the Period 3 elements and chlorine. A large difference in electronegativity leads to ionic bonding and a smaller difference leads to covalent bonding.

Table 3 and Fig. 13 show that the melting points gradually decrease across the Period. NaCl and $MgCl_2$ are ionic solids with a crystal lattice of ions attracting each other so they have high melting points. The electronegativity difference between Mg and Cl is smaller than that between Na and Cl, so there will be some covalency within the ionic bond, hence a lower melting point. There is a much smaller difference in electronegativity between Al and Cl than there is in NaCl and $MgCl_2$ so $AlCl_3$ is covalent and has a dimeric structure (Fig. 14).

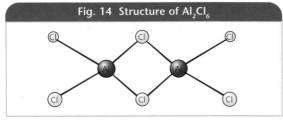

Fig. 14 Structure of Al_2Cl_6

There is a considerable decrease in melting point going from $MgCl_2$ to $AlCl_3$. $SiCl_4$ is a covalent molecule with a smaller electronegativity difference still, and it is a monomer, so it has a lower melting point than $AlCl_3$. PCl_3 has an even smaller electronegativity difference within the molecule, so the melting point decreases again. However, PCl_5 has a higher melting

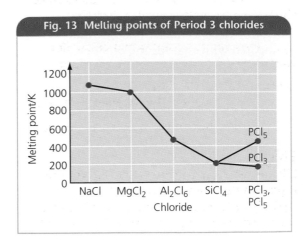

Fig. 13 Melting points of Period 3 chlorides

Melting point/K

PCl_5

PCl_3

NaCl $MgCl_2$ Al_2Cl_6 $SiCl_4$ PCl_3, PCl_5

Chloride

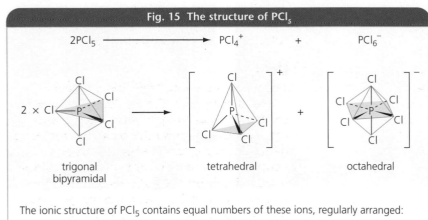

Fig. 15 The structure of PCl_5

$2PCl_5 \longrightarrow PCl_4^+ + PCl_6^-$

$2 \times$ Cl

trigonal bipyramidal

tetrahedral

octahedral

The ionic structure of PCl_5 contains equal numbers of these ions, regularly arranged:

point than $SiCl_4$. This is explained by the structure of PCl_5, which is not a simple molecule in the solid phase. The molecules form an ionic structure consisting of PCl_4^+ and PCl_6^- ions (see Fig. 15): Cl^- is donated from one PCl_5 molecule to the other. If the solid is heated, it forms discrete PCl_5 molecules.

The general trend for these chlorides indicates that the degree of covalent character in the ionic bond is increasing as the electronegativity difference decreases. As the electronegativity difference decreases, the element has a greater attraction for the electrons on the chorine. With the increased attraction, the charge cloud around the chloride ion distorts and produces covalent character. This gives a decrease in the melting points. The boiling points of the chlorides will also indicate the same trend.

Conductivity of the molten or dissolved compounds is an indication of ionic

character, and lack of conductivity is typical of covalent compounds. However, these covalent chlorides react with water and form ions, so conductivity in solution does not give reliable information about the original compounds. Table 3 shows only information about molten structures. This will indicate ionic character or covalency within the chlorides. The molten chlorides for sodium and magnesium show good conductivity. This indicates significant ionic bonding for sodium and magnesium. Poor conductivity for chlorides of aluminium, silicon and phosphorus indicates covalent bonding. Solid PCl_5 sublimes if heated, so does not exist in the molten state.

5 Aluminium chloride is a solid, yet it dissolves in non-polar solvents and does not conduct electricity when molten. Explain these properties in terms of its bonding.

13.5 Reactions of the Period 3 chlorides with water

NaCl
When sodium chloride dissolves in water, it forms solvated ions.

$NaCl(s) + aq \rightarrow Na^+(aq) + Cl^-(aq)$

The enthalpy of hydration of the Na^+ ions and the Cl^- ions supplies the energy needed to overcome the sodium chloride lattice dissociation enthalpy (see Chapter 12). For both $Na^+(aq)$ and $Cl^-(aq)$ ions the ion dipole attraction is weak because the size of the ion is large and the charge on the ion is small. This means that the attraction for the water molecules is sufficient to hydrate the ions but not sufficient to distort the water molecules and dissociate them into ions, so hydrolysis does not occur. The ions just dissolve and the solution remains neutral at pH 7.

$MgCl_2$
The magnesium ion, Mg^{2+}, has a higher charge density than the Na^+ ion because it is smaller and has a higher ionic charge with the same shielding, so the attraction for water molecules is greater. The higher charge density means there is some distortion of the charge clouds on the water molecules. This weakens the bond between the oxygen and the hydrogen and a small amount of

hydrolysis occurs forming a very slightly acidic solution.

$[Mg(H_2O)_6]^{2+}(aq) + H_2O(l)$
$\rightleftharpoons [Mg(H_2O)_5OH]^+(aq) + H_3O^+(l)$

This equilibrium lies well over to the left and the solution has a pH of 6.5.

$AlCl_3$
When aluminium chloride is added to water, there is a vigorous reaction to form hydrated ions $[Al(H_2O)_6]^{3+}$. The ion–dipole attraction is much more pronounced with aluminium because of the greater ionic charge and the smaller size of the Al^{3+} ion. The distortion of the water molecule is greater than that produced by Mg^{2+}, so there is even more hydrolysis and the equilibrium below lies further to the right than for magnesium.

$[Al(H_2O)_6]^{3+}(aq) + H_2O(l)$
$\rightleftharpoons [Al(H_2O)_5OH]^{2+}(aq) + H_3O^+(aq)$

The pH of this solution is approximately 3.

$SiCl_4$
Silicon has such a strong attraction for the dipole in the water molecules that it hydrolyses instantly. It forms $Si(OH)_4$ as an intermediate gel-type structure, but this will

break down into more stable solid oxides:

$$SiCl_4(l) + 2H_2O(l) \rightleftharpoons SiO_2(s) + 4H^+(aq) + 4Cl^-(aq)$$

The fuller reaction sequence is given in Fig. 16. The hydrolysis equilibrium lies well over to the right and the solution is strongly acidic with a pH of 1.

$$PCl_5(s) + 4H_2O(l) \rightarrow H_3PO_4(aq) + 5H^+(aq) + 5Cl^-(aq)$$

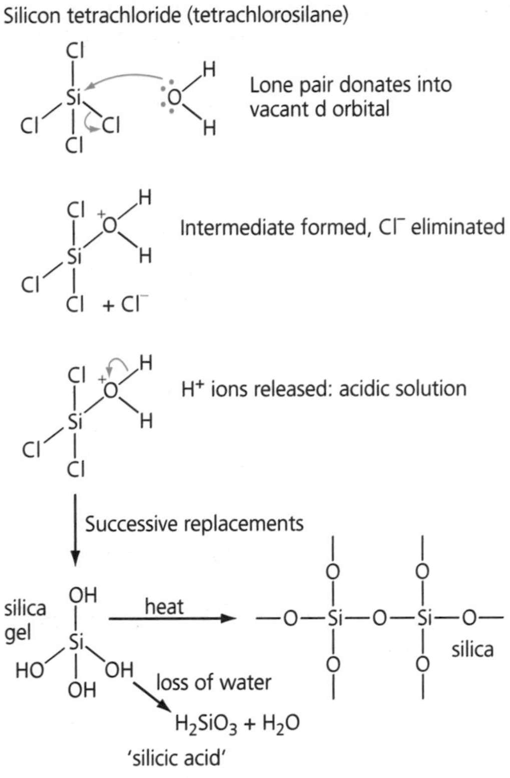

In these reactions, a large number of $H_3O^+(aq)$ ions are released and the solution has a very low pH (0–1), e.g. Fig. 17.

With the covalent chlorides, the attraction between the element and oxygen is so strong that $H_2O(l)$ donates an electron pair into a vacant d orbital and displaces the electrons in the X–Cl bond, which then forms a chloride ion. Loss of a proton follows, resulting in a substitution reaction. Successive substitutions occur to replace all the Cl^- ions, giving strongly acidic solutions with pH values of 0 to 1.

Fig. 16 Hydrolysis of SiCl₄

Silicon tetrachloride (tetrachlorosilane)

Lone pair donates into vacant d orbital

Intermediate formed, Cl⁻ eliminated

H⁺ ions released: acidic solution

Successive replacements

silica gel

heat

loss of water

silica

$H_2SiO_3 + H_2O$

'silicic acid'

Fig. 17 Hydrolysis of PCl₅

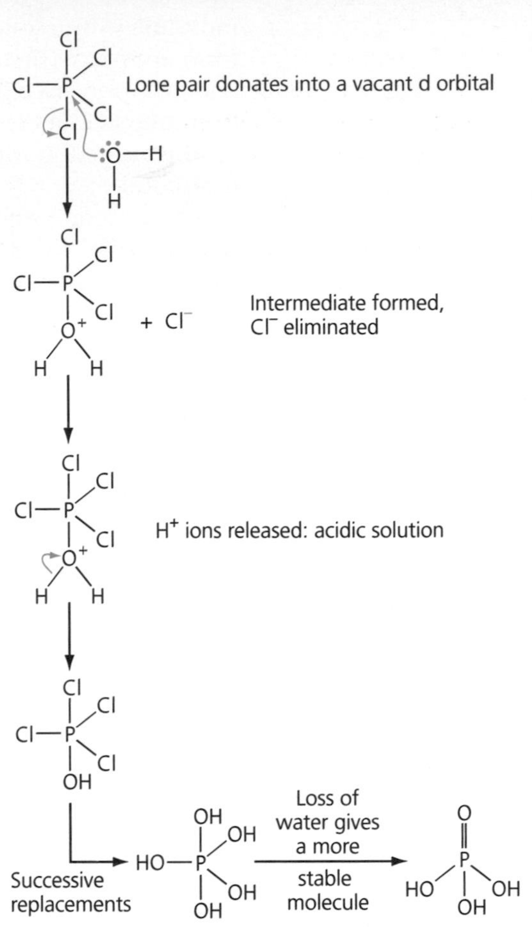

Lone pair donates into a vacant d orbital

Intermediate formed, Cl⁻ eliminated

H⁺ ions released: acidic solution

Successive replacements

Loss of water gives a more stable molecule

6 When aluminium chloride is mixed with aqueous sodium carbonate solution, bubbles are slowly formed and a white precipitate results. Explain these observations and write chemical equations for any reactions that are taking place.

The reaction of silicon (IV) chloride with water.

APPLICATION Davy's observations of the more reactive elements

Many of the elements and compounds mentioned in this chapter were discovered in the early 1800s. Although the scientists of the time could make the compounds and examine their properties, they could not explain them because they did not have the detailed knowledge of atoms that we have today. One of the scientists who did a lot of the work was Humphry Davy. He was a leading scientist of his day and his lectures at the Royal Institution (where the Children's Christmas Lectures are filmed) were very popular. In his work, he used electricity to isolate many of the more reactive elements. He kept notes of his observations, in his experiments. Some of them included inhaling gases and tasting solids, which would certainly not be recommended today.

The passages below are taken from Davy's laboratory notes.

Sodium

When thrown upon water it effervesces violently, but does not inflame, swims on the surface, gradually diminishes with great agitation and renders the water a solution of soda.

Potassium

When thrown upon water, it acts with great violence, swims on the surface and burns with a beautiful light, which is white mixed with red and violet; the water in which it burns is found alkaline, and contains a solution of potassa.

1 Why do both of these alkali metals float on water?
2 Write equations for the reactions when these metals react with water.
3 Give two reasons why the reaction between potassium and water seems to be more violent than that for sodium and water.

Sodium oxide

When a little water is added to it, there is a violent reaction between the two bodies.

Magnesium oxide

It is scarcely soluble in water, but produces heat when the water is mixed with it, and it absorbs a considerable portion of the fluid.

Aluminium oxide

Has no taste or smell, adheres strongly to the tongue, has no action on vegetable colours, is insoluble in water, is soluble in all the mineral acids and in hot solutions of ...alkalis

4 Why is sodium oxide soluble but magnesium and aluminium oxides are not?
5 Write an equation for the violent reaction between sodium oxide and water.

The experimental notes from Davy seemed to indicate that he was surprised about the reactions of aluminium oxide. *'is soluble with all the mineral acids and hot solutions of .....alkalis'* seemed unusual to him. He did not expect it to react with both acids and alkalis.

6 Write equations for a reaction of aluminium oxide with the *mineral acid* hydrochloric acid and with the *alkali* potassium hydroxide.
7 How can we explain these unusual properties of aluminium oxide?

Phosphorus oxide

Has no smell; its taste is intensely, but not disagreably acid. It dissolves in water, producing great heat; and its saturated solution is the consistence of syrup. It unites with alkalis.

Sulphur oxide

Is obtained when sulphur is burnt in oxygen with a beautiful violet flame. It reddens vegetable blues and gradually destroys most of them. It is absorbed by water; this fluid takes up about 30 times its bulk, and gains a nauseous subacid taste.

8 Write equations to explain how both these oxides react with water to give acidic solutions.
Explain why these oxides give acidic behaviour but sodium and magnesium oxides give basic behaviour.

Phosphorus chloride

Its properties are very peculiar. It is a snow-white substance. It is very volatile, and rises in a gaseous form at a temperature much below that of boiling water ... it acts violently upon water, which it decomposes. Its phosphorus reacts with the oxygen producing phosphoric acid and its chlorine with hydrogen to give muriatic [hydrochloric] acid.

9   What does *rises in a gaseous form at a temperature much below that of boiling water* indicate about the bonding in phosphorus pentachloride?
10 Explain why and how phosphorus pentachloride reacts with water to give phosphoric acid and hydrochloric acid.
11 When Davy was doing his experiments, aluminium chloride and silicon chloride had not been prepared. What sort of properties might he have been able to observe for these chlorides if he'd had samples of them?

1 Write equations to show what happens when each of the following substances is added to water and, in each case, suggest an approximate pH value for the resulting solution.

a Sodium
b Sodium chloride
c Sulphur dioxide
d Silicon tetrachloride (8)

AQA June 2000 CH01 Q6

2 Sketch a graph to show how the melting points of the elements vary across Period 3 from sodium to argon. Account for the shape of the graph in terms of the structure of, and the bonding in, the elements. (21)

AQA June 2000 CH01 Q7(part)

3 The table below contains electronegativity values for the Period 3 elements, except chlorine.

Element	Na	Mg	Al	Si	P	S	Cl	Ar
Electronegativity	0.9	1.2	1.5	1.8	2.1	2.5		no value

a Define the term *electronegativity*. (2)
b Explain why electronegativity increases across Period 3. (2)
c Predict values for the electronegativities of chlorine and of lithium. (2)
d State why argon has no electronegativity value. (1)
e How can electronegativity values be used to predict whether a given chloride is likely to be ionic or covalent? (2)
f **i)** State the type of bonding in anhydrous aluminium chloride.
  ii) Write an equation to show what happens when anhydrous aluminium chloride dissolves in water.
  iii) Suggest a value for the pH of aluminium chloride solution and give one reason why some H^+ ions are released into this solution. (4)
g State the type of bonding in sodium oxide. (1)
h Write an equation for the reaction of sodium oxide with water and suggest a value for the pH of the resulting solution. (2)
i State one piece of experimental evidence which suggests that lithium iodide has covalent character. (1)

AQA March 2000 CH01 Q5

4
a Why are the elements sodium to argon placed in Period 3 of the Periodic Table? Describe and explain the trends in electronegativity and atomic radius across Period 3 from sodium to sulphur. (7)
b Describe the trend in pH of the solutions formed when the oxides of the Period 3 elements, sodium to sulphur, are added separately to water. Explain this

trend by reference to the structure and bonding in the oxides and by writing equations for the reactions with water. (19)
c Describe and explain any differences in the thermal stabilities of the carbonates of Group I metals. (4)

AQA March 1999 CH01 Q6

5
a Name the shape of the module of $AlCl_3$ and give its bond angle. (2)
b Explain why a molecule of $AlCl_3$ is able to form a bond with a chloride ion and name the type of bond formed. (3)
c Write equations for the reactions of $AlCl_3$ and $SiCl_4$ with water. (2)
d Sketch the arrangement of oxygen atoms around silicon in the silicon-containing species formed by reaction of $SiCl_4$ with water. Indicate a value for one of the bond angles on your diagram. (2)

NEAB June 1998 CH01 Q4

6
a Explain the meaning of the term *periodic trend* when applied to trends in the Periodic Table. (2)
b Explain why atomic radius decreases across Period 2 from lithium to fluorine. (2)
c The table below shows the melting temperatures, T_m, of the Period 3 elements.

Element	Na	Mg	Al	Si	P	S	Cl	Ar
T_m/K	371	923	933	1680	317	392	172	84

Explain the following in terms of structure and bonding.
i) Magnesium has a higher melting temperature than sodium.
ii) Silicon has a very high melting temperature.
iii) Sulphur has a higher melting temperature than phosphorus.
iv) Argon has the lowest melting temperature in Period 3. (8)

NEAB June 1998 CH01 Q6

7
a Define the term *electronegativity*. (2)
b State and explain the trend in electronegativity across Period 3 from sodium to chlorine. (4)
c **i)** What is the trend in bond type in the oxides of the Period 3 elements from sodium to sulphur?
  ii) Explain how this trend is related to the differences in electronegativity between the Period 3 element and oxygen. (3)
d Write an equation for the reaction of phosphorus(V) oxide, P_4O_{10}, with water. (2)

NEAB March 1998 CH01 Q5

8

a Explain why the first ionisation energy of aluminium is less than the first ionisation energy of magnesium. (3)

b Explain why the first ionisation energy of aluminium is less than the first ionisation energy of silicon. (2)

c Explain why the second ionisation of energy of aluminium is greater than the first ionisation energy of aluminium. (2)

d Write an equation to illustrate the third ionisation energy of aluminium. (1)

e Explain why the third ionisation energy of aluminium is much less than the third ionisation energy of magnesium. (2)

NEAB March 1998 CH01 Q6

9 The table below shows the melting points of the Period 3 elements except for silicon.

Element	Na	Mg	Al	Si	P	S	Cl	Ar
m.p./K	371	923	933		317	392	172	84

a Explain in terms of bonding why the melting point of magnesium is higher than that of sodium. (3)

b State the type of bonding between atoms in the element silicon and name the type of structure which silicon forms. (2)

c Predict the approximate melting point of silicon. (1)

d Explain why chlorine has a lower melting point than sulphur. (2)

e Predict the approximate melting point of potassium and give one reason why it is different from that of sodium. (2)

NEAB JUNE 1997 CH01 Q5

10

a Write equations to show what happens when the following oxides are added to water and predict approximate values for the pH of the resulting solutions.
i) sodium oxide
ii) sulphur dioxide (4)

b What is the general relationship between bond type in the oxides of the Period 3 elements and the pH of the solutions which result from addition of the oxides to water? (2)

c Write equations to show what happens when the following chlorides are added to water and predict approximate values for the pH of the resulting solutions. (4)

NEAB June 1997 CH01 Q6

14 Redox equilibria

A 9 volt battery has six cells, each with a voltage of 1.5 V.

Many modern appliances need a portable source of electricity – for watches, personal stereos, mobile phones and lap-top computers. The list is endless.

Scientists regularly develop newer and better batteries, claiming that they give 'more energy', 'more power', are 'longer lasting', or are 'the best'. What determines how powerful and long lasting they are depends on the use they are put to, and particularly on their chemical construction.

Chemical changes take place inside batteries; they cause electrons to move round a circuit in the same direction. This is what we call the current. The battery, or 'cell', has one component which releases electrons – it is oxidised – and the other component accepts electrons – it is reduced. When we combine one of each component to make a cell, a voltage is produced, its size depending on the particular combination of components used.

The AA batteries we buy for use in most of our everyday items, such as stereos and torches, use zinc and carbon, and this produces a voltage of about 1.5 volts because of the particular reactivity of zinc and carbon. Watches often use lithium batteries because lithium is light and can give a high voltage, but they have to be replaced when they wear out. Mobile phones use NiCd (nickel cadmium) batteries and can be recharged by passing a current through them in the reverse direction. A car battery supplies only 12 volts, which you could obtain from eight AA batteries in series, but a car battery supplies a very large current – enough to give you a nasty shock.

The very first batteries invented were so poor that they hardly worked at all and were difficult to incorporate into equipment powered by them, but the chemical changes in them were exactly the same as in modern batteries. The chemical reactions determine the voltage, the engineering and design determine how efficiently they work, and how suitable they are for different uses.

The use of electric vehicles is proposed as a way to reduce pollution, but wide use of electric transport will depend upon when high voltage cells using low density materials are available. A sodium–sulphur cell, being developed, shows promise.

14.1 Oxidation states

A characteristic of transition elements is their **variable oxidation state**. The transition elements are all metals and they lose electrons when they react. We can represent this in a general equation:

$$M \rightarrow M^{n+} + ne^-$$

In this reaction, transition metal M is oxidised to M^{n+} by loss of n electrons. Remember OIL RIG from *AS Chemistry*:

Oxidation Is Loss of electrons
Reduction Is Gain of electrons

When the metal reacts with a non-metallic element, there is a **redox (reduction–oxidation) reaction** and electrons are transferred. In the reaction:

$$X + ne^- \rightarrow X^{n-}$$

X is reduced by gaining electrons. When these equations for loss and gain of electrons are written separately, they are called **half-equations** (see Chapter 8 in *AS Chemistry*).

Assigning oxidation states

The number of electrons lost or gained is called the **oxidation state**. You will have read about oxidation states in *AS Chemistry*, Chapter 8.

A metal forming a positive ion has a positive oxidation state. For example, when calcium metal loses two electrons, $Ca \rightarrow Ca^{2+} + 2e^-$, its oxidation state changes from 0 to +2. Note that the oxidation state and the charge on the ion have the same value, but for the oxidation state the sign (+) is put in front.

A non-metal combined with a metal will have a negative oxidation state. In the example, $Cl_2 + 2e^- \rightarrow 2Cl^-$, the oxidation state of chlorine changes from 0 to –1.

If two non-metals are joined to each other, the more *electronegative* element (see pages 30 and 64 in *AS Chemistry*) is assigned the negative oxidation state.

For compounds, this rule applies:

In a neutral compound, the sum of the positive and the negative oxidation states is zero.

So, for copper(II) chloride:

$$CuCl_2 \rightarrow Cu^{2+} + 2Cl^-$$

the oxidation state of Cu is +2, and the oxidation state of Cl is –1:

$$\begin{array}{ll} Cu & Cl_2 \\ +2 \quad + & (2 \times -1) = 0 \end{array}$$

Many of the s and p block elements have just one or perhaps two oxidation states that are common to nearly all their reactions. For example, oxygen is almost always –2, chlorine is almost always –1.

However, transition metals are special. They have electrons in the d orbital and, because the energy levels in the d orbital are similar, several oxidation states can exist with similar stabilities. For example, chromium has oxidation states of +2 (e.g. $CrCl_2$), +3 (e.g. $CrCl_3$) and +6 (e.g. CrO_2Cl_2).

Calculating oxidation states

Oxidation states in compounds

To work out the oxidation state of an element X in a compound, we write down the charges of all the other elements and, given that the overall charge of the compound is zero, we can then calculate the charge of X.

The oxidation state of iron in Fe_2O_3 can be found by knowing that the oxidation state of oxygen is –2, so we use its charge of 2–. Since Fe_2 is the only other species, we know that n must have a positive value:

$$Fe_2O_3: \quad \begin{aligned} 2 \times Fe^{n+} &= 3 \times O^{2-} \\ (2 \times n+) + (3 \times 2-) &= 2n+ + 6- = 0 \\ 2n+ &= 6+ \\ n &= 3+ \end{aligned}$$

The oxidation state of iron in Fe_2O_3 is +3.

A more complicated calculation, to find the oxidation state of manganese in $KMnO_4$, follows the same rules:

$$KMnO_4: \quad \begin{aligned} K &= 1+ \quad Mn = n \quad 4 \times O^{2-} = 4 \times 2- \\ & 1+ \qquad n \qquad \qquad 8- \\ & 1+ + n + (8-) = 0 \\ & n = 7+ \end{aligned}$$

The oxidation state of manganese in $KMnO_4$ is +7.

Oxidation states in ions

Transition elements form complex ions with a wide variety of ligands, groups arranged round the central ion of the transition element. Some complexes are shown in Table 1.

For a complex ion, the sum of the charges inside the bracket must equal the overall charge on the ion.

Table 1 Oxidation states for some complex ions			
Complex ion	Charge due to ligands	Charge on complex	Charge on central ion
$[Co(H_2O)_6]^{2+}$	0	2+	2+
$[Cr_2O_7]^{2-}$	$7 \times 2- = 14-$	2–	6+
$[Cu(NH_3)_4(H_2O)_2]^{2+}$	0	2+	2+
$[Ni(CN)_4]^{2-}$	$4 \times 1- = 4-$	2–	2+
$[Co(C_2O_4)_3]^{3-}$	$3 \times 2- = 6-$	3–	3+
$[Co(en)_3]^{3+}$ (en = ethane-1,2-diamine)	0	3+	3+

For the complex ion $[Fe(CN)_6]^{4-}$, the overall charge is 4–. To find out the oxidation state of iron, we need to know that the overall charge of each of the six –CN groups (ligands) is –1 (see $[Ni(CN)_4]^{2-}$ in Table 1).

$[Fe(CN)_6]^{4-}$:

Fe = n+ and CN = 1–
So n+ + (6 × 1–) = 4–
So n = 2+

The oxidation state of iron in $[Fe(CN)_6]^{4-}$ = +2.

1 Calculate the oxidation states of the chromium in each of the following:
a $Cr_2(SO_4)_3$
b $K_2Cr_2O_7$
c CrO_3
d $[Cr(H_2O)_4Cl_2]^-$

The changing oxidation states of transition metals

It can be relatively easy for the oxidation state of transition elements to be changed chemically, for example for Fe^{2+} to be changed to Fe^{3+}. In the change from one oxidation state to another, a redox reaction will occur, and often the colour changes as well.

In an aqueous solution, iron(II) ions (left) darken to iron(III) ions (right).

Take, for example, soluble iron(II) ions in water. The iron(II) ions are oxidised to iron(III) ions, $Fe^{2+}(aq) \rightarrow Fe^{3+}(aq)$, as iron(II) loses one electron. An **oxidant** will accept this electron and become reduced: in this case, oxygen dissolved in the water is reduced (gains electrons) to form hydroxide ions:

$$2H_2O(l) + O_2(aq) + 4e^- \rightarrow 4OH^-(aq)$$

Balancing full redox equations
Oxidation of iron(II) to iron(III)
To write a balanced overall equation for the oxidation of iron(II) to iron(III) described above, we carry out the following steps.

Step 1: Calculate the changes in oxidation state for the two ions:

$$Fe^{2+}(aq) \rightarrow Fe^{3+}(aq) + e^-$$

$$O_2(aq) + 4e^- \rightarrow 4OH^-(aq)$$

Step 2: Balance the O and H atoms by adding $H_2O(l)$ or $H^+(aq)$ (if needed) and complete the two half-equations:

$$Fe^{2+}(aq) \rightarrow Fe^{3+}(aq) + e^-$$
(no $H_2O(l)$ or $H^+(aq)$ needed here)

$$2H_2O(l) + O_2(aq) + 4e^- \rightarrow 4OH^-(aq)$$

Step 3: Balance the electron transfer (4 electrons) in the two half-equations:

$$4Fe^{2+}(aq) \rightarrow 4Fe^{3+}(aq) + 4e^-$$

$$2H_2O(l) + O_2(aq) + 4e^- \rightarrow 4OH^-(aq)$$

Step 4: Add the two half-reactions:

$$4Fe^{2+}(aq) \rightarrow 4Fe^{3+}(aq) + 4e^-$$
$$2H_2O(l) + O_2(aq) + 4e^- \rightarrow 4OH^-(aq)$$

$$4Fe^{2+}(aq) + 2H_2O(l) + O_2(aq) \rightarrow 4OH^-(aq) + 4Fe^{3+}(aq)$$

Step 5: Check the charges in the equation balance. In this method, H_2O sometimes needs cancelling out as well at a later stage.

Oxidation of sulphur dioxide by potassium chromate(VI)
For this reaction, $H_2O(l)$ and $H^+(aq)$ should be cancelled in the final equation.

Step 1: Calculate the changes in oxidation state for the two ions:

$$SO_2(aq) \rightarrow SO_4^{2-}(aq) + 2e^-$$
$$Cr_2O_7^{2-}(aq) + 6e^- \rightarrow 2Cr^{3+}(aq)$$

Step 2: Balance the O and H atoms by adding $H_2O(l)$ or $H^+(aq)$ (if needed) and complete the two half-equations:

$SO_2(aq) + 2H_2O(l) \rightarrow SO_4^{2-}(aq) + 2e^- + 4H^+(aq)$
$Cr_2O_7^{2-}(aq) + 6e^- + 14H^+(aq) \rightarrow 2Cr^{3+}(aq) + 7H_2O(l)$

Step 3: Balance the electron transfer in the two half-equations:

$3SO_2(aq) + 6H_2O(l) \rightarrow 3SO_4^{2-}(aq) + 6e^- + 12H^+(aq)$

$Cr_2O_7^{2-}(aq) + 6e^- + 14H^+(aq) \rightarrow 2Cr^{3+}(aq) + 7H_2O(l)$

Step 4: Add the two half-reactions:

$3SO_2(aq) + 6H_2O(l) \rightarrow 3SO_4^{2-}(aq) + 6e^- + 12H^+(aq)$
$Cr_2O_7^{2-}(aq) + 6e^- + 14H^+(aq) \rightarrow 2Cr^{3+}(aq) + 7H_2O(l)$

$3SO_2(aq) + 6H_2O(l) + Cr_2O_7^{2-}(aq) + 14H^+(aq) \rightarrow 3SO_4^{2-}(aq) + 12H^+(aq) + 2Cr^{3+}(aq) + 7H_2O(l)$

Here, some cancelling is needed for $H_2O(l)$ and $H^+(aq)$:

$\qquad\qquad\qquad 2H^+(aq) \qquad\qquad\qquad\qquad\qquad H_2O(l)$
$3SO_2(aq) + 6H_2O(l) + Cr_2O_7^{2-}(aq) + 14H^+(aq) \rightarrow 3SO_4^{2-}(aq) + 12H^+(aq) + 2Cr^{3+}(aq) + 7H_2O(l)$

The full equation is therefore:

$3SO_2(aq) + Cr_2O_7^{2-}(aq) + 2H^+(aq) \rightarrow 3SO_4^{2-}(aq) + 2Cr^{3+}(aq) + H_2O(l)$

Step 5: Check that the total of all the charges in the equation balance. In the last equation, the totals are 0 to the left and 0 to the right: the equation is balanced.

Oxidation of iron(II) by potassium manganate(VII)

Potassium manganate(VII) can be used to determine the concentration of iron(II) ions in an acidified solution. In the titration, the potassium manganate(VII) is the oxidising agent and it is reduced to $Mn^{2+}(aq)$. The Fe^{2+} is the reducing agent and is oxidised to $Fe^{3+}(aq)$.

Step 1: For the change in oxidation states, insert the charges:

$\quad$ Mn $\quad$ O$_4^-$ $\quad \rightarrow \quad$ Mn^{2+}
$\quad$ 7+ $\;+\;$ 8– = 1– $\; \rightarrow \quad$ 2+

The oxidation state of manganese has changed from +7 to +2: Mn has gained $5e^-$.
For the $Fe^{2+}(aq)$:

$MnO_4^- + 5e^- \rightarrow Mn^{2+}$

$\quad$ Fe^{2+}(aq) $\quad \rightarrow \quad$ Fe^{3+}(aq) $+ e^-$
$\quad$ 2+ $\qquad\qquad \rightarrow \quad$ 3+

Step 2: Balance the oxygen and hydrogen atoms by adding $H_2O(l)$ or $H^+(aq)$ (if needed) and complete the two half-equations:

$MnO_4^-(aq) + 5e^- + 8H^+(aq) \rightarrow Mn^{2+}(aq) + 4H_2O(l)$
$Fe^{2+}(aq) \rightarrow Fe^{3+}(aq) + e^-$

Step 3: Balance the electron transfer in the two half-equations:

$MnO_4^-(aq) + 5e^- + 8H^+(aq) \rightarrow Mn^{2+}(aq) + 4H_2O(l)$
$5Fe^{2+}(aq) \rightarrow 5Fe^{3+}(aq) + 5e^-$

Step 4: Add the two half-reactions:

$MnO_4^-(aq) + 5e^- + 8H^+(aq) + 5Fe^{2+}(aq) \rightarrow Mn^{2+}(aq) + 4H_2O(l) + 5Fe^{3+}(aq) + 5e^-$

$\quad$ 1– $\qquad$ 5– $\qquad$ 8+ $\qquad$ 5 × 2+ $\; \rightarrow \;$ 2+ $\qquad\qquad\qquad$ 5 × 3+ $\qquad$ 5–

Step 5: Check that the charges balance: 17+ on the left and 17+ on the right.
The full equation is:
$MnO_4^-(aq) + 5e^- + 8H^+(aq) + 5Fe^{2+}(aq) \rightarrow Mn^{2+}(aq) + 4H_2O(l) + 5Fe^{3+}(aq) + 5e^-$

14.2 Half-reactions

Batteries

When an electrical circuit containing a battery is connected, the transfer of electrons in redox reactions inside the battery produces a flow of electrons in an external circuit. The flow of charge in the system is the current, and so a component in the circuit, such as a bulb or watch mechanism, will be switched on.

Modern batteries are much more advanced than the first, crude, electrochemical cells constructed by Allessandro Volta in 1800, but the chemical principles are the same. Volta used silver and zinc discs separated by pasteboard soaked in salt solution to produce a voltage. Today's batteries use metal/metal ion solutions separated by an electrolyte.

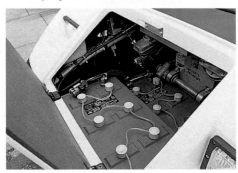

The batteries of this electric vehicle are being recharged using solar power. This not only cuts out urban pollution; it also eliminates the pollution caused when electricity is generated from non-renewable fuels.

Half-reactions

To understand how chemical reactions produce a voltage, we first look at what happens when a metal strip is dipped in a solution of one of its salts. Some of the metal atoms give up electrons and dissolve to form metal ions, leaving behind electrons on the metal strip (Fig. 1).

For zinc leaving the metal strip:

$$Zn(s) \rightarrow Zn^{2+}(aq) + 2e^- \qquad \text{half-equation (1)}$$

However, metal ions in the solution are free to recombine with the electrons on the metal to form metal atoms:

$$Zn^{2+}(aq) + 2e^- \rightarrow Zn(s) \qquad \text{half-equation (2)}$$

The greater the excess of charge on the metal strip, the more likely it is that metal ions will form atoms again. An equilibrium is set up, where the rate of formation of metal *ions* equals the rate of formation of metal *atoms*.

$$\xrightarrow{\text{zinc ions reduced}}$$
$$Zn^{2+}(aq) + 2e^- \rightleftharpoons Zn(s)$$
$$\xleftarrow{\text{zinc atoms oxidised}}$$

This is an example of a **dynamic equilibrium**. At equilibrium, the negative charge on the metal strip sets up a **potential difference** (voltage) between the metal and the solution. The greater the tendency of the metal to produce ions (and therefore hold a negative charge on the metal), the greater is

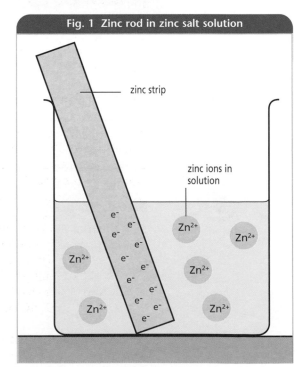

Fig. 1 Zinc rod in zinc salt solution

zinc strip

zinc ions in solution

e^- e^- e^- e^- e^- e^- e^- e^- e^-

Zn^{2+} Zn^{2+} Zn^{2+} Zn^{2+} Zn^{2+} Zn^{2+}

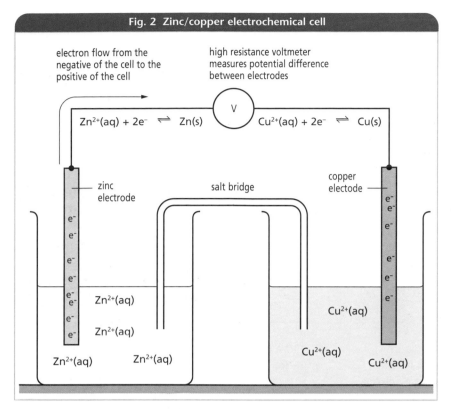

Fig. 2 Zinc/copper electrochemical cell

electron flow from the negative of the cell to the positive of the cell

high resistance voltmeter measures potential difference between electrodes

$Zn^{2+}(aq) + 2e^- \rightleftharpoons Zn(s)$ V $Cu^{2+}(aq) + 2e^- \rightleftharpoons Cu(s)$

zinc electrode

salt bridge

copper electode

$Zn^{2+}(aq)$

$Zn^{2+}(aq)$

$Zn^{2+}(aq)$ $Zn^{2+}(aq)$

$Cu^{2+}(aq)$

$Cu^{2+}(aq)$

$Cu^{2+}(aq)$

the potential difference at equilibrium.

Zinc tends to give up electrons and dissolve in a solution of its ions more readily than copper (Fig. 2). At equilibrium, the potential difference for the zinc/zinc ion system is greater than for the copper/copper ion system. For copper and zinc, very few metal ions actually end up in solution, and the equilibrium in each of the half-equations above lies well over to the right in both cases.

In summary, the half-reactions for metal/metal ion systems are described by half-equations.

Redox equilibria

We have seen that half-reactions are examples of redox reactions. By convention, standard half-reactions are written as reductions. This convention was set by the International Union of Pure and Applied Chemistry (IUPAC). Conventions such as these are important, as they enable information to be communicated consistently by chemists across the world, whether they are researchers working world-wide, or students studying in UK schools or colleges. For example, the half-reaction for the zinc/zinc ion system is written in the IUPAC convention as follows:

$Zn^{2+}(aq) + 2e^- \rightleftharpoons Zn(s)$
reduction of zinc ions to zinc atoms

2 Write out half-reactions for the following, so that they obey the IUPAC convention:

a magnesium metal (Mg) in equilibrium with magnesium ions (Mg^{2+});

b iron(II) ions in equilibrium with iron(III) ions;

c chlorine gas (Cl_2) in equilibrium with chloride (Cl^-) ions.

Electrochemical cells: Zn/Cu

Half-reactions in a metal/metal ion system produce an electrode potential. How is this used to generate useful electric current? If two metal/metal ion systems are linked (Fig. 2), the electrons on the metal strips can move around the new circuit. This set-up is called an **electrochemical cell**; it consists of two electrodes (metal conductors) immersed in electrolytes. The electrolyte can be a solution or a molten salt.

Electrons flow when two different half-reactions, such as Zn/Zn²⁺ with Cu/Cu²⁺ are connected as part of a complete circuit. The zinc metal (Fig. 2) has a greater build-up of negative charge at its surface than the copper does. This means that the zinc electrode has a more negative **electrode potential**. Relative to the zinc, the copper has a less negative electrode potential, so the copper is said to be the more positive electrode.

The difference in electrode potential, or electromotive force, **e.m.f.**, is a measure of the force that moves the electrons round the circuit. The bigger the difference in electrode potential and the more cells connected in series (see the Opener to this chapter), the greater is the e.m.f.

Batteries have a vast range of uses and can be made using different combinations of metals and numbers of cells, so they can have different electrode potentials appropriate to their uses.

The two half-reactions for the zinc/copper cell can be combined to give the overall equation. Electrons flow round the circuit from the *more* negative electrode to the *less* negative electrode. With the two half-reactions written according to the IUPAC convention, the overall equation is obtained by subtracting the more negative (zinc) half-equation from the less negative (copper) one:

$$Cu^{2+}(aq) + 2e^- \rightleftharpoons Cu(s)$$
$$- (Zn^{2+}(aq) + 2e^- \rightleftharpoons Zn(s))$$

$$Cu^{2+}(aq) + 2e^- - Zn^{2+}(aq) - 2e^- \rightleftharpoons Cu(s) - Zn(s)$$

Rearranging, this gives:

$$Zn(s) + Cu^{2+}(aq) \rightleftharpoons Cu(s) + Zn^{2+}(aq)$$

This is the same reaction as would be predicted from the electrochemical series (page 200), so that when zinc metal is put in copper sulphate solution:

$$Zn(s) + CuSO_4(aq) \rightarrow ZnSO_4(aq) + Cu(s)$$

zinc forms zinc ions, and copper ions form metallic copper. Conventionally, the zinc electrode is referred to as 'negative' and the copper electrode as 'positive'.

Notice in Fig. 2 that there is a **salt bridge** to complete the circuit. The two half-reactions need to be kept separate, yet they still need to be connected by a **conductor**. The salt bridge provides an ionic connection between the two ionic solutions. The ions are free to move in the bridge, so the charge is transferred through the bridge solution and keeps each compartment of the cell electrically neutral. The salt bridge in Fig. 2 allows electrons to flow from the negative zinc electrode to the positive (copper) electrode.

Typically, a salt bridge contains a solution of a salt such as potassium chloride or potassium nitrate. A salt bridge solution can be set in agar jelly and held in a glass tube with a porous plug at each end. In commercial batteries, the bridge jelly is held in an absorbent material.

KEY FACTS

- When a metal is placed in a solution of its ions, some of the metal atoms form ions and go into solution, resulting in a slight build-up of electrons on the metal.

- The build-up of charge on the metal electrode produces an electrode potential between the metal and its solution.

- The equilibrium between the metal and its ions is called a half-reaction. Half-reactions are examples of redox reactions.

- Half-reactions are written as reductions according to the IUPAC convention, e.g. $M^{n+}(aq) + ne^- \rightarrow M(s)$

- Two different half-reactions can be connected using a salt bridge to form an electrochemical cell. There is a potential difference (or e.m.f.) between the two electrodes of the cell.

14.3 Electrode potentials

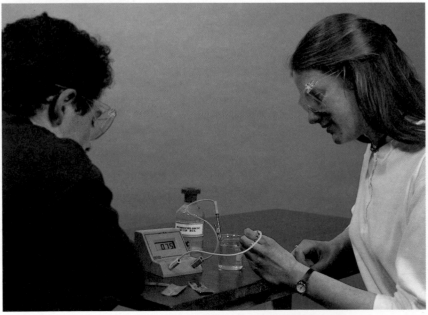

Measuring electrode potentials in the laboratory.

Measuring electrode potentials

The position of an equilibrium is affected by factors such as temperature and concentration of solution. Because the two half-reactions that make up a cell are both in equilibria, cell potentials can only be compared if they are measured under standard conditions, and against a standard half-reaction. Factors affecting cell potential are:

- concentration of ions in each half-reaction
- temperature
- pressure if gases form part of the cell
- cell current.

To produce standardised values, cell potentials are measured under standard conditions using a high-resistance voltmeter. The conditions for measuring are identified in Fig. 3, and the values measured in volts are called **standard cell (electrode) potentials**, symbol $E^{\ominus}$.

It is not possible to measure the potential of a single electrode. Only potential *differences* can be measured, so it is necessary to define a standard against which potential differences can be measured.

The standard hydrogen electrode (SHE)

All potentials are measured relative to the standard hydrogen electrode (SHE) operating under standard conditions. This is assigned an electrode potential, $E^{\ominus} = 0$ (Fig. 3).

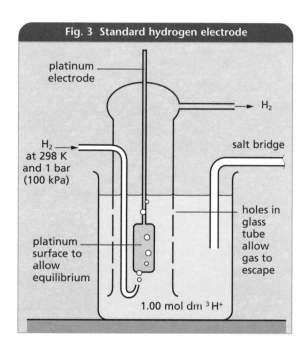

Fig. 3 Standard hydrogen electrode

The cell consists of a platinum electrode with hydrogen gas bubbling over its surface. The electrode is dipped in a solution containing hydrogen ions at a concentration of 1 mol dm⁻³. The purposes of the platinum electrode are to provide a non-reacting (inert) metal contact, and to act as a sink or source for electrons, thereby allowing hydrogen molecules to reach equilibrium with hydrogen ions. The half-reaction for this cell is:

$$2H^+(aq) + 2e^- \rightleftharpoons H_2(g)$$

Fig. 4 shows the conditions for measuring standard cell potentials, while Table 2 in the next section lists some combinations of different metal/metal ion half-reactions and the e.m.f.s they produce.

Convention for writing cells

Electrodes are written below in the conventional form, where the vertical bar | represents the boundary between two different phases. Two electrodes in a cell can be represented by putting the notation for the two electrodes together, and joining them with a salt bridge, symbol ||. The oxidation reaction (more reactive metal) is on the left and the reduction reaction (less reactive metal) on the right:

$$Zn(s)|Zn^{2+}(aq)||Cu^{2+}(aq)|Cu(s)$$

If a gas is present, as in the SHE, the platinum is included in the notation:

$$Pt(s)|H_2(g)|H^+(aq)$$

3 Using data from Table 2, write the correct notation for the following cells:
a A zinc electrode in zinc sulphate solution connected to a lead electrode in lead sulphate solution;
b A zinc electrode in zinc sulphate solution connected to the SHE.
c A copper electrode in copper(II) sulphate solution connected to a platinum electrode in a solution of iron(II) and iron(III) ions.

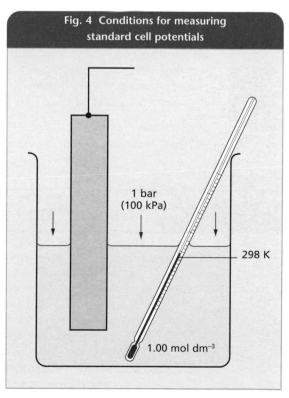

Fig. 4 Conditions for measuring standard cell potentials

The **standard electrode potential, $E^{\ominus}$, is the potential difference between the electrode under standard conditions and the standard hydrogen electrode.**

Some values for standard electrode potentials are given in Table 2. The series of standard electrode potentials in Table 2 forms part of the **electrochemical series**. Electrodes that have more negative potentials have a greater tendency to form positive ions.

Table 2 Standard electrode potentials	
Reduction half-equation	$E^{\ominus}$/V
$MnO_4^-(aq) + 8H^+(aq) + 5e^- \rightleftharpoons Mn^{2+}(aq) + 4H_2O(l)$	+1.51
$Cr_2O_7^{2-}(aq) + 14H^+(aq) + 6e^- \rightleftharpoons Cr^{3+}(aq) + 7H_2O(l)$	+1.33
$MnO_2(s) + 4H^+(aq) + 2e^- \rightleftharpoons Mn^{2+}(aq) + 2H_2O(l)$	+1.23
$Ag^+(aq) + e^- \rightleftharpoons Ag(s)$	+0.80
$Cu^{2+}(aq) + 2e^- \rightleftharpoons Cu(s)$	+0.34
$2H^+(aq) + 2e^- \rightleftharpoons H_2(g)$	0.00
$Pb^{2+}(aq) + 2e^- \rightleftharpoons Pb(s)$	−0.13
$Fe^{2+}(aq) + 2e^- \rightleftharpoons Fe(s)$	−0.44
$Zn^{2+}(aq) + 2e^- \rightleftharpoons Zn(s)$	−0.76
$Mg^{2+}(aq) + 2e^- \rightleftharpoons Mg(s)$	−2.37
$Na^+(aq) + e^- \rightleftharpoons Na(s)$	−2.71
$Ca^{2+}(aq) + 2e^- \rightleftharpoons Ca(s)$	−2.87
$K^+(aq) + e^- \rightleftharpoons K(s)$	−2.92
$Li^+(aq) + e^- \rightleftharpoons Li(s)$	−3.03

Secondary standards

The standard hydrogen electrode is cumbersome to use. It is easier to choose a secondary standard, calibrated against an SHE. The silver/silver chloride electrode has been used for these purposes, but the most common secondary standard is the calomel electrode. Its half-reaction (written as a reduction) is:

$Hg_2Cl_2(s) + 2e^- \rightleftharpoons 2Hg(l) + 2Cl^-(aq)$

The electrode is represented as:

$Pt(s)|Hg(l)|Hg_2Cl_2(sat).KCl\ (1.00\ mol\ dm^{-3})$

$E^{\ominus} = +0.27$ V

Summary

- The strongest reducing agents lose electrons easily and have more negative potentials.
- The strongest oxidising agents accept electrons easily and have more positive potentials.
- Half-reactions with more negative potentials correspond to electron loss (oxidation) reactions and go readily from right to left.
- Half-reactions with more positive potentials correspond to electron gain (reduction) reactions and go readily from left to right.

 4 How would you convert a standard electrode potential measured against a calomel electrode to one measured against the SHE?

14.4 Calculating cell potentials

Standard electrode potentials can be used to calculate standard cell potentials (or e.m.f.s) across electrochemical cells. We can determine standard cell potentials using the overall equation for the cell reaction written as a spontaneous change. For example, in the zinc/copper cell, the zinc is a more reactive metal than the copper. Its electrode potential is therefore more negative. The cell is written:

$Zn(s)|Zn^{2+}(aq)||Cu^{2+}(aq)|Cu(s)$

more reactive   less reactive
more negative   less negative/more positive

The cell potential is calculated by subtracting the *left-hand* electrode potential from the *right-hand* one, so that:

$E^{\ominus}_{cell} = E^{\ominus}_{right} - E^{\ominus}_{left}$

For the cell reaction to be *spontaneous*, $E^{\ominus}_{cell}$ must be positive. For the zinc/copper cell, $E^{\ominus}_{cell}$ will be positive if the standard electrode potential for the $Zn^{2+}|Zn$ electrode is subtracted from the standard electrode potential of the $Cu^{2+}|Cu$ electrode (using figures from Table 2):

$$E^{\ominus}_{cell} = +0.34 \text{ V} - (-0.76 \text{ V})$$

$$E^{\ominus}_{cell} = +1.10 \text{ V}$$

Fig. 5 gives a diagrammatic treatment of this calculation.

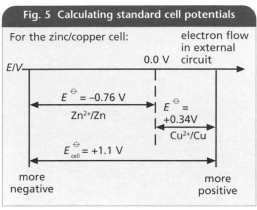

Fig. 5 Calculating standard cell potentials

The calculation above is for a positive value with a negative value. Figs. 6(a) and (b) show calculations of cell potentials for two negative values and for two positive values.

5 If you were choosing two metal electrodes for a cell to give a high voltage, which ones would you select from Table 2? Why?

6 What will happen to the zinc/zinc ions equilibrium if the concentration of ions in the solution is increased?

7

a Draw a diagram like Fig. 5 for the zinc/silver cell.

b Calculate the standard cell potential for this cell.

c Use Table 2 to calculate the cell e.m.f.s using different combinations of the electrodes Mg^{2+}/Mg, Zn^{2+}/Zn and Ag^+/Ag.

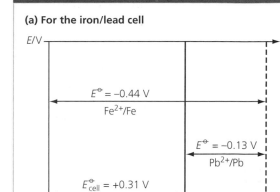

Fig. 6 Calculating standard cell potentials for Fe/Pb and Ag/Cu

(a) For the iron/lead cell

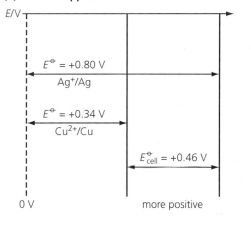

(b) For the copper/silver cell

14.5 Predicting reaction direction

The direction of redox reactions can be predicted by using standard electrode potentials. In Table 2, the standard electrode potential is a measure of how readily electrons will be accepted for the half-reaction. The oxidised species in a half-reaction with a more positive potential will accept electrons more readily than will the oxidised species in a half-reaction with a more negative potential.

When two half-reactions are put together:

• the more positive system will gain electrons (be reduced), and
• the more negative system will lose electrons (be oxidised).

This is an important rule to remember. For example, the half-reaction:

$$Cl_2(g) + 2e^- \rightleftharpoons 2Cl^-(aq) \qquad E^{\ominus} = +1.36 \text{ V}$$

has a more positive electrode potential than:

$$Br_2(l) + 2e^- \rightleftharpoons 2Br^-(aq) \qquad E^{\ominus} = +1.07 \text{ V}$$

so Cl_2 will accept electrons from Br^- (Fig. 7), and the Cl_2 will oxidise Br^- ions to Br_2 and will itself form Cl^- ions when reduced.

Fig. 7 shows how the direction of redox reactions can be predicted using potentials of half-reactions from the electrochemical series. Table 3 shows standard potentials for non-metals, and Fig. 8 shows how to write redox equations for cells.

Fig. 8 Writing redox equations using standard potentials

To write an overall equation for a reaction, we can use the following steps:

1 Write the more negative (less positive) half-reaction as an oxidation:
$$2Br^-(aq) \rightleftharpoons Br_2(l) + 2e^-$$
2 Write the more positive half-reaction as a reduction:
$$Cl_2(g) + 2e^- \rightleftharpoons 2Cl^-(aq)$$
3 Balance the number of electrons being transferred in each equation if necessary.
4 Combine ('add together') the two half-reactions:

$$2Br^-(aq) \rightleftharpoons Br_2(l) + 2e^-$$
$$Cl_2(g) + 2e^- \rightleftharpoons 2Cl^-(aq)$$

$$2Br^-(aq) + Cl_2(g) + 2e^- \rightleftharpoons Br_2(l) + 2e^- + 2Cl^-(aq)$$

The two electrons on each side of the new equation balance out, so they can be struck from the equation (if this is not the case you may need to go back to step 3):
$$2Br^-(aq) + Cl_2(g) + 2\cancel{e^-} \rightleftharpoons Br_2(l) + 2\cancel{e^-} + 2Cl^-(aq)$$

The redox equation for the combination of chlorine gas and bromine ions is therefore:
$$2Br^-(aq) + Cl_2(g) \rightleftharpoons Br_2(l) + 2Cl^-(aq)$$

(This is borne out in the laboratory.)

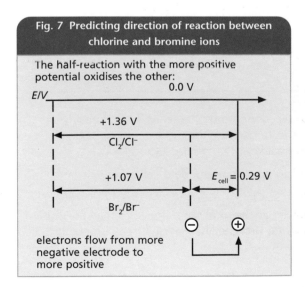

Fig. 7 Predicting direction of reaction between chlorine and bromine ions

The half-reaction with the more positive potential oxidises the other:

electrons flow from more negative electrode to more positive

Table 3 Standard electrode potentials for non-metals

Reduction half-equation	$E^{\ominus}$/V
$F_2(g) + 2e^- \rightleftharpoons 2F^-(aq)$	+2.87
$H_2O_2(aq) + 2H^+(aq) + 2e^- \rightleftharpoons 2H_2O(l)$	+1.78
$Cl_2(g) + 2e^- \rightleftharpoons 2Cl^-(aq)$	+1.36
$Br_2(l) + 2e^- \rightleftharpoons 2Br^-(aq)$	+1.07
$I_2(s) + 2e^- \rightleftharpoons 2I^-(aq)$	+0.54
$2SO_2(aq) + 2e^- \rightleftharpoons S_2O_4^{2-}(aq)$	+0.40
$O_2(g) + 2H_2O(l) + 4e^- \rightleftharpoons 4OH^-(aq)$	+0.40
$2H^+(aq) + 2e^- \rightleftharpoons H_2(g)$	0.00

9 Use electrode potentials to explain why iodine will not displace bromine from a solution containing bromide ions. Use Fig. 7 to help you.

10

a Use electrode potentials in Table 2 to determine the e.m.f.s for the lithium cell (Li/MnO_2).

b Predict the reaction for the Li/MnO_2 cell.

11 Use electrode potentials in Tables 2 and 3 to answer the following, and explain each of your answers:

a Which metal will displace hydrogen from acid: zinc or copper?

b Will bromine oxidise iron(II) to iron(III)?

c Will iodine oxidise iron(II) to iron(III)?

d A common redox titration is acidified potassium manganate(VII) with iron(II) solution. Will acidified manganate(VII) oxidise iron(II)?

Predicting made easy

The electrochemical series and the standard electrode potentials of the corresponding half-reactions can be used to predict the direction of chemical reactions. The series forms into an 'electrode potential chart' and, by following simple rules (Fig. 9), the direction of reactions and cell potential can be determined. Cell potential is given by the difference in $E^{\ominus}$ between the two half-equations.

Fig. 9 Predicting reaction direction

Electrode half-equation	$E^{\ominus}$ / V
$MnO_4^-(aq) + 8H^+(aq) + 5e^- \rightleftharpoons Mn^{2+}(aq) + 4H_2O(l)$	+1.51
$Cl_2(g) + 2e^- \rightleftharpoons 2Cl^-(aq)$	+1.36
$Cr_2O_7^{2-} + 14H^+(aq) + 6e^- \rightleftharpoons Cr^{3+}(aq) + 7H_2O(l)$	+1.33
$Br_2(l) + 2e^- \rightleftharpoons 2Br^-(aq)$	+1.07
$Ag^+(aq) + e^- \rightleftharpoons Ag(s)$	+0.80
$Fe^{3+}(aq) + e^- \rightleftharpoons Fe^{2+}(aq)$	+0.77
$I_2(s) + 2e^- \rightleftharpoons 2I^-(aq)$	+0.54
$Cu^{2+}(aq) + 2e^- \rightleftharpoons Cu(s)$	+0.34
$2H^+(aq) + e^- \rightleftharpoons H_2(g)$	0.00
$Pb^{2+}(aq) + 2e^- \rightleftharpoons Pb(s)$	−0.13
$Fe^{2+}(aq) + 2e^- \rightleftharpoons Fe(s)$	−0.44
$Zn^{2+}(aq) + 2e^- \rightleftharpoons Zn(s)$	−0.76
$Mg^{2+}(aq) 2e^- \rightleftharpoons Mg(s)$	−2.37
$Na^+(aq) + e^- \rightleftharpoons Na(s)$	−2.71
$Ca^{2+}(aq) + 2e^- \rightleftharpoons Ca(s)$	−2.87
$K^+(aq) + e^- \rightleftharpoons K(s)$	−2.92

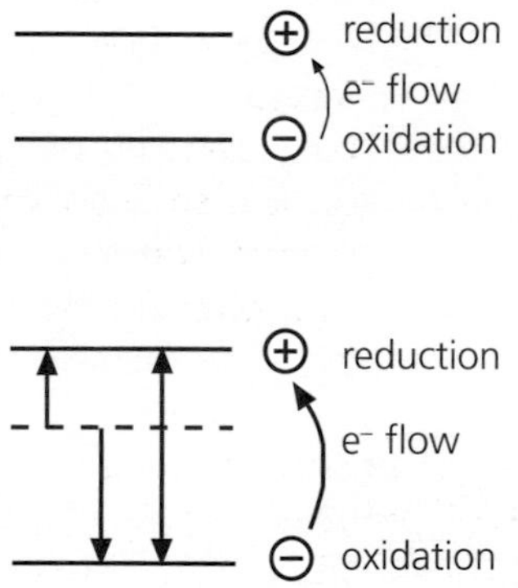

Steps

1 Draw horizontal lines against the two half-equations you are interested in.

2 Mark with a minus sign the one which is more negative.

3 Mark with a plus sign the one which is more positive.

4 Mark the direction of 'electron flow' (from the minus sign to the plus sign).

5 The 'electron flow' will produce a reduction reaction in the half-reaction marked with the plus sign, and an oxidation reaction in the half-reaction marked with the minus sign.

A solar rechargeable aircraft

The Pathfinder prototype being prepared for take-off. The craft has a take-off distance of just 24 metres; less than its own wing span!

The 'flying wing' has reached an altitude of 15 385 m, and can travel at 24 kilometres per hour.

AeroVironment, an American company based in California, has been developing prototypes for solar rechargeable aircraft (SRAs) since the early 1980s. Pathfinder is a recent addition to a line of prototypes for use in monitoring weather and relaying data transmitted through the atmosphere (in mobile communications). Pathfinder can also act like a **geostationary satellite**, but much closer to Earth.

The aircraft is solar powered during the day, but at night, until a technological solution is found, an SRA is still dependent to some extent on electrochemical cells, for its navigational lights, for example. Electrochemical cells were used during aircraft tests, to save on building expensive solar cell arrays.

According to an engineer working on the research, 'When we started work on Pathfinder, we did some calculations with $E^{\ominus}$ values to get an idea of the cells that would give us a large voltage. But potential isn't everything. We needed to consider how much energy could be stored per kg of cell – energy density, how quickly it could be released safely – power density, and the cell had to be rechargeable.'

During the developmental stages, engineers investigated electrode potentials (Tables 2 and 3) for a variety of cells, including Ag/Zn, Li/MnO₂ and Li/SO₂ cells.

1 Researchers world-wide need to use a standard notation system. Write out half-reactions that obey the IUPAC convention for the following:
a zinc metal (Zn) in equilibrium with zinc ions (Zn^{2+});
b manganese(II) ions in equilibrium with manganese(IV) ions in acidic solution;
c oxygen gas (O_2) in equilibrium with oxide (OH^-) ions in aqueous solution.
d explain what would happen to the electrode potential of the $Zn^{2+}|Zn$ electrode if $[Zn^{2+}]$ was increased.

2 Write the correct notation for electrodes that have the following half-reactions:
a $Zn^{2+}(aq) + 2e^- \rightleftharpoons Zn(s)$;
b $Li^+(aq) + e^- \rightleftharpoons Li(s)$;
c $H^+(aq) + e^- \rightleftharpoons \frac{1}{2}H_2(g)$.
d What special purpose has the reaction in **c**?
e What conditions are used for standard electrode potentials?

3 Write the correct notation for the following cells:
a A zinc electrode in zinc sulphate solution connected to a silver electrode in silver sulphate solution;
b A zinc electrode in zinc sulphate solution connected to the SHE. The SHE is represented by $Pt(s)|H_2(g)|H^+(aq)$.

4
a The cell used in the Pathfinder prototype for test flights was a zinc/silver cell. Give one reason why this cell was chosen in preference to the zinc/copper cell. (Use Table 2 to help you.)
b Draw a diagram like Fig. 5 for the zinc/silver cell.
c Calculate the standard cell potential for this cell.
d The potential for the Li/SO₂ cell is +3.43 V. Give two advantages of this cell over the Zn/Ag cell.

5 Predict what would happen (if anything) in the following:
(i) Zn(s) with $Ag_2SO_4(aq)$ **(ii)** Ag(s) with $ZnSO_4(aq)$
(iii) $I_2(g)$ with $Br^-(aq)$ **(iv)** $Br_2(g)$ with $I^-(aq)$
Explain your reasoning.

1

a The following reaction occurs in aqueous solution.

$$5S_2O_8^{2-} + Br_2 + 6H_2O \rightarrow 2BrO_3^- + 12H^+ + 10SO_4^{2-}$$

Identify the reducing agent in this reaction and write a half-equation for its action. (2)

b The electrode potential for the half-equation:

$$Co^{2+}(aq) + 2e^- \rightarrow Co(s)$$

is measured by reference to a standard hydrogen electrode.

i) State the temperature at which the standard electrode potential $E^{\ominus}$ is measured, and give the concentration of $Co^{2+}(aq)$ that must be used.

ii) Electrode potentials are usually measured by reference to a secondary standard electrode. Identify a secondary standard electrode and give a reason why it is used rather than standard hydrogen electrode.

c Cobalt in oxidation states +2 and +3 forms complex ions with water, ammonia and cyanide ligands. Use, where appropriate, the data given below to answer the questions which follow.

$$[Co(H_2O)_6]^{3+}(aq) + e^- \rightarrow [Co(H_2O)_6]^{2+}(aq) \quad E^{\ominus} = +1.81 \text{ V}$$
$$\tfrac{1}{2}O_2(g) + 2H^+(aq) + 2e^- \rightarrow H_2O(l) \quad E^{\ominus} = +1.23 \text{ V}$$
$$[Co(NH_3)_6]^{3+}(aq) + e^- \rightarrow [Co(NH_3)_6]^{2+}(aq) \quad E^{\ominus} = +0.10 \text{ V}$$
$$2H^+(aq) + 2e^- \rightarrow H_2(g) \quad E^{\ominus} = 0.00 \text{ V}$$
$$[Co(CN)_6]^{3-}(aq) + e^- \rightarrow [Co(CN)_6]^{4-}(aq) \quad E^{\ominus} = -0.80 \text{ V}$$

i) Which of the six cobalt species shown above is the most powerful oxidising agent?

ii) Identify a cobalt(II) species which cannot be oxidised by gaseous oxygen.

iii) Hydrogen is evolved when a salt containing the cobalt species $[Co(CN)_6]^{4-}(aq)$ is reacted with a dilute acid. Use the electrode potentials given above to explain the formation of hydrogen gas. (4)
AQA June 2000 CH04 Q1

2

a Name the standard reference electrode against which electrode potentials are measured and, for this electrode, state the conditions to which the term standard refers. (4)

b The standard electrode potentials for two electrode reactions are given below.

$$S_2O_8^{2-}(aq) + 2e^- \rightarrow 2SO_4^{2-}(aq) \quad E^{\ominus} = +2.01 \text{ V}$$
$$Ag^+(aq) + e^- \rightarrow Ag(s) \quad E^{\ominus} = +0.80 \text{ V}$$

i) A cell is produced when these two half-cells are connected. Deduce the cell potential, $E^{\ominus}$, for this cell nd write an equation for the spontaneous reaction.

ii) State how, if at all, the electrode potential of the $S_2O_8^{2-}/SO_4^{2-}$ equilibrium would change if the concentration of SO_4^{2-} ions was increased. Explain your answer. (6)
AQA March 2000 CH04 Q1

3 Use the data below to answer the questions that follow.

Reaction at 298 K	$E^{\ominus}$/V
$Ag^+(aq) + e^- \rightarrow Ag(s)$	+0.80
$AgF(s) + e^- \rightarrow Ag(s) + F^-(aq)$	+0.78
$AgCl(s) + e^- \rightarrow Ag(s) + Cl^-(aq)$	+0.22
$AgBr(s) + e^- \rightarrow Ag(s) + Br^-(aq)$	+0.07
$H^+(aq) + e^- \rightarrow \tfrac{1}{2}H_2(g)$	0.00
$D^+(aq) + e^- \rightarrow \tfrac{1}{2}D_2(g)$	-0.004
$AgI(s) + e^- \rightarrow Ag(s) + I^-(aq)$	-0.15

The symbol D denotes deuterium, which is heavy hydrogen, ^2_1H.

a By considering electron transfer, state what is meant by the term *oxidising agent*. (1)

b State which of the two ions, $H^+(aq)$ or $D^+(aq)$, is the more powerful oxidising agent. Write an equation for the spontaneous reaction which occurs when a mixture of aqueous H^+ ions and D^+ ions are in contact with a mixture of hydogen and deuterium gas. Deduce the e.m.f. of the cell in which this reaction would occur spontaneously. (3)

c Write an equation for the spontaneous reaction which occurs when aqueous F^- ions and Cl^- ions are in contact with a mixture of solid AgF and solid AgCl. Deduce the e.m.f. of the cell in which this reaction would occur spontaneously. (2)

d Silver does not usually react with dilute solutions of strong acids to liberate hydrogen.

i) State why this is so.

ii) Suggest a hydrogen halide which might react with silver to liberate hydrogen in aqueous solution. Write an equation for the reaction and deduce the e.m.f. of the cell in which this reaction would occur spontaneously. (4)
AQA June 1999 CH04 Q4

15 Transition metals and their compounds

The simplest glazes that potters use are colourless, and include the oxides of silicon, sodium, calcium, aluminium and boron. To apply them to pottery, these compounds are finely powdered and suspended in water. Potters also want to create colourful designs in intricate patterns, and so need a range of colours that can be painted on and yet will withstand the heat of the kiln. Simple dyestuffs cannot be used because they are organic compounds and will decompose at such high temperatures. Compounds of transition metals are used because they are stable at high temperatures: some produce the required colour only after firing.

The colour of glazes is produced because of the way the glaze compound (pigment) affects light. It absorbs radiation from part of the visible spectrum, and we see as a colour the part of the visible spectrum that the pigment reflects. In this chapter we learn about the arrangement of electrons within the transition metal atoms and how it gives rise to colour.

The Chinese used iron oxides in glazes as long ago as 200 BC and could produce a whole range of shades for their pottery. Iron(II) oxide, FeO, has the outer electron configuration $4s^0 3d^6$ formed by loss of the two 4s electrons. Iron can also lose a d electron and become iron(III), as in iron(III) oxide, Fe_2O_3, where the iron has an electron configuration of $4s^0 3d^5$. FeO is black and Fe_2O_3 is reddish-brown, so the change in oxidation state has produced a change in colour. By changing the firing conditions, the final oxidation state of the glaze can be controlled: in oxidising conditions it forms Fe(III), while in reducing conditions it forms Fe(II).

A common compound for producing a rich blue glaze is cobalt(II) oxide, CoO. Like iron, cobalt is one of the d block elements of the Periodic Table.

The acidity or alkalinity of the glaze also affects the colour. This is because transition metal ions have other molecules or ions attached to them, called ligands, which alter when conditions change from acid to alkaline, or vice versa.

The potters of the ancient world did not understand the theory of ligand or redox chemistry, nor did they understand how colour is produced by absorbing various parts of the visible spectrum, and yet they used these properties to make some of the greatest artistic treasures of the world.

15.1 Transition metals

Transition metals are used in a variety of alloys to add strength, hardness and resistance to corrosion. Centre: copper. Clockwise from upper left: nickel–chrome ore, nickel bars, titanium bars, iron–nickel ore, niobium bars, chromium granules.

The transition metal elements are found in the central d block of the Periodic Table (see *Data section*). The first transition series, in Period 4 of the Periodic Table, contains the elements from scandium to copper. The Period 4 elements have common properties:

- they are generally hard metals;
- they form complex ions;
- their ions are coloured;
- they take part in catalytic activity;
- their oxidation states are variable.

A transition element is an element having an incomplete d (or f) shell either in the element or in one of its ions.

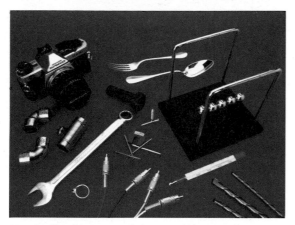

Many familiar objects are made from transition metal elements or alloys of them.

the values of the energies of the sub-shells are similar to each other.

The orbitals for Period 3 (Na to Ar) are filled according to a regular pattern of s followed by p, but the way the d orbitals in transition metals fill is different and a little more complex, as Table 1 shows. (Note that the table shows the order of filling, but not the order in which a first electron is removed on ionisation.)

The energies of the 4s and 3d levels are similar (see *AS Chemistry*, page 15), and the energy levels change as new electrons are added.

As the nuclear charge increases, so does its attraction for electrons, the sub-shells move

---

Fig. 1 The Periodic Table (Periods 1 to 6) with the transition metals of d block highlighted

s block

p block

H 1																	He 2
Li 3	Be 4											B 5	C 6	N 7	O 8	F 9	Ne 10
Na 11	Mg 12		d block transition elements									Al 13	Si 14	P 15	S 16	Cl 17	Ar 18
K 19	Ca 20	Sc 21	Ti 22	V 23	Cr 24	Mn 25	Fe 26	Co 27	Ni 28	Cu 29	Zn 30	Ga 31	Ge 32	As 33	Se 34	Br 35	Kr 36
Rb 37	Sr 38	Y 39	Zr 40	Nb 41	Mo 42	Tc 43	Ru 44	Rh 45	Pd 46	Ag 47	Cd 48	In 49	Sn 50	Sb 51	Te 52	I 53	Xe 54
Cs 55	Ba 56	La 57	Hf 72	Ta 73	W 74	Re 75	Os 76	Ir 77	Pt 78	Au 79	Hg 80	Tl 81	Pb 82	Bi 83	Po 84	At 85	Rn 86

The properties of transition elements are directly related to the electronic structures of the atoms.

A selection of the Period 4 (3d block) properties is given in Table 1.

The order of filling the sub-shells follows the pattern shown in the third row of Table 1. The pattern is not completely regular because

closer to the nucleus, and so their energies are lowered. Each type of sub-shell is affected differently by the attraction of the nucleus and by repulsion between electrons in the occupied orbitals.

When d sub-shells in Period 4 are empty, the 4s sub-shell is at a lower energy than 3d, but when the electrons occupy the d sub-

---

Table 1 Some properties of Period 4 elements										
	Sc	**Ti**	**V**	**Cr**	**Mn**	**Fe**	**Co**	**Ni**	**Cu**	**Zn**
Melting point/°C	1539	1675	1900	1890	1244	1535	1495	1453	1083	420
First ionisation energy/kJ mol⁻¹	633	659	650	653	717	762	759	736	745	906
Outer electron configuration	$3d^14s^2$	$3d^24s^2$	$3d^34s^2$	$3d^54s^0$	$3d^54s^2$	$3d^64s^2$	$3d^74s^2$	$3d^84s^2$	$3d^{10}4s^1$	$3d^{10}4s^2$
Atomic radius/nm	0.164	0.147	0.135	0.129	0.137	0.126	0.125	0.125	0.128	0.137

shells, the 4s sub-shell is at a higher energy level than the 3d sub-shell. This means that when transition metal atoms react, they lose the higher energy 4s electrons first, acquiring a +2 oxidation state. This is one of the reasons why the transition metals have very similar chemical properties to each other and why +2 is the common oxidation state for metals.

As an atom, copper has the electron configuration $[Ar]3d^{10}4s^1$. When it reacts, it can lose $4s^1$ to form Cu(I) compounds which have a full 3d shell. In addition, however, it can lose the 4s plus one of its d electrons when it forms Cu(II).

Compare this with zinc. The d orbitals in zinc are so low in energy that they are not involved in bonding at all. For this reason, and since transition metals are defined as those elements which have an unfilled d orbital in their atoms or ions, zinc is not normally regarded as a transition metal.

Now consider scandium. The normal ion formed in scandium compounds is Sc(III). It has no electrons in the d orbital – its electron configuration is $[Ar]3d^04s^0$ – so scandium compounds do not show the typical properties of transition metal compounds. But, as the element has the electron configuration $[Ar]3d^14s^2$, it is classed as a transition element.

 Write out the electron configuration for Fe^0, Fe^{2+}, Fe^{3+}.

KEY FACTS

■ Transition metals have several properties in common.

■ Transition metal compounds have partially filled 3d sub-shells.

■ When transition metals react, the 4s electrons are lost first.

15.2 Complex ions

Transition metal *ions* form links with specific numbers of molecules or ions which surround the central metal ion. Likewise, many transition metal *compounds* have a central transition metal ion bonded by **co-ordinate** (covalent) bonds to atoms, groups of atoms or ions called **ligands**. Ligands can donate a pair of electrons to the metal ion. The bond is called co-ordinate because both the electrons in the bond are donated by the same atom. These bonds are often represented by arrows (see Fig. 15.2). When ions such as the one shown in Fig. 15.2 are dissolved in water, they will become hydrated and be aqueous ions.

These transition metal ions with ligands attached are the species involved in transition metal chemistry and they are called **complex ions**. The whole complex is an ion because it carries a charge, but the bonds within the complex are co-ordinate covalent bonds. The number of ligands associated with the central metal ion is called its **co-ordination number**. For example, $[Ag(H_2O)_2]^+$ has a co-ordination number of 2 (2 water ligands attached), and $[Cu(H_2O)_6]^{2+}$ has a co-ordination number of 6.

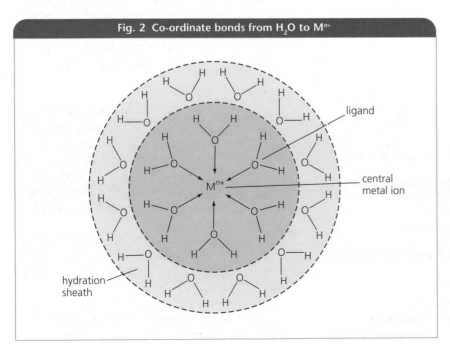

Fig. 2 Co-ordinate bonds from H_2O to M^{n+}

ligand

central metal ion

M^{n+}

hydration sheath

2 What are the co-ordination numbers of the copper and nickel ions in Fig. 3?

$[Cu(NH_3)_4(H_2O)_2]^{2+}$ is often described as a square planar complex, but it also has two water ligands at a greater distance than the ammine ligands, forming a distorted octahedral structure.

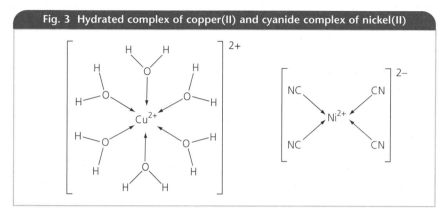

Fig. 3 Hydrated complex of copper(II) and cyanide complex of nickel(II)

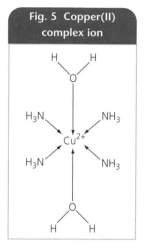

Fig. 5 Copper(II) complex ion

Shapes of complex ions

Complexes with a co-ordination number of 6, whether they are ions or electrically neutral, are usually octahedrally shaped. Those with a co-ordination number of 4 are usually tetrahedral (Fig. 4).

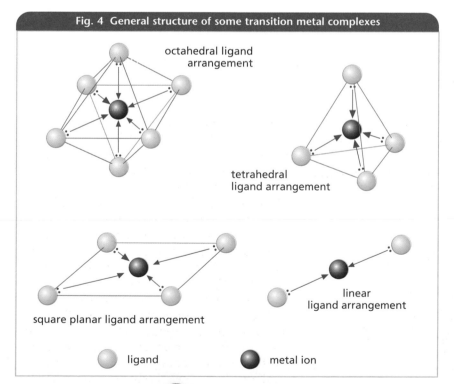

Fig. 4 General structure of some transition metal complexes

octahedral ligand arrangement

tetrahedral ligand arrangement

square planar ligand arrangement

linear ligand arrangement

⚪ ligand   ⚫ metal ion

Ligands

Ligands may be molecules or ions. Water – called 'aqua' when it is part of a complex – is the most common ligand. Many others exist: ammonia (amine), the hydroxide ion (hydroxo), the chloride ion (chloro) and cyanide ion (cyano) all take part in complex formation. The size of the ligand has an effect on the shape of the complex ion. Water, ammonia and cyanide are small, and six ligands can fit round the central metal ion. With larger ions such as chloride, only four ligands can fit round the central metal ion. Some transition elements form complexes with the co-ordination number of 2. These are linear complexes. Examples include Ag(I) complexes and Cu(I) complexes, such as $[Ag(NH_3)_2]^+$, $[Ag(CN)_2]^-$ and $[Ag(S_2O_3)_2]^{3-}$.

Types of ligand

Each water molecule in diaquasilver(I) ions, $[Ag(H_2O)_2]^+$, forms *one* co-ordinate bond with the central silver ion, because there is only *one* atom (oxygen) in the water molecule with a pair of electrons available for co-ordinate bonding. The ligand is said to be **unidentate** as it is only able to donate *one* lone pair of electrons (Fig. 6).

Bidentate ligands can form *two* bonds. An example is the ethanedioate ion. A lone pair of electrons on each of the two oxygen atoms can form co-ordinate bonds.

Multidentate ligands form several bonds with the central metal ion because they have a number of atoms with lone pairs of

3 Write the formula for the copper complex in Fig. 5.

4
a Give the bond angles in:
octahedrally shaped complexes,
b tetrahedrally shaped complexes.

Table 2 Names of common ligands			
Ion/molecule	**Formula**	**Ligand name**	**Type of ligand**
chloride	Cl^-	chloro	unidentate
hydroxide	OH^-	hydroxo	unidentate
water	H_2O	aqua	unidentate
ammonia	NH_3	ammine	unidentate
cyanide	CN^-	cyano	unidentate
ethane-1,2-diamine	$H_2NCH_2CH_2NH_2$ or (en)	ethane-1,2-diamine or (en)	bidentate
ethanedioate	$(COO)_2^{2-}$	ethanedioate	bidentate
ethylenediaminetetraacetic acid	H_4EDTA	$EDTA^{4-}$	hexadentate
ethylenediaminetetraacetate EDTA			

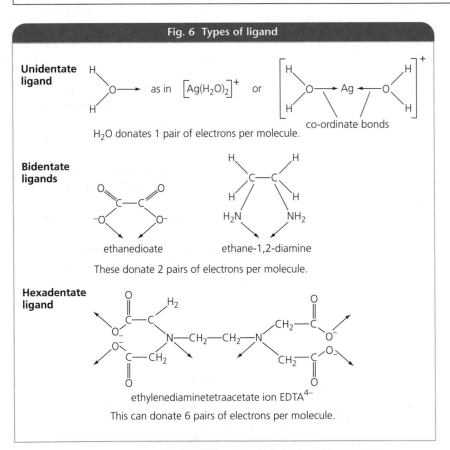

Fig. 6 Types of ligand

Unidentate ligand

as in $\left[Ag(H_2O)_2\right]^+$ or

co-ordinate bonds

H_2O donates 1 pair of electrons per molecule.

Bidentate ligands

ethanedioate   ethane-1,2-diamine

These donate 2 pairs of electrons per molecule.

Hexadentate ligand

ethylenediaminetetraacetate ion $EDTA^{4-}$

This can donate 6 pairs of electrons per molecule.

electrons available for bonding (see Table 2 and Fig. 6). A hexadentate ligand has six donor atoms with lone electron pairs. $EDTA^{4-}$ is a hexadentate ligand.

The lone pairs in EDTA are on the two nitrogen atoms and on four of the oxygen atoms. EDTA forms one-to-one complexes with metal(II) atoms, such as $[Cu(EDTA)]^{2-}$ (Fig. 7). The complexes formed by polydentate ligands are very stable, because the central cation is firmly held by many co-ordinate bonds. They are sometimes called **chelating** ligands (chelate is derived from the Greek *khele* = claw).

Sometimes, bidentate ligands can form similar structures, e.g. $[Cu(en)_3]^{2+}$ (see Fig. 7).

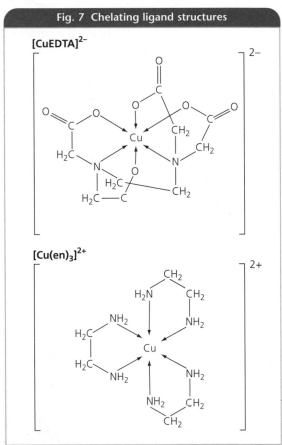

Fig. 7 Chelating ligand structures

$[CuEDTA]^{2-}$

$[Cu(en)_3]^{2+}$

The charge on complex ions

In the complex ion $[Cr(H_2O)_6]^{3+}$, the water ligands are neutral molecules, so the overall charge on the complex is the charge on the transition metal ion. In complexes such as $[CoCl_4]^{2-}$, the ligands are ions, so the overall charge on the complex is the sum of the charges on the central metal ion and the ligands. Some examples of complexes and ligands with their charges are shown in Table 3.

Table 3 Charges for some complex ions and ligands				
Complex	**Ligand**	**Charge due to ligands**	**Overall charge**	**Charge on metal**
$[CoCl_4]^{2-}$	Cl^-	4−	2−	2+
$[CoEDTA]^{2-}$	$EDTA^{4-}$	4−	2−	2+
$[Ag(CN)_2]^-$	CN^-	2−	1−	1+
$[Cu(H_2O)_6]^{2+}$	H_2O	0	2+	2+

Haemoglobin

One of the most important compounds in the body is haemoglobin. Haemoglobin is the oxygen-carrying protein in the blood. It transports oxygen from the lungs to every other part of the body. Haemoglobin consists of a haem molecule (Fig. 8) bonded to a molecule of a protein called globin. At the centre of the haem molecule is an iron(II) ion, surrounded by a **porphyrin** structure. The iron(II) complex is octahedrally co-ordinated. The porphyrin ligand provides four nitrogen atoms, each with a lone pair of electrons. All four are bonded to the iron(II) ion, so that porphyrin is a **tetradentate ligand**. The bonds form the square planar part of the octahedral structure. The haem is embedded in a globin protein, which has a nitrogen atom occupying the fifth position, and an oxygen forming the sixth bond (Fig. 9).

5 Look at the paragraph and diagrams relating to haemoglobin and answer the following questions.

a What is the overall co-ordination number of the iron ion?

b What is the oxidation number of the iron?

c Is the porphyrin a neutral or negative ligand?

d Why is porphyrin called a tetradentate ligand?

e Draw the arrangement of the nitrogens linking to the iron in a square planar pattern.

Fig. 8 Haem structure

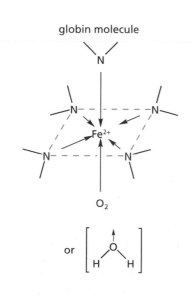

Fig. 9 The haemoglobin molecule

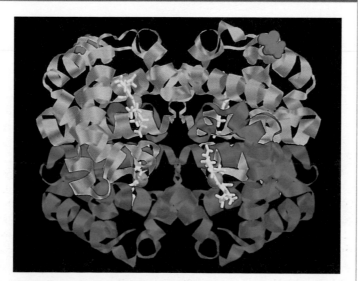

Computer graphic of haemoglobin. It has four globin groups (shown as yellow and blue) with a haem group (white) in each.

- Co-ordination number is the number of ligands surrounding the central metal ion.

- Shapes of complex ions can be linear, square planar, tetrahedral, octahedral.

- Types of ligands are unidentate, bidentate, tetradentate, hexadentate (multidentate).

- The overall charge on the complex ion is the sum of the charges on the ligands and the central metal ion.

15.3 Colour

A typical property of transition metal compounds is that they are coloured. Colour is produced when some parts of the visible spectrum are absorbed and others are reflected. The haem complex in haemoglobin is red and is responsible for the red colour of blood. The colour of these complexes is related to the 3d electron configuration of the transition metal and to the nature and number of the ligands. If the oxygen attached to the haemoglobin (making oxyhaemoglobin) is replaced by carbon monoxide, carboxyhaemoglobin is formed, which has a darker red colour, a sign of carbon monoxide poisoning.

The electron configuration for any element can be worked out using the Periodic Table. So, from the Periodic Table, the s, p, d notation for chromium is:

$1s^2 2s^2 2p^6 3s^2 3p^6 3d^5 4s^1$ (and $4p^0$)

This can also be written as $[Ar]3d^5 4s^1$.

6 What is the electron configuration of iron (atomic number 26)?

Paint pigments made from transition metal compounds are coloured because they each absorb a particular range of the frequencies of light in the visible spectrum, and reflect others.

How does the absorption produce colour?

Chemical compounds we call pigments exhibit an enormously wide range of colours. Light is electromagnetic radiation in the visible spectrum, and when it falls on pigments, they absorb and reflect different wavelengths of its energy. The wavelengths of visible (white) light range from 400 to 700 nm, made up of all the colours of the rainbow. Violet has the shortest (highest energy) wavelengths and red the longest (lowest energy) wavelengths of light.

If an electron absorbs wavelengths within that range, it will reflect a colour (which we see) made up of all the other visible wavelengths minus the wavelengths it absorbs. Only electrons in d energy levels do this, so if an ion or compound has atoms with no d electrons, or if they have a full (d^{10}) configuration, the ion or compound will absorb outside the visible region and appear colourless or white.

In 1900, Max Planck, working on his quantum theory, proposed that energy can only be absorbed (or emitted) in definite amounts (called quanta) and that each of these amounts of energy corresponds to a particular **frequency** of radiation (symbol v, pronounced nu). Frequency is the number of wavelengths of light passing a particular point in a second. Planck summed up his findings in his equation:

$$\Delta E = hv$$

where ΔE is an energy difference and h is Planck's constant.

If an outer electron of atom X absorbs the right amount of energy, it can move from a lower to a higher energy level: it makes a **transition**. When the electron moves to the higher energy level, it absorbs an amount of energy exactly equal to the energy difference (ΔE) between the two allowed energy levels for that electron in atom X. The energy absorbed can cover a range of frequencies. If the frequencies are in the visible spectrum, colour will be observed.

If the transition energy is larger, the radiation may fall in the ultraviolet region (sunscreens absorb harmful ultraviolet radiation); if it is smaller, it may be in the infrared region.

Van Gogh used a chromium compound to provide his yellow colours (*chroma* means colour).

In a transition metal *atom*, all the electrons in the d orbitals are at the same energy level. When the atom forms part of a complex ion, however, the ligand 'splits' the d orbitals into two groups (Fig. 10).

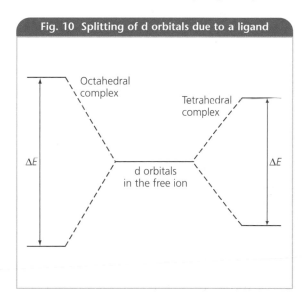

Fig. 10 Splitting of d orbitals due to a ligand

Octahedral complex

Tetrahedral complex

ΔE

d orbitals in the free ion

ΔE

Identifying transition metals

Each of the transition elements is different, so the electrons of each element have their own set of values for allowed energy levels, their own absorption frequencies and so reflect particular colours. Different transition elements will have different nuclear charges and therefore different attractive forces for the electrons in their orbitals. A larger nuclear charge will have a stronger force of attraction, so more energy will be needed to move the electron from one d energy level to another. The absorption is therefore at a different frequency, and a different colour is produced. This property can be used to identify different transition metal compounds.

When white light passes through a solution and some of the frequencies are absorbed, the light that emerges will be different. If the red part of the spectrum is absorbed, the emerging light will have lost this part of the spectrum and will be bluer. An example of this is $[Cu(H_2O)_6]^{2+}$, which absorbs red light and so appears blue (Fig. 11).

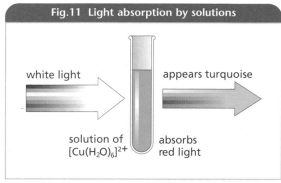

Fig.11 Light absorption by solutions

white light

appears turquoise

solution of $[Cu(H_2O)_6]^{2+}$

absorbs red light

If the blue part of the spectrum (different energy) is absorbed, the emerging light will have a colour which is more yellow or red. The blue end of the visible spectrum is of higher energy, so ions which absorb blue light have a larger ΔE value (Fig. 12).

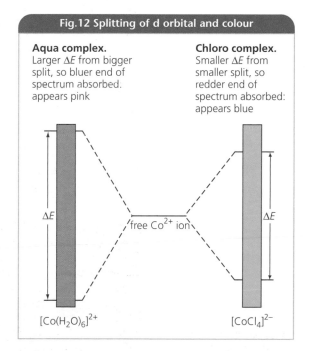

Fig.12 Splitting of d orbital and colour

Aqua complex.
Larger ΔE from bigger split, so bluer end of spectrum absorbed. appears pink

Chloro complex.
Smaller ΔE from smaller split, so redder end of spectrum absorbed: appears blue

ΔE

free Co^{2+} ion

ΔE

$[Co(H_2O)_6]^{2+}$

$[CoCl_4]^{2-}$

Changing colours

The factors which affect the value of ΔE are:

- the ligand
- the size of the ligand
- the strength of the ligand–metal bond in the complex
- the shape of the complex
- the co-ordination number
- the oxidation state

Changing the ligands

If a transition metal ion is dissolved in water, the lone pairs on the water molecules are attracted to the positive metal ion. The best way for these ligands to fit round the ion is in an octahedral arrangement (Fig. 13). If the ligand is changed, the value of ΔE is also changed.

Fig. 13 $[Co(H_2O)_6]^{2+}$ (octahedral) changing to $[CoCl_4]^{2-}$ (tetrahedral)

Changing the oxidation state

Changing the oxidation state of a given element changes the number of electrons in the d levels. This will alter the value of ΔE.

Table 4 summarises some examples.

> **7** Which end of the spectrum is more likely to be absorbed by an ion with a large energy split?

Water and chloride ions cause ligand and co-ordination number to change, but with Co^{2+} ions there is no change in oxidation state. With cobalt chloride paper, this colour change is used to detect the presence of water:

water present	water absent
$[Co(H_2O)_6]^{2+}$	$[CoCl_4]^{2-}$
pink	blue

The aqua complex absorbs blue, green and yellow so it is pink. The chloro-complex absorbs orange, red and pink, so it is blue.

Cobalt chloride paper before and after the addition of water

> **8** Which ligand, Cl^- or H_2O, is producing a larger d orbital split?
>
> **9** Why does the colour change if the ligand is changed?
>
> **10** If $[Cu(H_2O)_6]^{2+}$ is treated with excess CN^- ions it will go through a series of reactions and eventually form the complex ion $[Cu(CN)_4]^{3-}$.
>
> **a** Give the oxidation state for the copper in each of these complexes.
>
> **b** Write the electronic configuration for the copper in each of these complexes.
>
> **c** The aqua complex is blue and the cyano complex is colourless. Explain this change in terms of electronic configuration.

Table 4 Examples of factors affecting colour					
Changing the ligand		**Changing the co-ordination number**		**Changing the oxidation state**	
$[Cr(H_2O)_6]^{3+}$	$[Cr(NH_3)_6]^{3+}$	$[Ni(H_2O)_6]^{2+}$	$[Ni(CN)_4]^{2-}$	$[Fe(H_2O)_6]^{2+}$	$[Fe(H_2O)_6]^{3+}$
red-violet	purple	green	orange-red	green	pale violet
octahedral	octahedral	octahedral	square planar	octahedral	octahedral
$[Cu(H_2O)_6]^{2+}$	$[Cu(NH_3)_4(H_2O)_2]^{2+}$	$[Cu(H_2O)_6]^{2+}$	$[CuCl_4]^{2-}$	$[Cr(H_2O)_6]^{3+}$	$[Cr(NH_3)_6]^{2+}$
blue	blue-violet	blue	yellow	red-violet	blue
octahedral	octahedral	octahedral	tetrahedral	octahedral	octahedral

Analysing for colour

Absorption of light

The amount of light absorbed by a solution depends upon the number of ions it interacts with, so it will be affected by:

- the nature of the ion (relates to ΔE)
- the concentration of the solution
- the distance the light has to travel through the solution (path length)

The more light that is absorbed, the less is transmitted (Fig. 14).

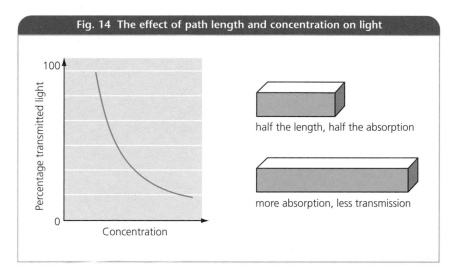

Fig. 14 The effect of path length and concentration on light

half the length, half the absorption

more absorption, less transmission

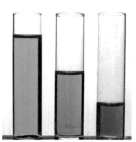

A more concentrated solution will allow less light to be transmitted.

If a solution is more concentrated, more ions are present in a given volume, more will interact, so the colour will be more intense. If the light has to travel a greater distance, there will again be more ions to interact with the light.

UV/visible spectrophotometry

The instrument used to find the concentration of metal ions in solution is called a **spectrophotometer** (Fig. 15). It uses visible or ultraviolet radiation to shine through the sample. Some radiation is absorbed and the remainder passes through to the detector. The amount absorbed is displayed as a meter reading or as a graph. If the nature of the ion and the path length are fixed, we can find concentration in a sample by comparing it with standard solutions.

Procedure

Because aqueous complexes have only very weak absorptions, a suitable ligand is first added to the ion to give an intense colour. The visible range is scanned to find the most intense absorption frequency and then a suitable range of frequencies is passed through a solution of fixed path length.

A typical spectrophotometer has two radiation sources, to provide frequencies both in the ultraviolet and visible regions. The various frequencies are separated by using a prism (or more commonly a diffraction grating), which rotates very slowly. As the prism rotates, the different wavelengths pass, in turn, through a very narrow slit, producing a series of very precise wavelengths of light, each referred to as **monochromatic radiation**. As the radiation passes through the sample, particular frequencies are absorbed, giving a broad band of absorption characteristic of the ion. This absorption pattern is displayed either as a spectrum drawn on a paper chart, a visual display unit (VDU) of the spectrum, or as a printout of absorbance at particular wavelengths. The height of the peak relates to the concentration of the solution (Fig. 16).

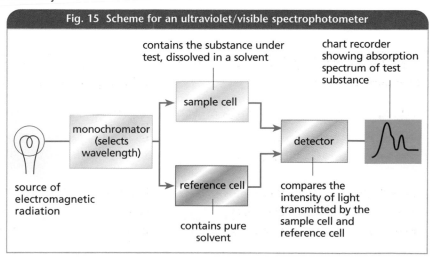

Fig. 15 Scheme for an ultraviolet/visible spectrophotometer

source of electromagnetic radiation

monochromator (selects wavelength)

sample cell — contains the substance under test, dissolved in a solvent

reference cell — contains pure solvent

detector — compares the intensity of light transmitted by the sample cell and reference cell

chart recorder showing absorption spectrum of test substance

SCN⁻ gives an intense red colour with Fe^{3+}, so it is used frequently for analysing Fe^{3+} compounds.

Colorimeter

A simpler instrument which can be used to measure the absorption of radiation is the **colorimeter**. This uses a lamp as a source of white light which passes through a filter to produce light of one colour. This colour will be the one that the sample will absorb the most and is called its **complementary colour**. For example, a blue solution will absorb red, and therefore a red filter is used so that only red light passes through the solution and maximum absorption occurs.

Usually, a set of standard solutions is prepared to give a calibration graph for the instrument, and then the unknown concentration can be determined from the calibration graph, using the colorimeter reading for the unknown.

11 Why does the colour of a solution become paler if water is added?

12 What is the wavelength of the peak absorption in Fig. 16?

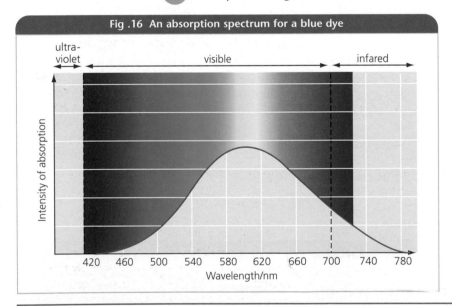

Fig .16 An absorption spectrum for a blue dye

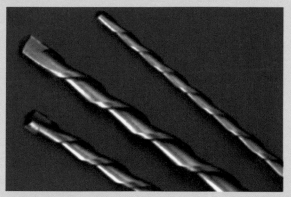

APPLICATION

Absorption and concentration

High-speed steel is used for making drill bits for light engineering manufacturing processes. The composition of alloys they contain is carefully controlled so that it matches the specifications of the manufacturer.

Drill bits must maintain their sharpness during repeated use.

One of the alloy metals is cobalt. The quality control laboratory regularly analyses samples of the steel, determining the amount of cobalt in it colorimetrically, using a spectrophotometer (see Fig. 15). The steel is dissolved in a mixture of concentrated acids, the iron is removed by precipitation, and the remaining solution diluted in water. The colour of the aqueous cobalt ion solution is pale pink.

1 Write the electronic configuration of Co metal.

2 Write the electronic configuration of the Co in the complex ion $[Co(H_2O)_6]^{2+}$.

3 Draw the structure of the aqua species in question 2 and state its co-ordination number.

4 The solution is pale pink because radiation from the visible spectrum is being absorbed. How does the electronic structure of Co^{2+} explain this?

In the spectrophotometer, visible or ultraviolet radiation shines through the sample. Some radiation is absorbed and the remainder passes through to the detector. The amount absorbed is displayed as a meter reading or graph. The amount of radiation absorbed changes with wavelength (see Fig. 17).

5 If thiocyanate ions SCN^- are added to the cobalt solution, a tetrahedral deep blue complex is formed, with maximum absorbance at 613 nm.

a Draw the complex (the lone pair on the nitrogen donates electrons to the Co^{2+} ion), and state the co-ordination number.

b Explain why the colour changes when thiocyanate ions are added.

6 If we want to measure the concentration of cobalt ions in solution in a spectrophotometer (Fig. 15), the thiocyanate complex is used rather than the aqua complex. Why is the thiocyanate complex preferred?

If the solution is more concentrated, more light will be absorbed and the solution will appear darker.

Lambert in 1760 and Beer in 1852 studied the effects of concentration and path length and formulated their respective laws about transmission of colour. Now, *absorbance* is used rather than transmission. The combined Beer–Lambert law is used when measuring colour or ultraviolet absorption.

$$A = \varepsilon lc$$

A = absorbance
ε = the molar absorption (extinction) coefficient, a constant for how strongly the molecule absorbs at a 1 molar solution, measured in $mol^{-1}\ dm^3\ cm^{-1}$
l = path length/cm
c = concentration of solution/mol dm^{-3}

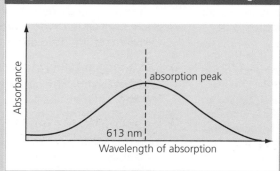

Fig.17 Variation of absorbance with wavelength

Absorbance

613 nm

absorption peak

Wavelength of absorption

APPLICATION

Absorption and concentration (continued)

The sample is poured into a rectangular cell called a cuvette (Fig. 18). The cell is of a standard size and its surfaces are made very accurately, so that they are *optically flat*. This minimises the reflection or scattering from the surfaces, because scattering would reduce the amount of light being transmitted. Also, the cells must be handled very carefully.

Fig.18 Diagram of a cuvette

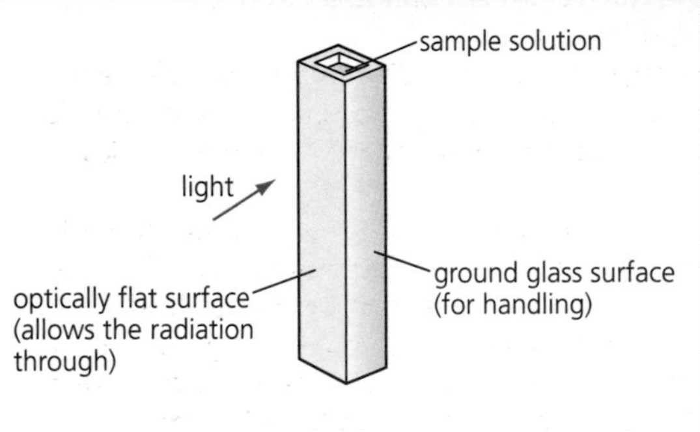

sample solution

light

optically flat surface (allows the radiation through)

ground glass surface (for handling)

7 Why should the optically flat surfaces not be touched?

8 Why is it essential to know the size of the cells?

To determine the concentration of the cobalt solution, a calibration curve is plotted (Fig. 19).

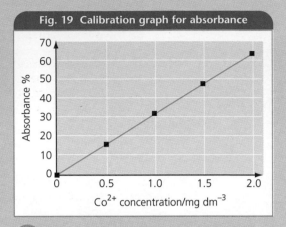

Fig. 19 Calibration graph for absorbance

9 The reading from the test solution was 37%. Determine the concentration of cobalt in the solution.

1

a The ethanedioate ion, $C_2O_4^{2-}$, acts as a *bidentate ligand*.

i) What is meant by the term *bidentate ligand*?

ii) This ligand forms an octahedral complex with iron(III) ions. Deduce the formula of this complex and draw its structure showing all the atoms present.

b i) Give the name of a naturally–occurring complex compound which contains iron.

ii) What is an important function of this complex compound? (2)

AQA June 2000 CH05, Q4(a,b)

2

a Scandium and zinc both appear in the d block of the Periodic Table but their common compounds do not show the characteristic properties associated with transition element compounds.
Give the electronic configurations of each of these elements and deduce the oxidation state each has in its common compounds.

b When concentrated hydrochloric acid is added to an aqueous solution of a cobalt(II) salt there is a change in colour. Account for this observation by giving the two cobalt species involved, their colours, their shapes and the co-ordination numbers of the cobalt

they contain. Write an equation for the reaction and explain why the addition of dilute hydrochloric acid does not cause the same colour change. (11)

NEAB June 1998 CH02 Q6(a,b)

3

a i) Explain what is meant by a *complex ion*.

ii) Describe the bonding in a complex ion.

iii) Explain what is meant by the term *bidentate ligand* and give the formula of one compound or ion that can act as a bidentate ligand. (6)

b Give the formula of a complex ion that is: octahedral; tetrahedral; linear. (3)

c Give, by name or formula, a transition metal complex which:

i) occurs in the body;

ii) is used in medicine. (2)

NEAB June 1997 CH05, Q1

4

a Anhydrous cobalt(II) chloride is a blue solid. Write an equation to show the reaction which occurs when cobalt(II) chloride dissolves in water.

b When cobalt(II) chloride is added to concentrated hydrochloric acid, a blue solution is formed. Give the formula and shape of the cobalt–containing species present in the blue solution. (3)

c When aqueous cobalt(II) chloride is treated with an excess of aqueous ammonia, in the presence of hydrogen peroxide, a mixture of complexes is formed. One complex which can be isolated from this mixture is $[Co(NH_3)_6]Cl_3$.

i) State the oxidation state and co-ordination number of cobalt in $[Co(NH_3)_6]Cl_3$. (2)

ii) Another complex which can be isolated from this mixture has the formula $CoCl_3.5NH_3$. This complex contains cobalt in the same oxidation state and with the same co-ordination number as the complex in part i) above. Suggest a structural formula for $CoCl_3.5NH_3$ which indicates the number of ligands around the cobalt ion. (1)

NEAB February 1997 CH05 Q3(a,b,d)

5

a Use the Periodic Table to complete the electronic configurations of the following atoms. (2)
Ge: [Ar]
Cu: [Ar]

b Explain why copper is considered to be a transition metal. (2)

c In an experiment on the properties of aqueous transition metal ions, a solution containing the pink species **A** turned blue when an excess of concentrated hydrochloric acid was added, due to the formation of species **B**.

i) Before any experiments were carried out, what is the evidence which indicates that **A** was a transition metal species? (1)

ii) Give the formulae and shapes of the species **A** and **B**, state the oxidation state of the metal in each species and write an equation for the conversion of **A** into **B**.

d i) State a feature of the molecules NH_3 and H_2O which enables them to act as ligands.
ii) Explain the term co-ordination number as applied to transition metal complexes. (3)

NEAB February 1997 CH02 Q4

6

a Give the formula, colour, shape and co-ordination number of the complex ions formed by cobalt(II) ions with water and with chloride ions. Include in your answer an explanation of the terms *ligand* and *co-ordination number*. (11)

b Explain why the copper species, $[CuCl_2]^-$, is colourless. (3)

NEAB June 1997 CH02 Q7

7

a What is meant by the terms *ligand* and *complex ion*? (2)

b Give the **full** electronic configuration of the copper(II) ion. (1)

c When anhydrous $CuCl_2$ is dissolved in water a blue solution is formed. Identify the species responsible for the blue colour and state the shape of this species. (2)

d When anhydrous $CuCl_2$ is dissolved in concentrated hydrochloric acid, a yellow-green solution is formed due to the presence of the copper species **X**. If sulphur dioxide is bubbled through this yellow-green solution in the presence of an excess of hydrochloric acid, the colourless species $[CuCl_2]^-$ is formed together with SO_4^{2-} ions.

i) Identify the yellow–green copper species, **X**, state its shape and give the oxidation state of copper in this series.

ii) State the role of sulphur dioxide in the conversion of species **X** into $[CuCl_2]^-$.

iii) Explain, in terms of electronic configuration, why $[CuCl_2]^-$ is colourless.

iv) When the solution containing the yellow-green copper species **X** is added to water, a blue solution is obtained. Write an ionic equation for this reaction. (6)

NEAB March 1998 CH02 Q1

16 Variable oxidation states, and catalysts

Charlotte has been working hard for her first-year exams at university, and she hasn't been sleeping too well. She complained of feeling very tired. Her friends thought she looked too pale so they persuaded her to see a doctor. The doctor sent a sample of Charlotte's blood for analysis, and the result confirmed her friends' suspicions that Charlotte is anaemic.

Charlotte has experienced difficulty because of a lack of iron in her body system, and she will probably have to take nutritional supplements to make up the deficiency. In this chapter you will see how important transition metals are in assisting chemical reactions. Iron is a common element in the body, which forms a haemoglobin complex with oxygen (a ligand) in red blood cells. The iron in haem must be in its Fe(II) oxidation state, and iron supplements can be analysed in the laboratory to ensure the correct dosage by titrating with manganate(VII) – an oxidising agent. The iron(II) is oxidised to iron(III) using a known amount of manganate(VII).

Enzymes are biological catalysts and they facilitate body processes, but they are quite specific in their activity. Certain transition metals, e.g. nickel and platinum can catalyse a range of reactions, and transition metal ions can act as catalysts by changing from one oxidation state to another. When they do this, there is a transfer of electrons, producing a redox reaction.

16.1 Variable oxidation states

One of the properties of transition metals listed in Chapter 15 is that of variable oxidation state.

When metals react, they lose electrons and the higher energy electrons will be lost first. For transition metals these are the 4s electrons. This gives a commonly occurring oxidation state for metals of +2. However, transition metals can form compounds in which the transition metal has other oxidation states, e.g. iron can form Fe^{2+} or Fe^{3+} compounds.

16.2 Changing oxidation states

When a transition metal or its ion is oxidised (loses electrons) it forms an ion with a higher oxidation state, the reacting species must be reduced (gains electrons) and forms a lower oxidation state. These are redox reactions. Metallic zinc is a good reducing agent. It reacts with dilute acid to form Zn^{2+} ions and releases electrons for the reduction:

$$Zn \rightarrow Zn^{2+} + 2e^-$$

We can see this process in the reactions of many transition elements. For example, when you shake an orange solution of VO_2^+ (a vanadium(V) compound) with zinc and dilute hydrochloric acid, it changes colour gradually from yellow through green to blue, then green and finally to violet. These colours are due to the different oxidation states of vanadium (see Table 1). As we saw in Chapter 15 (page 212), changing the oxidation state, changes ΔE, the energy of the d orbital splitting, and therefore the colour of the complex.

Table 1 Oxidation states of vanadium			
Oxidation state	+5	+4	+3	+2
Colour	yellow	blue	green	mauve
Ion	VO_2^+(aq)	VO^{2+}(aq)	V^{3+}(aq)	V^{2+}(aq)
Complex ion	variable structure	$[VO(H_2O)_5]^{2+}$	$[VCl_2(H_2O)_4]^+$	$[V(H_2O)_6]^{2+}$

If sulphuric acid is used instead of hydrochloric acid, the V^{3+} species will be $[V(H_2O)_6]^{3+}$ which is a dull, grey-blue colour. This is an example of the ligand changing the colour of the complex. The final V^{2+} species is easily oxidised, so air must be excluded from the reaction mixture.

The equation for the overall redox reaction can be shown as follows.

Writing the half-equations for

$VO_2^+(aq) + 3e^- \rightarrow V^{2+}(aq)$ and $Zn(s) \rightarrow Zn^{2+}(aq) + 2e^-$ gives:

$VO_2^+(aq) + 3e^- + 4H^+(aq) \rightarrow V^{2+}(aq) + 2H_2O(l)$
$Zn(s) \rightarrow Zn^{2+}(aq) + 2e^-$

Balancing the electron transfer gives:

$2VO_2^+(aq) + 6e^- + 8H^+(aq) \rightarrow 2V^{2+}(aq) + 4H_2O(l)$
$3Zn(s) \rightarrow 3Zn^{2+}(aq) + 6e^-$

Adding these two half-equations gives:

$2VO_2^+(aq) + 6e^- + 8H^+(aq) \rightarrow 2V^{2+}(aq) + 4H_2O(l)$
$3Zn(s) \rightarrow 3Zn^{2+}(aq) + 6e^-$

$2VO_2^+(aq) + 8H^+(aq) + 3Zn(s) \rightarrow 2V^{2+}(aq) + 4H_2O(l) + 3Zn^{2+}(aq)$

If granulated zinc and sulphuric acid (1 mol dm^{-3}) are added to an alkaline solution of ammonium vanadate(V) (NH_4VO_3), the vanadium is reduced and the solution changes colour from yellow to blue to green to mauve.

If you take the highest oxidation state of chromium Cr(VI), you can form lower oxidation states using zinc and hydrochloric acid as a reducing agent. As with V^{2+} species, air must be excluded from the Cr^{2+} species otherwise oxidation will occur giving Cr^{3+} (see Fig. 1).

Fig. 1 Oxidation states of chromium			
The $Cr_2O_7^{2-}$ ion can be reduced using zinc and hydrochloric acid			
Oxidation state	+6	+3	+2
Chromium ion	$Cr_2O_7^{2-}$	$[CrCl_2(H_2O)_4]^+$	$[Cr(H_2O)_6]^{2+}$
Colour in aqueous solution	orange	green	blue

The oxidation states of chromium change in the presence of HCl as acid, therefore in Fig.1 $[CrCl_2(H_2O)_4]^+$ (green) is the complex, not $[Cr(H_2O)_6]^{3+}$.

1
a Write a balanced redox equation for Zn reducing Cr(VI) to Cr(III).
b Write a balanced redox equation for Zn reducing Cr(III) to Cr(II).
c Write a balanced redox equation for $O_2(g)$ oxidising Cr(II) to Cr(III).

When hydrochloric acid is used for the reduction, the Cl^- ion will act as a ligand donating to the Cr^{3+} ion. With sulphuric acid, the complex ion is still green because both SO_4^{2-} and the water act as ligands. $[Cr(H_2O)_6]^{3+}$, which is red-violet, is not formed. Isomers of $CrCl_3(H_2O)_6$ are shown in Table 2.

Table 2 Isomers of $CrCl_3(H_2O)_6$		
Molecular formula	**Structural formula**	**Colour**
$CrCl_3(H_2O)_6$	$[Cr(H_2O)_6]^{3+}$ $3Cl^-$	violet
$CrCl_3(H_2O)_6$	$[Cr(H_2O)_5Cl]^{2+}$ $2Cl^-$ H_2O	light green
$CrCl_3(H_2O)_6$	$[Cr(H_2O)_4Cl_2]^+$ Cl^- $2H_2O$	dark green

16.3 Changing redox properties

The effect of the ligands on the ease of oxidation can be shown with cobalt. In aqueous solution, the oxidation of $[Co(H_2O)_6]^{2+}$ to $[Co(H_2O)_6]^{3+}$ is not normally possible. However, in alkaline solution the precipitate $Co(OH)_2(aq)$ is formed initially and this is readily oxidised to $Co(OH)_3$ by any air present (at the surface or dissolved in the solution). An oxidising agent such as H_2O_2 will also do this. Similarly, if an excess of concentrated ammonia solution is added to the initial precipitate of $Co(OH)_2(aq)$, it dissolves to form the pale brown complex of $[Co(NH_3)_6]^{2+}$, which is again readily oxidised by air or by H_2O_2. This reaction produces the yellow complex hexaammine cobalt(III).

Another ion that can be oxidised by this method is Cr(III). First, the chromium(III) ions are reacted with sodium hydroxide to give $Cr(OH)_3(aq)$:

$$[Cr(H_2O)_6]^{3+} + 3OH^- \rightarrow [Cr(H_2O)_3(OH)_3]$$
$$\text{green}$$

This precipitate will dissolve in an excess of sodium hydroxide solution giving a deep green solution of $[Cr(OH)_6]^{3-}$:

$$[Cr(H_2O)_3(OH)_3] + 3OH^- \rightarrow [Cr(OH)_6]^{3-}$$

When H_2O_2 is added, $[Cr(OH)_6]^{3-}$ is oxidised to CrO_4^{2-}:

$$2[Cr(OH)_6]^{3-} + 3H_2O_2$$
$$\rightarrow 2CrO_4^{2-} + 2OH^- + 8H_2O$$

The yellow tetraoxochromate(VI) ion forms.

Chromate(VI) can be converted to dichromate(VI) by simply adding acid:

$$2CrO_4^-(aq) + 2H^+(aq) \rightleftharpoons Cr_2O_7^{2-}(aq) + H_2O(l)$$

Chromium has its highest oxidation state of VI in both the complexes CrO_4^{2-} and $Cr_2O_7^{2-}$. The position of equilibrium:

$$2CrO_4^{2-} + 2H^+ \rightleftharpoons Cr_2O_7^{2-} + H_2O$$

depends upon the pH of the solution. In acidic solution the equilibrium lies to the right, so if H^+ is added, the equilibrium as written will shift from left to right. In high pH, e.g. if OH^- is added to $Cr_2O_7^{2-}$ the equilibrium as written below shifts to the right.

$$Cr_2O_7^{2-} + 2OH^- \rightleftharpoons 2CrO_4^{2-} + H_2O$$

2 Use a calculation to show that there is no change in oxidation states in converting $CrO_4^-(aq)$ to $Cr_2O_7^{2-}(aq)$

Another example of where a transition metal ion in a low oxidation state, in an alkaline solution, is readily oxidised to a higher oxidation state is iron(II) being oxidised to iron(III). Aqueous iron(II) ions must be stored as an acidified solution, because any air present will oxidise the solution to iron(III). This process is very rapid if alkali is added to the solution. If all air is excluded from the

---

General rules

$[M(H_2O)_6]^{2+}$ acid solution harder to oxidise	$[M(H_2O)_4(OH)_2]$ neutral	$[M(OH)_4]^{2-}$ alkaline solution easier to oxidise
High oxidation state $\rightarrow$ e.g. MnO_4^-	low oxidation state $[Mn(H_2O)_6]^{2+}$	use acid + reducing agent
Need to change O^{2-} to H_2O Need to change Mn(VII) to Mn(II)		
Low oxidation state $\rightarrow$ e.g. $[Mn(H_2O)_6]^{2+}$	high oxidation state MnO_4^-	use alkali + oxidising agent
Need to change H_2O to O^{2-} Need to change Mn(II) to Mn(VII)		

solution (including dissolved oxygen in the solution), a white precipitate is formed.

Normally, when OH^- is normally added to Fe^{2+} ions, a 'dirty' green solution is formed, because oxidation is taking place, giving a complex mixture of compounds. If the solution is left standing, the precipitate will go darker and a brown precipitate of hydrated Fe_2O_3 is formed, especially at the surface of the solution, where it is in contact with the atmosphere.

16.4 Redox titrations

Potassium manganate(VII) is a strong oxidising agent and it can be used in titrations to estimate the amount of iron(II) ions in solution.

Titration of iron(II) with manganate(VII) ions

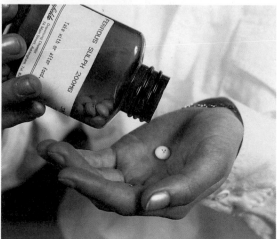

Iron supplement

In the titration the potassium manganate(VII) is the oxidising agent and it is reduced to $Mn^{2+}(aq)$. The iron is the reducing agent and it is oxidised to $Fe^{3+}(aq)$. The reaction must be acidified and excess acid is added to the iron(II) ions before the reaction begins. If insufficient acid is present, the brown solid MnO_2 will be formed and MnO_4^- to Mn^{4+} is only a $3e^-$ change, and has different stoichiometry. The acid used must not react with the manganate(VII) ions, so the acid normally used is dilute sulphuric acid. Acids such as hydrochloric acid are not used because the chloride ions in the solution can be oxidised to chlorine.
The acid must:
- be a strong acid because a high concentration of hydrogen ions is needed
- not be an oxidising agent, because it may react with the reductant and affect the titration results
- not be a reducing agent, because it may be oxidised by the manganate(VII) ions and affect the titration results.

Choose dilute sulphuric acid because it is a strong acid and the dilute acid does not oxidise under these conditions.

Don't choose:
- hydrochloric acid because it can be oxidised to chlorine by the manganate(VII) ions
- nitric acid because it is an oxidising agent and may react with the reductant
- concentrated sulphuric acid because it is an oxidising agent
- ethanoic acid because it is a weak acid, so the concentration of hydrogen ions will be insufficient.

To find the stoichiometry for the reaction, follow the usual steps.
Complete the two half-equations:

$MnO_4^-(aq) + 5e^- + 8H^+(aq) \rightarrow Mn^{2+}(aq) + 4H_2O(l)$
$Fe^{2+}(aq) \rightarrow Fe^{3+}(aq) + e^-$

Balance the electrons:

$MnO_4^-(aq) + 5e^- + 8H^+(aq) \rightarrow Mn^{2+}(aq) + 4H_2O(l)$
$5Fe^{2+}(aq) \rightarrow 5Fe^{3+}(aq) + 5e^-$

Add the two half-equations:

$MnO_4^-(aq) + \cancel{5e^-} + 8H^+(aq) \rightarrow Mn^{2+}(aq) + 4H_2O(l)$
$5Fe^{2+}(aq) \rightarrow 5Fe^{3+}(aq) + \cancel{5e^-}$
_____

$MnO_4^-(aq) + 8H^+(aq) + 5Fe^{2+}(aq) \rightarrow Mn^{2+}(aq) + 4H_2O(l) + 5Fe^{3+}(aq)$

Check the charges balance: 17+ on LHS and 17+ on RHS of the equation.

Potassium manganate(VII) acts as its own indicator. As the purple potassium manganate(VII) solution is added to the titration flask from a burette, it reacts rapidly with the $Fe^{2+}(aq)$. The manganese(II) ions have a very pale pink colour, but they are present in such low concentrations that the solution looks colourless. As soon as all the iron(II) ions have reacted with the added manganate(VII) ions, a pink tinge appears in the flask due to an excess of manganate(VII).

Calculating the amount of iron(II) in an iron tablet

An iron tablet weighing 0.850 g is dissolved in dilute sulphuric acid and the whole of this solution is titrated with 0.02 mol dm^{-3} potassium manganate(VII) solution. A titre of 17.4 cm^3 potassium manganate(VII) solution is needed to reach the endpoint. The equation (as above) for the reaction is:

$MnO_4^-(aq) + 8H^+(aq) + 5Fe^{2+}(aq) \rightarrow Mn^{2+}(aq) + 4H_2O(l) + 5Fe^{3+}(aq)$

Using the equation for the number of moles in a solution:

$$\text{Number of moles} = \frac{\text{volume} \times \text{molar concentration}}{1000}$$

We can work out the number of moles of manganate(VII) used in the titration:

$$\text{Moles of } MnO_4^- \text{ used} = \frac{17.4 \times 0.02}{1000}$$
$$= 3.48 \times 10^{-4} \text{ moles}$$

Since 5 moles of iron(II) are required for every mole of manganate(VII) ion,

$$\text{moles of iron(II)} = 5 \times 3.48 \times 10^{-4} \text{ moles}$$
$$= 1.74 \times 10^{-3} \text{ moles}$$

$$\text{Number of moles} = \frac{\text{mass}}{\text{relative atomic mass}}$$

So mass = number of moles × relative atomic mass

Mass of iron(II) $= 1.74 \times 10^{-3} \times 56$ g
$= 0.0974$ g

The percentage by mass of iron in the tablet is given by the expression:

$$= \frac{0.0974 \times 100}{0.850}$$

$$= 11.5\%$$

In many titrations 250 cm^3 of the iron(II) solution will be made up in a volumetric flask. Usually, 25.0 cm^3 of iron(II) solution is pipetted into a titration flask and manganate(VII) solution is titrated with this aliquot. When doing the calculations, you must remember to put this dilution factor into your calculation.

Manganate(VII) is in the burette and iron(II) is being pipetted from the volumetric flask into the conical flask.

At the endpoint, the contents of the flask start to turn pink.

Titrating with potassium dichromate(VI) solution

Like the manganate(VII) ion, the dichromate(VI) ion can be used to find the amount of iron in iron tablets. Dichromate(VI) is a powerful oxidising agent when it changes from the +6 oxidation state in the dichromate ion, $Cr_2O_7^{2-}$, to the +3 oxidation state in Cr^{3+}:

$$Cr_2O_7^{2-}(aq) + 14H^+(aq) + 6e^- \rightarrow 2Cr^{3+}(aq) + 7H_2O(l)$$

And this reacts with iron(II) ions:

$$6Fe^{2+}(aq) \rightarrow 6Fe^{3+}(aq) + 6e^-$$

The colour change for this titration is orange to bluish-green. To give a more visible endpoint, the indicator sodium diphenylaminesulphonate is used. This turns from colourless to purple at the endpoint.

A different brand of iron tablet, weighing 0.960 g, was dissolved in dilute sulphuric acid. A titre of 28.5 cm^3 of 0.0180 mol dm^{-3} potassium dichromate(VI) solution was needed to reach the endpoint. What is the percentage by mass of iron in the tablet? Again, using the correct equation for the redox reaction and the relevant formulae, this can be calculated.

One mole of dichromate ions oxidises six moles of iron(II) ions and the overall equation is:

$$6Fe^{2+}(aq) + Cr_2O_7^{2-}(aq) + 14H^+ \rightarrow 6Fe^{3+}(aq) + 2Cr^{3+}(aq) + 7H_2O(l)$$

Moles of $Cr_2O_7^{2-}$ used in titration to reach endpoint:

$$= \frac{28.5 \times 0.0180}{1000}$$

$$= 5.13 \times 10^{-4}$$

Moles of iron(II) $= 6 \times 5.13 \times 10^{-4}$ moles
$$= 3.078 \times 10^{-3} \text{ moles}$$

Mass of iron(II) $= 56 \times 3.078 \times 10^{-3}$
$$= 0.1724 \text{ g}$$

The percentage by mass of iron in the tablet

$$= \frac{0.1724 \times 100}{0.960}$$

$$= 18.0\%$$

3 An iron tablet weighing 0.780 g was dissolved in dilute sulphuric acid and titrated with 0.018 mol dm^{-3} potassium manganate(VII). A titre of 21.5 cm^3 of this solution was needed to reach the endpoint. What is the percentage by mass of iron in the tablet?

4 The balanced redox equation for the reaction of $Cr_2O_7^{2-}(aq)$ with $Fe^{2+}(aq)$ is:

$$Cr_2O_7^{2-}(aq) + 14H^+(aq) + 6Fe^{2+}(aq) \rightarrow 2Cr^{3+}(aq) + 6Fe^{3+}(aq) + 7H_2O(l)$$

Use your knowledge of oxidation states to show that this is correct.

5 2.225 g of a sample of iron wire was dissolved in dilute sulphuric acid to give a solution containing Fe(II) ions and was made up to 250 cm^3 in a volumetric flask. 25.0 cm^3 of this solution were acidified and titrated against a 0.0185 mol dm^{-3} solution of potassium dichromate(VI). The iron solution needed 31.00 cm^3 of dichromate(VI). Use the following steps to calculate the percentage of iron metal in the iron wire.

a Calculate the number of moles of dichromate(VI) in 31.0 cm^3 of a 0.0185 mol dm^{-3} solution.

b From the redox equation write down the ratio of moles of iron(II) reacting with 1 mole of dichromate(VI).

c Calculate the number of moles of iron(II) in the 25.0 cm^3 sample.

d Calculate the number of moles of iron(II) in the flask.

e Calculate the mass of iron(II) in the flask (made from 2.225 g of iron wire).
A_r Fe = 56.

f Calculate the percentage of iron in the iron wire.

This sequence can be used for all redox titrations when the reductant and its oxidation product are known.

Step 1 Write the half-equations for oxidant and reductant.
Step 2 Deduce the equation for the overall reaction.
Step 3 Calculate the number of moles of manganate(VII) or dichromate(VI) used.
Step 4 Calculate the ratio of moles of **oxidant : reductant** from the redox equation.
Step 5 Calculate the number of moles in the sample solution of reductant.
Step 6 Calculate the number of moles in the original solution of reductant.
Step 7 Determine either the concentration of the original solution and/or the percentage of reductant in a known quantity of sample.

6 25.0 cm^3 of a sodium sulphite solution needs 45.0 cm^3 of 0.0200 mol dm^{-3} potassium manganate(VII) solution. What is the concentration of the sodium sulphite solution?

7 2.145 g of ethanedioic acid-2-water was dissolved in water and diluted to 250 cm³ in a volumetric flask. 25.0 cm³ of this solution was titrated with potassium manganate(VII) solution and needed 35.0 cm³ to oxidise the acid solution. Calculate the concentration of the potassium manganate(VII) solution.

KEY FACTS

■ For calculations in redox titrations: (a) write half-equations; (b) balance electron transfer; (c) add these balanced half-equations.

■ The commonly used oxidants, MnO_4^- and $Cr_2O_7^{2-}$ must be acidified.

■ For solutions, number of moles = molar concentration x volume/1000 (for volumes in cm³).

■ For masses, number of moles

$$= \frac{mass}{M_r} \quad or \quad \frac{mass}{A_r}$$

16.5 Catalysts

Transition metals and their compounds are widely used as catalysts. Catalysts alter the rate of chemical reactions by providing an alternative reaction pathway, which has a lower activation energy (Fig. 2), but they do not alter the equilibrium position because at equilibrium both the forward and reverse reactions are speeded up equally. In most cases, a catalyst will be used to speed up a chemical reaction. However, some catalysts, called inhibitors or negative catalysts, slow down chemical reactions.

In heterogeneous catalysis, at least one of the reactants binds onto the metal's surface using available orbitals by a mechanism similar to ligand donation. This process is called **adsorption**. It is in this new 'state' that the reactant is more likely to undergo reaction. Adsorption of a reactant onto a metal surface can speed up a reaction for the following reasons.

● It can weaken bonds within the reactant molecule, reducing the activation energy.
● It can cause a reactant molecule to break up into more reactive fragments, reducing the activation energy.
● It can hold a reactant in a position, increasing the chance of a favourable collision.
● It can give a higher concentration of one reactant on the catalyst surface, increasing the chance of a favourable collision with the other reactant.

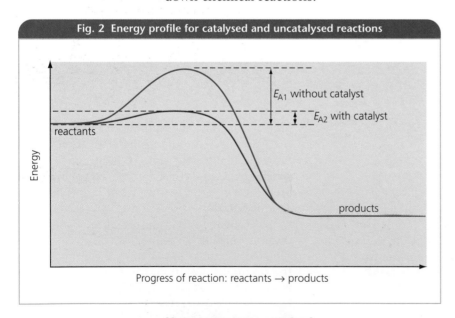

Fig. 2 Energy profile for catalysed and uncatalysed reactions

E_{A1} without catalyst
E_{A2} with catalyst
reactants
Energy
products
Progress of reaction: reactants → products

The strength of the bonding between the metal and the reactant is important in determining the suitability of catalysts for a particular reaction. Metal atoms on the surface of the catalyst need to form bonds with the reactant gas molecules that are strong enough to hold the reactant in position while it reacts with other molecules (Figs. 3 and 4). However, if the bond is too strong, the product molecule will not be released by the metal catalyst and the reaction will not proceed. (The opposite of adsorption is desorption.) Tungsten forms very strong bonds with some molecules, so it is not often used as a catalyst.

Heterogeneous catalysis

Metals used in catalysis are heterogeneous catalysts, because they are solids and in a different phase to the reactants, which will be liquids or gases. Heterogeneous catalysis occurs at the surface of the metal.

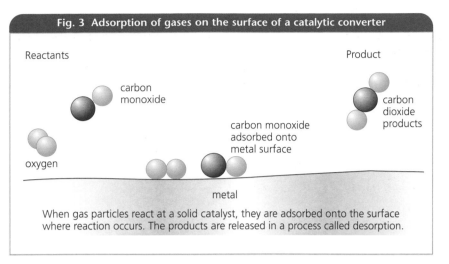

Fig. 3 Adsorption of gases on the surface of a catalytic converter

Reactants

carbon monoxide

oxygen

carbon monoxide adsorbed onto metal surface

Product

carbon dioxide products

metal

When gas particles react at a solid catalyst, they are adsorbed onto the surface where reaction occurs. The products are released in a process called desorption.

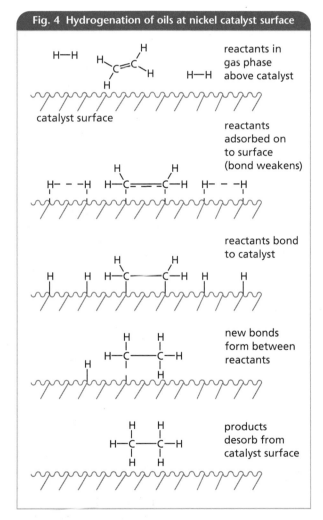

Fig. 4 Hydrogenation of oils at nickel catalyst surface

reactants in gas phase above catalyst

catalyst surface

reactants adsorbed on to surface (bond weakens)

reactants bond to catalyst

new bonds form between reactants

products desorb from catalyst surface

Weak adsorption also results in poor catalysis. Silver forms very weak bonds with molecules, so it is unable to hold onto the gas long enough for a reaction to take place, although it is used in the manufacture of epoxyethane.

Good catalysts include palladium, rhodium and platinum, but unfortunately they can be expensive. To minimise costs, every last microgram of precious metal should be available for catalysis. It is more cost-effective to include a cheap support medium for the catalyst and coat it with catalytic material. (Remember: the reactions to be catalysed take place only on the surface of the catalyst.) The metal catalyst provides a series of active sites. These are places on the surface of the catalyst where the reactions take place. The catalyst coating is not a smooth surface – it is applied in such a way that it has an irregular surface (Fig. 5).

Another way of maximising the surface area of the catalyst in contact with the reactants is to use a fluidised bed (Fig. 6). When gases are blown through a very fine powder, the powder 'floats' on the gas separating the catalyst particles, so that all the surface of the catalyst is available for reactions.

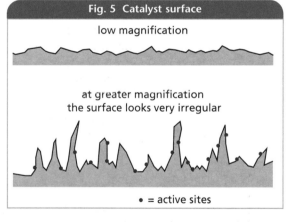

Fig. 5 Catalyst surface

low magnification

at greater magnification the surface looks very irregular

● = active sites

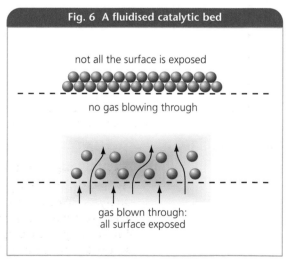

Fig. 6 A fluidised catalytic bed

not all the surface is exposed

no gas blowing through

gas blown through: all surface exposed

227

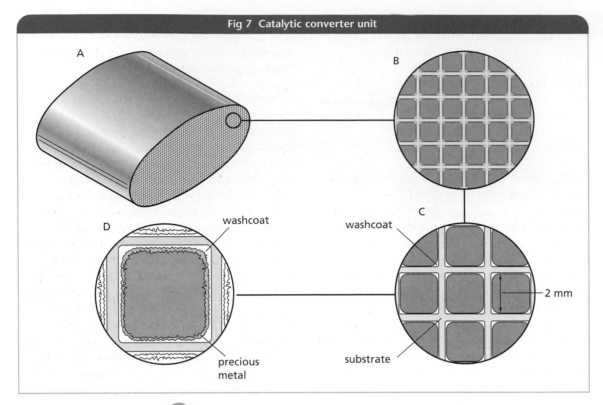

Fig 7 Catalytic converter unit

8 Why do you think the coating is deliberately applied so that its surface is not smooth?

Catalyst poisoning

Sometimes unwanted gases adsorb very strongly on to the catalyst. These 'block' the active sites on the catalyst, reducing the catalyst's effectiveness. This is called poisoning. During the Haber process it is vital that traces of sulphur and carbon monoxide are removed, because these will poison the iron catalyst. Catalytic converters on cars contain a mixture of palladium, platinum and rhodium metals, coated on to cerium oxide (Fig. 7). These will remove unwanted products of combustion. If the exhaust emissions contain sulphur dioxide, this gas will also adsorb on to the catalyst. The sulphur dioxide is adsorbed very strongly and it 'blocks' the active sites on the catalyst, reducing its effectiveness. All petrol contains some sulphur, but the lower the sulphur content achieved during the refining process, the better.

To help overcome sulphur dioxide poisoning, manufacturers of catalytic converters mix in aluminium oxide with the catalyst. The aluminium oxide 'stores' the sulphur oxides under oxidising (normal running) conditions, and converts them to hydrogen sulphide under reducing (accelerating) conditions. This removes the sulphur periodically, so that it does not permanently poison the catalyst. This is why the emissions from catalytic converters occasionally smell of hydrogen sulphide.

Other metals are also capable of bonding permanently to the active sites of catalysts. Lead is a particular problem, and this is why it is absolutely vital that only unleaded fuel is used in cars that are fitted with a catalytic converter. The use of leaded fuel would completely poison the catalyst and catalytic converters are very expensive to replace.

16.6 Industrial processes

Many industrial processes involve the use of catalysts. Two of the most important industrial processes in the economy of any industrialised nation are the manufacture of sulphuric acid and the manufacture of ammonia. Both of these processes involve equilibria, and for each process it is vital to obtain the maximum possible output, in the shortest possible time, at the lowest possible cost. A catalyst is used to increase the rate of reaction, allowing a much lower temperature to be used, yet maintaining sufficient output for the process to be cost-effective.

Haber process

In the Haber process an iron catalyst is used:

$$N_2(g) + 3H_2(g) \overset{Fe}{\rightleftharpoons} 2NH_3(g)$$

The iron is mixed with aluminium oxide and potassium oxide, which act as promoters. The promoter is present in smaller amounts than the iron, and improves the efficiency of the catalyst. The hydrogen for this reaction is obtained from methane, so trace amounts of steam and carbon monoxide are found as impurities in the products. If the carbon monoxide was allowed to remain in the gas mixture, it would poison the catalyst by forming an iron carbonyl compound, so it must be removed. The carbon monoxide is heated with an excess of steam in a two-stage process, first with an iron oxide catalyst and then with a zinc/copper catalyst. The products are carbon dioxide which is easier to remove from the mixture, and hydrogen which is used in the process.

Contact process

The main step in the manufacture of sulphuric acid is the oxidation of sulphur dioxide gas to sulphur trioxide:

$$2SO_2(g) + O_2(g) \rightleftharpoons SO_3(g)$$

The reactant mixture is passed over a catalyst of vanadium(V) oxide (V_2O_5). This catalyst has replaced platinum, despite the fact that platinum produces a faster reaction rate, reducing the time needed to reach equilibrium, because it is cheaper and less prone to poisoning by impurities, e.g. any traces of compounds produced from the roasting of metal ores. When it behaves as a catalyst, it shows variable oxidation states. The catalyst, V_2O_5, has an oxidation state is +5. This reacts with the SO_2 and is reduced to the +4 oxidation state and the sulphur is oxidised to +6. The oxygen present then oxidises the +4 vanadium back to +5, so that it can now oxidise another molecule of SO_2. The vanadium(V) oxide catalyst takes part in the reaction, but is unchanged at the end.

1    $\underset{+5}{V_2O_5(s)} + SO_2(g) \rightarrow \underset{+4}{V_2O_4(s)} + SO_3(g)$

2    $\underset{+4}{V_2O_4(s)} + O_2(g) \rightarrow \underset{+5}{2V_2O_5(s)}$

KEY FACTS

- A heterogeneous catalyst is in a different phase to the reactants, e.g. a *metal* catalyst and *gaseous* reactants.

- In heterogeneous catalysis, reaction occurs at the surface of the catalyst.

- Many transition metals are good catalysts.

- The strength of the bond formed between reactant and catalyst must allow both adsorption

- of reactant molecules and desorption of product molecules to occur, if the catalyst is to be effective.

- Different metal atoms form bonds of different strengths with reactants.

- Catalysts are poisoned if active sites are blocked.

- Catalysts are unchanged in their chemical composition at the end of the reaction.

Homogeneous catalysts

These are in the same phase as the reactants. The ability of transition metal ions to act as catalysts is often linked to a transition metal's ability to form ions in different oxidation states. Because these are ions in solution with: the reactants, they are in the same phase, hence they are homogeneous catalysts. An ion that behaves as an homogeneous catalyst is the hexaaquacobalt(II) ion

$[Co(H_2O)_6]^{2+}(aq)$. Under suitable reaction conditions the Co^{2+} ion is easily oxidised to Co^{3+}, and this transition can be used to form temporary intermediates in the reaction between hydrogen peroxide and sodium potassium tartrate $Na^{+-}OOCCH(OH)CH(OH)COO^{-+}K$.

Cobalt(II) ions form a pink complex with water molecules. Fig. 8 plots the reduction potentials for $M^{3+}(aq)/M^{2+}(aq)$ systems for transition metals in aqueous solution and shows that the reduction Co(III) to Co(II) is favoured. With H_2O as ligands, reduction from Co(III) to Co(II) can occur spontaneously, but not the oxidation Co(II) to Co(III). A powerful oxidising agent would be needed to overcome this reduction potential. However, some ligands can lower the reduction potential sufficiently for a mild oxidising agent to produce the change Co(II) to Co(III). Here is an example.

If cobalt(II) ions are added to a solution of sodium potassium tartrate and hydrogen peroxide, the cobalt ions form an intermediate complex with the tartrate. In this intermediate the cobalt(II) ions are more easily oxidised to cobalt(III) by the hydrogen peroxide. (Changing the ligands changes the electrode potential for oxidation of Co(II) to Co(III).) This means that a green cobalt(III)-tartrate complex forms. The cobalt(III) ions in the complex are then able to oxidise the tartrate, to produce carbon dioxide gas and a carbonyl compound (Fig. 9), and the Co(III) is reduced back to the pink Co(II) complex $[Co(H_2O)_6]^{2+}$.

Colour changes often accompany a change in oxidation state in transition metals, and the colour change in this reaction reveals the catalytic ability of cobalt in this reaction.

Another reaction that involves homogeneous catalysis is the oxidation of iodide ions by peroxodisulphate(VI) ions.

$$S_2O_8^{2-} + 2I^- \rightarrow I_2 + 2SO_4^{2-}$$

The reaction is quite slow even though it is energetically favourable. Both ions are negatively charged, so they are unlikely to make fruitful collisions with each other. However, if iron(II) ions are added to the reaction, the rate is much quicker. The positive Fe^{2+} ions will make effective collisions with the negative peroxodisulphate ions; and the iron(II) is oxidised to iron(III), which will rapidly oxidise the iodide ions to iodine. In this redox reaction the iron(III) is reduced to iron(II). This will then continue the catalysis.

$$2Fe^{2+} + S_2O_8^{2-} \rightarrow 2Fe^{3+} + 2SO_4^{2-}$$
$$\underline{2Fe^{3+} + 2I^- \rightarrow 2Fe^{2+} + I_2}$$
$$S_2O_8^{2-} + 2I^- \rightarrow I_2 + 2SO_4^{2-}$$

The Fe^{2+} ions do not appear in the final equation because they are catalysts.

Autocatalysis

When manganate(VII) is titrated with a warmed acidified solution of ethanedioate ions $C_2O_4^{2-}$, it is quite slow initially and, for the first addition of manganate(VII), the purple colour is slow to decolorise. When more manganate(VII) is added, the solution immediately turns colourless until the endpoint is reached. The reaction is catalysed by the Mn^{2+} ion, which is formed during the reaction.

To balance the electrons, $2MnO_4^-(aq)$ will react with $5C_2O_4^{2-}(aq)$. Then adding the two half-equations and cancelling gives:

$$2MnO_4^-(aq) + 8H^+(aq) + 5C_2O_4^{2-}(aq)$$
$$\rightarrow 2Mn^{2+}(aq) + 10CO_2(g) + 4H_2O(l)$$

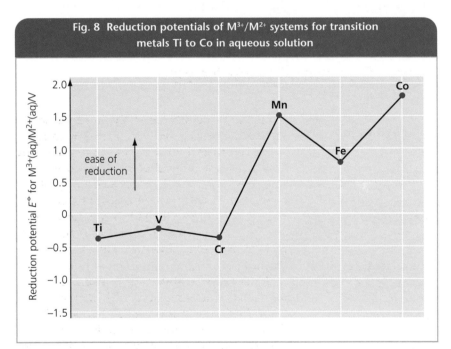

Fig. 8 Reduction potentials of M^{3+}/M^{2+} systems for transition metals Ti to Co in aqueous solution

Reduction potential $E^\ominus$ for $M^{3+}(aq)/M^{2+}(aq)$/V

ease of reduction

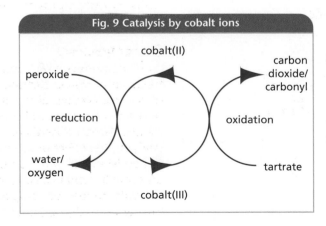

Fig. 9 Catalysis by cobalt ions

cobalt(II)

peroxide

carbon dioxide/ carbonyl

reduction

oxidation

water/ oxygen

tartrate

cobalt(III)

The reaction does not speed up until some Mn^{2+} ions have been formed. The catalysis of a reaction by one of its products is called **autocatalysis**.

This reaction is slow initially because both the oxidant and reductant are negatively charged, so they are unlikely to make fruitful collisions. However, the positive Mn^{2+} ions can react with the MnO_4^- ions to form Mn^{3+} ions.

$$4Mn^{2+} + MnO_4^- + 8H^+ \rightarrow 5Mn^{3+} + 4H_2O$$

The Mn^{3+} ions can then react with $C_2O_4^{2-}$ ions to liberate CO_2 and re-form Mn^{2+} ions, which can then continue the autocatalysis.

$$2Mn^{3+} + C_2O_4^{2-} \rightarrow 2Mn^{2+} + 2CO_2$$

9 Write the half-equation for the reaction Mn(VII) to Mn(II).
Write the half-equation for the reaction $C_2O_4^{2-}$ to $2CO_2$.
Why is the manganate(VII)/ethanedioate reaction an example of homogeneous catalysis?

KEY FACTS

- Homogeneous catalysis proceeds via an intermediate (or intermediates).
- Transition metal ions taking part in homogeneous catalysis usually go through a temporary change in oxidation state.
- Autocatalysis occurs when one of the products catalyses the reaction.

16.7 Other important transition metal complexes

Haemoglobin

As mentioned in Chapter 15, haemoglobin carries oxygen around the body. At the centre of this molecule is the haem complex which contains the Fe(II) ion. This ion will accept a pair of electrons from oxygen, which is behaving as a ligand. The oxygen is released in the cells for respiration and exchanged for carbon dioxide, which is carried back to the lungs for exhalation. Carbon monoxide is another ligand that will react with the haem molecule. The carbon monoxide molecule has a lone pair on the carbon atom, which can donate into the orbitals to the iron. Carbon monoxide forms a very strong link with the haem, stronger than the bond to oxygen. Any carbon monoxide in the blood, e.g. from smoking, will form the carboxyhaemoglobin complex and will prevent oxygen from linking with the Fe(II). This interferes with the transport of oxygen round the body and this is why carbon monoxide is toxic (Fig. 10).

Fig. 10 Formation of carboxyhaemoglobin

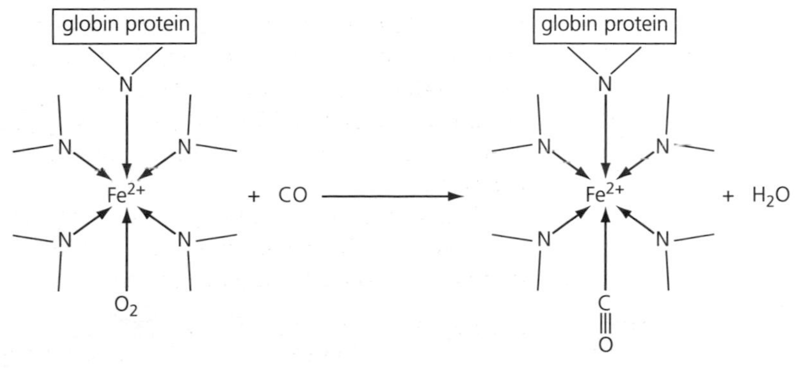

Most of the examples of complexes discussed so far have been with elements from the first transition series, Period 4, but there are some important complexes in Periods 5 and 6, especially those of silver and platinum. Platinum metal is used as a catalyst in the oxidation of ammonia by oxygen, which is a step in the manufacture of nitric acid.

Cancer treatment

One platinum complex ion has been at the forefront of cancer treatment. It was discovered accidentally, but is an example of how scientists need to be alert to unusual results. Barnett Rosenberg, a biophysicist, was studying the effect of alternating current on cell division and was testing his equipment on some bacterial cells. To his amazement, the cells stopped dividing and, after some further research, he found the platinum electrodes he was using in the electrolysis (which he thought would not react) reacted with the solution forming a complex ion. Eventually, with the help of a chemist, Andy Thompson, he found that the ion producing this effect was the square planar complex, *cis*[Pt(NH$_3$)$_2$Cl$_2$], *cis*diamminedichloroplatinum(II), more commonly called cisplatin (Fig. 11).

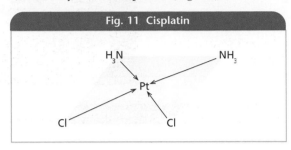

Fig. 11 Cisplatin

They then had the idea that this compound might be effective in the treatment of cancer cells, because these cells divide in an uncontrolled way. The complex exchanges its chloro ligands for the nitrogen atoms on the DNA, this distorts its structure, inhibits its function and eventually kills the cell. Cisplatin and some of its derivatives are now major drugs in cancer treatment. Although cisplatin is used as part of combined chemotherapy for a range of cancers, it has been found to be particularly effective in treating testicular and breast cancers. These are hormone-related cancers but, as yet, scientists do not understand why there should be this link.

Electrolysis

The cyano ligand is often used to complex with precious metals such as silver and gold in their extraction, their purification or their electroplating.

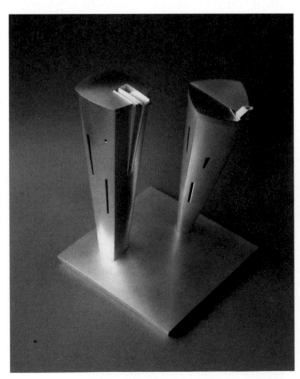

This designer silver-lined vinegar and oil set was exhibited at the V&A in London.

The metals are generally unreactive. Solutions containing Ag$^+$(aq) ions form the complex cyano ions such as dicyanoargentate(I) [Ag(CN)$_2$]$^-$ on addition of aqueous potassium cyanide. This complex is decomposed during electrolysis and is used in the electroplating of silver on to the surface of other metals. The object to be plated is made the cathode and the electrolyte contains the [Ag(CN)$_2$]$^-$ ions. The Ag$^+$ ion accepts electrons from the cathode forming metallic silver.

Test for aldehydes and ketones

The diammine complex of silver is diamminesilver(I), [Ag(NH$_3$)$_2$]$^+$. This will react with compounds that are easily oxidised to form metallic silver. This property is used in the silver mirror test to distinguish between ketones and aldehydes. Dilute ammonia is added to silver nitrate solution until the precipitate of Ag$_2$O, which is formed initially, just re-dissolves. The solution formed, called Tollens reagent, contains the soluble [Ag(NH$_3$)$_2$]$^+$ ion. When Tollens reagent is warmed with an aldehyde, the aldehyde is oxidised to an acid and the silver(I) ion is reduced to silver, which forms a silver mirror on the surface of the container. Ketones are not so easily oxidised, do not form a silver mirror on the surface of the container, and so can be readily distinguished from aldehydes.

$$CH_3CHO + H_2O + 2Ag^+ \rightarrow$$
$$CH_3COOH + 2H^+ + 2Ag$$

In the presence of aldehydes, Tollens reagent is reduced to metallic silver.

Test for silver halides

This $[Ag(NH_3)_2]^+$ complex is also responsible for the observation that a precipitate of AgCl and AgBr will dissolve in aqueous ammonia.

- AgCl will dissolve in dilute ammonia solution.
- AgBr will dissolve in concentrated ammonia solution.
- AgI will not dissolve in any ammonia solution.

This property can be used to confirm and/or distinguish between these halides.

The positive and negative image

Photography

For about 250 years scientists have known that silver halides are sensitive to light, but only when it was discovered that they would form a soluble complex with thiosulphate ions did the modern photographic process start. The photographic film is coated with an emulsion containing insoluble silver bromide. When the film is exposed to light, atoms of silver form and this process is enhanced by the developer. When the image has developed to the correct level, a solution of sodium thiosulphate is added. This forms the soluble complex ion $[Ag(S_2O_3)_2]^{3-}$, which removes any remaining light-sensitive silver bromide. This stops the development process and 'fixes' the film (Fig. 12).

$$AgBr + 2S_2O_3^{2-} \rightarrow [Ag(S_2O_3)_2]^{3-} + Br^-$$

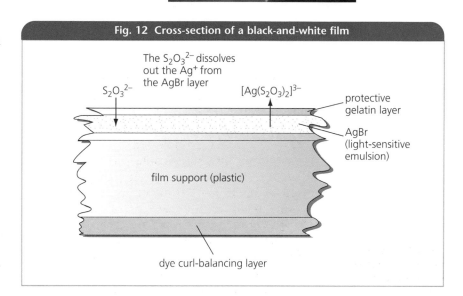

Fig. 12 Cross-section of a black-and-white film

The $S_2O_3^{2-}$ dissolves out the Ag^+ from the AgBr layer

$S_2O_3^{2-}$

$[Ag(S_2O_3)_2]^{3-}$

protective gelatin layer

AgBr (light-sensitive emulsion)

film support (plastic)

dye curl-balancing layer

APPLICATION

Catalysts

Catalysts have a wide variety of uses and are very important in an industrialised society as well as being very important for our own body's well-being. In our bodies we use biological catalysts called enzymes to make sure our bodies function properly. Iron is an important element for us, some is used in enzymes but mostly it is found in haemoglobin and is used to transport oxygen round the body. Sometimes through illness we need to increase our iron intake and it is possible to take iron supplements.

Iron is an example of one of several trace elements our bodies need to function properly. The trace elements Co, Cr, Cu and Zn are often used in conjunction with enzymes.

1 Why are Co, Cr and Cu described as transition metals and Zn is not?

Fig 13. Vitamin B$_{12}$ structure

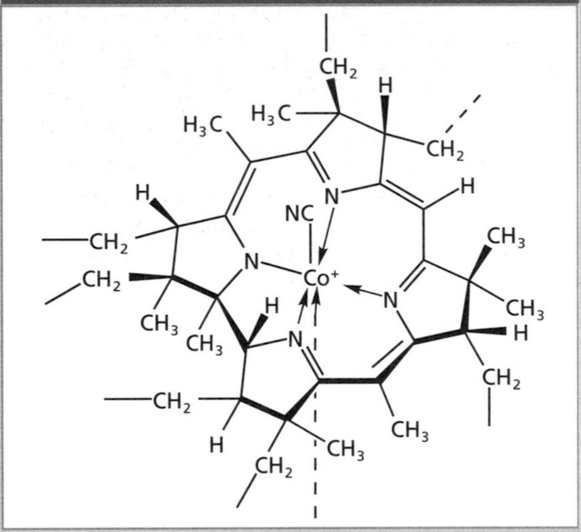

Chromium is needed for insulin production to control sugar levels and cobalt is at the centre of a porphyrin structure found in vitamin B$_{12}$ essential for building proteins for muscles and red blood cells (Fig. 13). In aqueous solution, the complex $[Cr(H_2O)_6]^{3+}$ is not normally formed.

2 Write the formula of the Cr(III) complex ion formed when an aqua ligand has been replaced by a hydroxo ligand.

Acids can act as catalysts, transition metals act as catalysts, and enzymes are biological catalysts.
Fats are made from glycerol $CH_2OHCH(OH)CH_2OH$ and three molecules of a fatty acid RCOOH. H$^+$(aq) catalyses this reaction by protonating the oxygen in the OH bond of the acid, and at a later stage H$^+$(aq) is released again.

$$
\begin{array}{llll}
CH_2OH + & HOOCR & & CH_2OOCR \\
| & & & | \\
CHOH + & HOOCR & \rightarrow & CHOOCR \\
| & & & | \\
CH_2OH + & HOOCR & & CH_2OOCR \\
\text{Glycerol} & \text{fatty acids} & & \text{fat}
\end{array}
$$

3 What is a catalyst?
4 Why is H$^+$(aq) considered to be (a) a catalyst (b) a homogeneous catalyst?

Food production

The food industry makes significant use of catalysts, especially nickel. This is particularly used for hydrogenating oils to make fats. Often, these catalysts catalyse reactions which stop at a very precise stage of hydrogenation, before taste or texture, or both, are ruined.

Vegetable oils contain long chains of carbon–carbon double bonds (unsaturated groups), and oils contain a higher proportion than fats, and are healthier. Vegetable products are generally healthier than animal fats. Oils can be hydrogenated to make them solids for use in foods such as chocolate by converting the double bonds to single bonds. The catalyst is very finely divided nickel, deposited on small particles of an inert carrier material, which can be filtered out.

5 Why is nickel called a heterogeneous catalyst?
6 Some chocolate is made from hydrogenated vegetable oils. The catalyst is nickel and it is mixed with the oil in the reactor. Why is it important that the nickel is finely divided?

Iron is used as an industrial catalyst in the Haber process $N_2(g) + 3H_2(g) \rightarrow 2NH_3(g)$

7 Explain why tungsten and silver are not used as catalysts.
8 Impurities such as carbon monoxide can poison the catalyst.
What does poisoning mean?
Explain how impurities poison the catalyst.

APPLICATION

Enzymes as catalysts

Enzymes usually catalyse a specific reaction, lipase breaks down fat but not carbohydrate.

9 Use Fig. 14 to explain how enzymes are so specific in their reactions. Use Fig. 15 to explain why enzymes are only effective over a narrow temperature range.

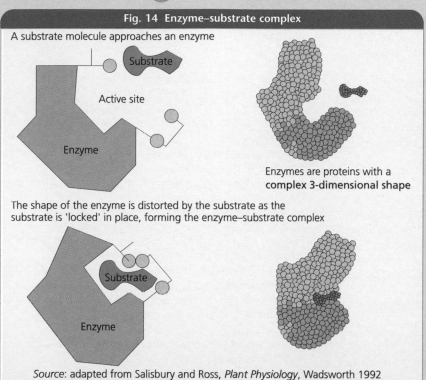

Fig. 14 Enzyme–substrate complex

A substrate molecule approaches an enzyme

Active site

Substrate

Enzyme

Enzymes are proteins with a **complex 3-dimensional shape**

The shape of the enzyme is distorted by the substrate as the substrate is 'locked' in place, forming the enzyme–substrate complex

Substrate

Enzyme

Source: adapted from Salisbury and Ross, *Plant Physiology*, Wadsworth 1992

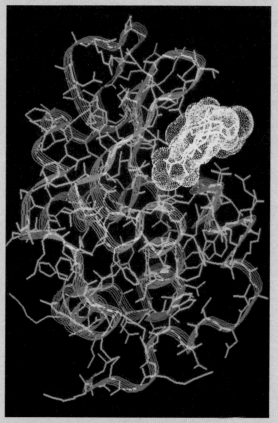

Computer graphic of lysozyme, an important catalyst in the breaking down of sugar structures in the body. The lysozyme (blue) fits with the substrate molecule (yellow) at the active site. The 'backbone' of the protein is shown in pink.

Fig. 15 Relationship between reaction rate and temperature for enzyme catalysts

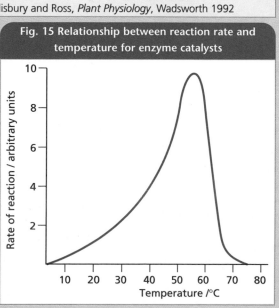

1

a i) On the axes below, draw a Maxwell–Boltzmann distribution curve for the molecular energies in a gas at fixed temperature. E_a is the activation energy for a chemical reaction in this gas.

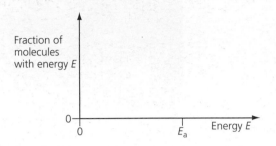

ii) Indicate on the graph the proportion of molecules which can react.

iii) A catalyst which increases the rate of reaction is added to the gas. What effect does this have on the Maxwell–Boltzmann distribution curve and on the proportion of molecules which can react? (6)

b The catalyst added is a heterogeneous catalyst.

i) Explain the term *heterogeneous*.

ii) Give an example of a heterogeneous catalyst and name a process in which it is used. (3)

c When transition metals are used as catalysts, some are more effective than others. Give **two** examples of transition metals that, for different reasons, are poor catalysts and, in each case, give an explanation for their poor catalytic activity. (6)

NEAB March 1998 CH04 Q1

2 Chemical reactions can be affected by homogeneous or by heterogeneous catalysts.

a Explain what is meant by the term *homogeneous* and suggest the most important feature in the mechanism of this type of catalysis when carried out by a transition–metal compound. (2)

b In aqueous solution, $S_2O_8^{2-}$ ions can be reduced to SO_4^{2-} ions by I^- ions.

i) Write an equation for this reaction.

ii) Suggest why the reaction has a high activation energy, making it slow in the absence of a catalyst.

iii) Iron salts can catalyse this reaction. Write two equations to show the role of the catalyst in this reaction. (4)

c Left is a sketch showing typical catalytic efficiencies of transition metals from Period 5 (Rb to Xe) and Period 6 (Cs to Rn) when used in heterogeneous catalysis.

Catalytic efficiency — Period 6 — Period 5 — Successive transition metals

i) Identify two metals which lie at opposite ends of these curves and explain why they show rather low catalytic efficiency.

ii) Suggest why these curves pass through a maximum. (5)

d In catalytic converters which clean up petrol engine exhaust gases, a catalyst promotes the reduction of nitrogen oxides using another polluting gas as reductant. State a suitable catalyst for this task, identify the reductant, and write an equation for the reaction that results. (3)

NEAB June 1998 CH04 Q1

3

a An aqueous solution contained both $Fe^{2+}(aq)$ and $Fe^{3+}(aq)$. 25.0 cm^3 of this solution were acidified with dilute sulphuric acid and titrated with a 0.0100 M solution of potassium manganate(VII). 15.0 cm^3 were required. The equation for the reaction occurring is given below.

$$5Fe^{2+} + 8H^+ + MnO_4^- \rightarrow 5Fe^{3+} + Mn^{2+} + 4H_2O$$

i) Explain why ethanoic acid would be an unsuitable acid for use in this analysis.

ii) Calculate the concentration of $Fe^{2+}(aq)$ in mol dm⁻³. (6)

b Another 25.0 cm^3 sample of the same solution was treated with an excess of zinc metal in the presence of dilute sulphuric acid. This converted the $Fe^{3+}(aq)$ into $Fe^{2+}(aq)$ and the zinc was converted into $Zn^{2+}(aq)$. After filtration to remove excess zinc, the solution was titrated as above. It now reacted with 17.6 cm^3 of the potassium manganate(VII) solution.

i) Write the equation for the reaction between $Fe^{3+}(aq)$ and zinc.

ii) Explain why it is necessary to remove excess of zinc from the solution. (2)

c Use the information given in parts a) and b) above to calculate the percentage of iron present as $Fe^{2+}(aq)$ in the original sample. (2)

NEAB March 1995 CH02 Q6

4

a State the difference between homogeneous and heterogeneous catalysts. (1)

b State two reasons why an inert medium is sometimes used as a support for a heterogeneous catalyst. (2)

c Soluble iron salts catalyse the following reaction:

$$2I^-(aq) + S_2O_8^{2-}(aq) \rightarrow I_2(aq) + 2SO_4^{2-}(aq)$$

Suggest a reason why the reaction is likely to be slow in the absence of a catalyst. Write two equations to illustrate the action of the catalyst and explain why both iron(II) salts and iron(III) salts can catalyse this reaction. (4)

EXAMINATION QUESTIONS

d i) State three important features in the mechanism of the heterogeneous catalysis.
ii) Write equations for **two** reactions which undergo heterogeneous catalysis by a transition metal or one of its compounds. (5)
AQA March 1999 CH04 Q2

5

a Describe an experiment to show that vanadium has several oxidation states.

b Describe the essential features of catalysis and explain what is meant by the term *heterogeneous catalyst*. State an industrial process which uses vanadium(V) oxide as a heterogeneous catalyst and explain, with the help of equations, the mode of action of the catalyst.
AQA June 2000 CH05 Q5

6

a Transition metals and their compounds can act as heterogeneous catalysts. Explain what is meant by both *heterogeneous* and by *catalyst*. (2)

b State one feature of transition metals which makes them able to act as catalysts. (1)

c Write an equation for the reaction which is heterogeneously catalysed by a transition metal or one of its compounds. State the catalyst used. (2)

d When an acidified solution of ethanedioate ions is titrated with a solution of potassium manganate(VII), the colour of the manganate(VII) ions disappears slowly at first but then more rapidly as the end-point is reached. Explain this observation. (3)
NEAB February 1997 CH05 Q2

7

a A blue solution is formed when chromium is dissolved in hydrochloric acid in the absence of air. This solution turns green in the presence of air. From your knowledge of the reduction of dichromate(VI) ions by zinc in the presence of hydrochloric acid, deduce the formulae of the chromium species present in the blue and in the green solutions.

b A 0.500 g sample of steel was dissolved in hydrochloric acid to produce a solution of iron(II) ions. When this was titrated with 0.0500 M potassium dichromate(VI) solution, 29.7 cm^3 were required to oxidise all the iron(II) to iron(III).

i) Write the equation for the reaction between iron dichromate(VI) ions in acid solution.

ii) Calculate the percentage, by mass, of iron in the steel.
NEAB February 1997 CH05 Q1(b,c)

8

a Starting from chromium(III) sulphate in both cases, describe briefly how you would obtain a solution containing chromate(VI) ions and also how you would obtain a solution containing chromium(II) ions. (Equations are not required.)

b In order to determine the chromium content of a sample of the industrially prepared metal, all the chromium in the sample was converted into aqueous dichromate(VI) ions. The amount of chromium present as dichromate(VI) was then determined by a titration involving a solution of iron(II) ions. It was found that a 0.106 g sample of this chromium gave a solution of dichromate(VI) ions which was equivalent to 30.0 cm^3 of a 0.200 M iron(II) sulphate solution. Calculate the percentage by mass of chromium in the sample. (15)
NEAB June 1997 CH05 Q7(b)

All living things need an energy source for all their life processes, and to build molecules such as proteins in order to grow healthily. The main elements required are carbon, nitrogen, hydrogen and oxygen. Both humans and plants need these for energy, growth and repair. The plants get carbon and oxygen from the air and water from the soil. But all living things need trace metal elements as well.

Plants need magnesium, which is the atom at the centre of the chlorophyll molecule, vital for photosynthesis. Chlorophyll (Fig. 1) has many similarities to the haemoglobin molecule (Chapter 15, Fig. 9), which we need for transporting oxygen around the body.

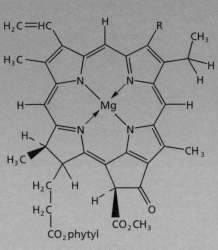

$$\left[phytyl = C_{29}H_{39} \right]$$

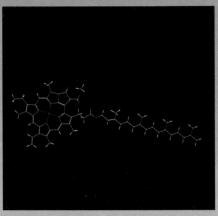

Computer graphic of a chlorophyll molecule; the magnesium atom is at the centre of a porphyrin ring structure and is bonded to it via four nitrogen atoms (see the diagram).

Chlorophyll structure

As a complex molecule, chlorophyll illustrates how atoms (in this case, nitrogen) with lone pairs of electrons can donate into the empty sub-shells available on the central atom, forming co-ordinate bonds. Two of the nitrogen atoms around the magnesium do this: the bonds are represented by arrows and the two nitrogen atoms are called ligands.

As well as magnesium, plants need iron. This element transfers an electron as part of the energy exchange from light to chemical processes. Iron does this by changing its oxidation state.

To carry out all the cellular processes, plants need other elements such as cobalt and manganese. All these elements are available if the soil (or growing medium) is fertilised properly. Because the plants need to extract the nutrients from the soil, equilibria are set up between the aqueous solutions in the plant and the aqueous solutions in the soil. The position of these equilibria will depend upon the ligands attached to the central metal ions and, very importantly, to the pH of the soil.

Changing the soil pH will affect any water ligands attached to the metal ions and this will affect how well the plant can take up the nutrients. An acid soil makes the complex ions more soluble, and they will be washed away by rain water.

Some of the reactions of photosynthesis rely on molecules which contain a central atom of either magnesium or iron. Other biochemical reactions require cobalt, manganese and other transition metal ions in trace quantities. Ligands bond with all these elements.

Growers often **sequester** the nutrients in fertiliser, which means the complex ion formed is very stable. This slows down the release of the nutrients to the plant, but the advantage is that it is not very soluble so that it is not washed away.

In this chapter you will see how transition metal ions behave in different aqueous solutions, and how they interact in solution with different ligands.

17.1   Aqueous reactions of inorganic compounds

The important properties of the transition and other metal compounds are affected by the ligands which attach themselves to the central metal ion when a complex ion forms. A complex ion is formed when a ligand donates a pair of electrons. This leads to complexes being described in terms of acids and bases.

In 1938, G. N. Lewis defined acids and bases as follows:

- A **Lewis acid** is an electron pair acceptor (see Fig. 1).
- A **Lewis base** is an electron pair donor.

Fig. 1 Donation of a lone pair from H_2O to HCl

$$\begin{array}{ccc} H & & H \\ \overset{|}{\underset{|}{\text{O}}} \cdots H{-}Cl & \longrightarrow & \left[\begin{array}{c} H \quad H \\ \overset{|}{\underset{|}{\text{O}}} \\ H \end{array} \right]^{+} + Cl^{-} \end{array}$$

1   Which are the Lewis acids and the Lewis bases in the following reactions?
a $H_2O + H^+ \rightarrow H_3O^+$
b $BF_3 + NH_3 \rightarrow F_3B{\leftarrow}NH_3$

When a Lewis acid reacts with a Lewis base, a co-ordinate bond is formed between the two species and this is represented by an arrow. The arrow goes from the donor to the acceptor. The donation of lone pairs means that a Lewis base is a ligand and a nucleophile (see Chapter 3).

In the dicyanoargentate(l) ion $[Ag(CN)_2]^-$,

$$[NC{\rightarrow}Ag{\leftarrow}CN]^-$$

the silver ion is accepting two electron pairs (one pair from each cyanide ion) and is acting as a Lewis acid. Each cyanide ligand is thus acting as a Lewis base. In all metal complexes, the central metal atom is a Lewis acid and all the ligands are Lewis bases.

2   Which of the following are Lewis acids?
i SO_3,   **ii** HCl,   **iii** H_2SO_4,   **iv** Ag^+,
v C_2H_5COOH,   **vi** Fe^{2+}.

17.2   Acidity reactions of metal ions in water

The acid–base definitions given above can be used to describe metal ions in water.

Oxidation state +2

When a transition metal ion is in water, the lone electron pairs on the water molecules (Lewis base) form co-ordinate bonds with the metal ions (Lewis acid). Complex ions are produced. Often, there is a striking colour change if the ligand is changed. Anhydrous cobalt chloride is blue, but the hexaaquacobalt(II) ion is pink.

$$CoCl_2 + 6H_2O \rightarrow [Co(H_2O)_6]^{2+} + 2Cl^-$$
$$\text{blue} \qquad\qquad\qquad \text{pink}$$

Copper(II) and iron(II) form similar aqua ions. Anhydrous copper(II) sulphate is white, but $[Cu(H_2O)_6]^{2+}$ is blue (see photo on page 240). Anhydrous iron(II) sulphate is almost white, but $[Fe(H_2O)_6]^{2+}$ is green.

Blue cobalt chloride paper before and after the addition of water

When water is added to white anhydrous $CuSO_4$, blue $[Cu(H_2O)_6]^{2+}$ is formed.

The reaction in which aqua ions are formed is called **hydration**. The hydration enthalpy of the ions present in the solid usually provides enough energy to break down the crystal lattice (to overcome the lattice energy), so the solid is water soluble. The mixture gets very hot.

The co-ordinate bond between the metal ion and the water can be strong enough to survive evaporation. For example, $FeSO_4.7H_2O$ is green and contains the complex $[Fe(H_2O)_6]^{2+}$ formed when a warm, saturated solution of iron(II) sulphate is cooled. Many metal ion complexes are surrounded by six ligands, i.e. they have a co-ordination number of 6. These ions will have an octahedral shape. Other hydrated salts of transition metals are:

- $CoCl_2.6H_2O$ which is pink and contains the complex $[Co(H_2O)_6]^{2+}$
- $Fe(NO_3)_3.9H_2O$ which is pale violet and contains the complex $[Fe(H_2O)_6]^{3+}$

Oxidation state +3

Metal(III) ions form similar aqua complexes. Anhydrous aluminium chloride, for example, dissolves in water to produce hexaaquaaluminium(III) ions:

$$AlCl_3 + 6H_2O \rightarrow [Al(H_2O)_6]^{3+} + 3Cl^-$$

When the oxidation state of the metal forming the complex is +3, the reaction is often very exothermic. Anhydrous aluminium chloride reacts quite violently with water, giving out heat and fuming strongly.

3 State the shapes and write the formulae for the metal-aqua complex ions formed with:
a vanadium(II);
b chromium(III);
c iron(III).

17.3 Reactions involving complex ions

We have seen in Chapter 16 that transition metals can take part in redox reactions by gain or loss of electrons and this can cause a colour change. The electron transfer can take place *without a ligand change*.

$$[M(H_2O)_6]^{2+} \rightarrow [M(H_2O)_6]^{3+} + e^-$$

If the co-ordinate bond between the metal ion and the water molecule is broken and replaced by another ligand, this is a **ligand substitution reaction**.

4 Why do these reactions result in a colour change?

If the O–H bond in a co-ordinated water molecule is broken, the reaction is called a hydrolysis or **acidity reaction**. All metal–aqua complex ions formed by metal ions in an oxidation state of +3 can take part in hydrolysis (acidity) reactions, and change the pH of the water. In water, the following equilibria are established:

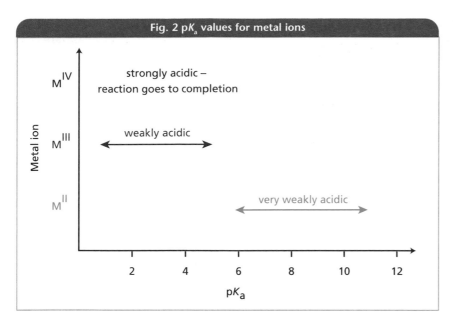

Fig. 2 pK_a values for metal ions

M^{IV}

M^{III}

M^{II}

Metal ion

strongly acidic – reaction goes to completion

weakly acidic

very weakly acidic

2   4   6   8   10   12

pK_a

$$[M(H_2O)_6]^{2+} + H_2O \rightleftharpoons [M(H_2O)_5(OH)]^+ + H_3O^+$$
or
$$[M(H_2O)_6]^{3+} + H_2O \rightleftharpoons [M(H_2O)_5(OH)]^{2+} + H_3O^+$$

This is called a hydrolysis reaction because the ligand water molecule has split into H$^+$ and OH$^-$ ions and it is an acidity reaction because H$_3$O$^+$ ions are formed as a result.

The degree of acidity of the solution depends on how many H$_3$O$^+$ ions are formed. This equilibrium can be quantified by an equilibrium constant K_a, the **dissociation** or **acidity constant** (Chapter 3). For the first of these equilibria (see Chapter 3):

$$K_a = \frac{[[M(H_2O)_5(OH)]^+] \,\{H_3O^+\}}{[[M(H_2O)_6]^{2+}] \,[H_2O]}$$

This expression can be simplified:

$$K_a \approx \frac{[H^+]^2}{[[M(H_2O)_6]^{2+}]}$$

Usually, it is easier to work with pK_a than with K_a, where pK_a is given by the expression:

$$pK_a = -\log_{10} K_a$$

The pH of a weakly acidic solution can be calculated using the equations for K_a and pH:

$$[H^+]^2 = [[M(H_2O)_6]^{2+}] \times K_a$$
and $-\log_{10} [H^+] = $ pH

See Chapter 3 for more details.

For metal(II) ions, K_a varies between 10^{-6} and 10^{-11}, so pK_a varies between 6 and 11. For metal(III) ions the equilibrium:

$$[M(H_2O)_6]^{3+} + H_2O$$
$$\rightleftharpoons [M(H_2O)_5(OH)]^{2+} + H_3O^+$$

lies much further to the right, and K_a varies between 10^{-2} and 10^{-5}.
(pK_a = 2 to 5)

For metal(IV) ions, the hydrolysis equilibrium lies so much further to the right that the reaction goes virtually to completion: the pK_a < 1. pK_a values for metal ions are shown in Fig. 2.

These equations represent the acidity produced by the exchange of one aqua ligand. You should be aware that, if the complex ion has six ligands, there will be an equilibrium for each aqua ligand being replaced. There is an overall equilibrium constant, which will be the sum of all these successive, individual equilibria, and this dictates the overall acidity of the solution. The equilibrium constant for each successive removal is lower than for the previous step, and since the K_a for subsequent equilibria will be so small, the steps may be insignificant.

Generally, the pH of metal ion solutions is:

$[M(H_2O)_6]^{2+}$	$[M(H_2O)_6]^{3+}$
very weak acids	weak acids
pH ≈ 6	pH ≈ 3

17.4 Effect of charge density on K_a values

The features of the metal ions that affect degree of acidity in solutions are:

• the charge on the metal ion (Fig. 3);
• the size of the metal ion.

Aluminium ions, Al^{3+}, are small, the combination of a high charge and a small radius means that Al^{3+} ions have a high

charge density (see Fig. 4). In the $[Al(H_2O)_6]^{3+}$ complex, the Al^{3+} ions attract the oxygen electrons very strongly:

electrons are attracted towards the Al^{3+} part of the aluminium–aqua ion

241

The attraction of the metal ion particle for the electrons in a co-ordinated water molecule is called its polarising power, the ion distorts or polarises the water ligand's electron cloud. This attraction of 2+ and 3+ ions weakens the O–H bonds so that H^+ ions can break away to form H_3O^+.

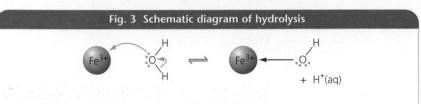

Fig. 3 Schematic diagram of hydrolysis

$$Fe^{3+} \quad \rightleftharpoons \quad Fe^{3+} \qquad + H^+(aq)$$

The highly charged cation attracts electrons from the water molecule, weakening an O–H bond. H^+ is released and becomes H_3O^+ with water, increasing the acidity.

When we compare a 2+ ion to a 3+ ion, we can see that the 3+ ion has a greater charge density and therefore a greater polarising power. The greater attraction for oxygen weakens the O–H bond more, so more OH bonds break and more H^+ ions are released. This results in a greater hydrogen ion concentration $[H^+(aq)]$ and so a lower pH.

The presence of aluminium ions in solution gives a pH usually between 3 and 6 (slightly acidic depending on the concentration of $Al^{3+}(aq)$ ions).

The 3+ charge on the ion is the difference between the number of protons and the number of electrons. For the elements Na^+, Mg^{2+} and Al^{3+}, the nuclear charge is increasing, the number of electrons is constant and the size of the ions is decreasing (see Fig. 4).

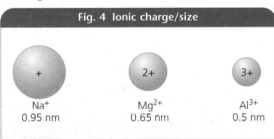

Fig. 4 Ionic charge/size

+	2+	3+
Na^+	Mg^{2+}	Al^{3+}
0.95 nm	0.65 nm	0.5 nm

For simplicity, we just refer to the charge difference, so we talk about a 3+ ion for aluminium. When combined with size, this

gives us the idea of charge density (ratio of charge : ionic radius). The ionic charge is more significant than the size.

5
a A solution contains chromium(III) ions. Write an equation to show the first equilibrium of Cr^{3+} in water.
b What pH value would you expect for this system?
c Explain your reasons for the pH you have chosen

Even for 3+ ions, the first equilibrium lies well over to the left:

$$[M(H_2O)_6]^{3+} + H_2O \rightleftharpoons [M(H_2O)_5(OH)]^{2+} + H_3O^+$$

If the pH of this solution is 3, then $[H_3O^+]$ is 10^{-3}, so the ratio of:

$$[M(H_2O)_6]^{3+} : [M(H_2O)_5(OH)]^{2+} \text{ is } 1000 : 1$$

For 2+ ions, the ratio is about 100 000 : 1.

If these salts are added to water, they will dissolve, but precipitates will not normally form. To produce a precipitate, an alkali such as sodium hydroxide must be added.

If an alkali, such as aqueous sodium hydroxide, is added to a solution of hexaaqua ions, H_3O^+ ions are removed from the solution and the equilibrium moves to the right, to replace the H_3O^+ ions removed. If an acid is added, the equilibrium will move to the left, to remove H_3O^+ ions from the system, and more metal–hexaaqua ions are produced.

Crystals of $Fe(NO_3)_3.9H_2O$ are pale violet in colour. When they are added to water, the $[Fe(H_2O_6)]^{3+}$ ions present in the violet crystals react with the water to give hydrolysis products:

$$[Fe(H_2O)_6]^{3+} + H_2O \rightleftharpoons [Fe(H_2O)_5(OH)]^{2+} + H_3O^+$$
pale violet   brown

Although the concentration of $[Fe(H_2O)_5(OH)]^{2+}$ ions is very low and $[Fe(H_2O)_6]^{3+}$ is still the main species, the intensity of the brown colour is so great compared with the pale violet of $[Fe(H_2O)_6]^{3+}$ that the solution appears brown.

If nitric acid is added to the brown solution, the equilibrium will shift to the left and the brown colour will disappear. If an alkali is added, the equilibrium shifts to the right and the brown colour is intensified.

Table 1 Charge: size ratios			
Element	**Charge**	**Ionic radius**	**Charge/size ratio**
Na	1+	0.95	1.1
Mg	2+	0.65	4.7
Al	3+	0.5	6.0

■ 2+ and 3+ metal-aqua ions take part in acidity or hydrolysis reactions.

■ The pH of a solution of M(III) aqua ions is lower than the pH of a solution of M(II) aqua ions of the same concentration.

■ The acidity of solutions of metal ions is decided by the charge and size of the metal ion.

17.5 Reactions of alkalis with aqua ions

Sodium hydroxide (strong base)

If a stronger base than water is added to a solution of the transition metal salt, the equilibrium will move even further to the right and an insoluble, neutral metal hydroxide can be formed. All transition metal salts form a metal hydroxide precipitate when aqueous sodium hydroxide is added to a solution containing the transition metal ion mixed with an alkali. The precipitate can be removed by filtration. Alkalis can be used to precipitate metal hydroxides. When sodium hydroxide is added to water containing copper(II) ions, the following reactions take place:

$$[Cu(H_2O)_6]^{2+} + H_2O \rightleftharpoons [Cu(H_2O)_5(OH)]^+ + H_3O^+$$

In water, this equilibrium lies well over to the left, but on addition of alkali the equilibrium is displaced.
The hydroxide ions react with H_3O^+ ions:

$$OH^- + H_3O^+ \rightarrow 2H_2O$$

This removes H_3O^+ ions and moves the equilibrium completely to the right.
If more OH^- is added, the second equilibrium:

$$[Cu(H_2O)_5(OH)]^+ + H_2O \rightleftharpoons [Cu(H_2O)_4(OH)_2] + H_3O^+$$

is also displaced to the right by OH^- ions reacting with the H_3O^+ formed and $[Cu(H_2O)_4(OH)_2]$, blue copper(II) hydroxide, is precipitated. We can write the overall equation as:

$$[Cu(H_2O)_6]^{2+} + 2OH^- \rightleftharpoons [Cu(OH)_2(H_2O)_4] + 2H_2O$$

This equilibrium can be reversed by adding a strong acid, such as nitric acid, to the metal hydroxide so that metal–aqua ions are formed again.

Ca^{2+} Mg^{2+} Cu^{2+} Fe^{2+} Fe^{3+} Co^{2+} Ni^{2+} Mn^{2+} Cr^{3+} Ag^+ Zn^{2+} Pb^{2+} Al^{3+}

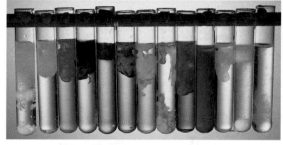

Aqueous sodium hydroxide can be used in qualitative analysis, because it gives characteristic coloured solutions and precipitates (lower) with aqueous metal ions.

Most metal hydroxides also react with excess hydroxide, OH^-, to form anionic complexes, i.e. negatively charged. Some, however, require a solution of sodium hydroxide which is more concentrated than is normally used as a laboratory solution. When sodium hydroxide solution is added to an aqueous solution of an aluminium salt, for instance, a white precipitate of aluminium hydroxide $[Al(H_2O)_3(OH)_3]$ is formed initially. As more sodium hydroxide is added, the precipitate dissolves to give a colourless solution containing tetrahydroxoaluminate(III) ions $[Al(OH)_4]^-$:

$$[Al(H_2O)_6]^{3+} + 3OH^- \rightleftharpoons [Al(H_2O)_3(OH)_3] + 3H_2O$$
white precipitate

$$[Al(H_2O)_3(OH)_3] + OH^- \rightleftharpoons [Al(OH)_4]^- + 3H_2O$$
colourless solution

When dilute acid is added to $[Al(OH)_4]^-$, first there is a white precipitate of $[Al(H_2O)_3(OH)_3]$ and then this re-dissolves to give a colourless solution of $[Al(H_2O)_6]^{3+}$.

Metal hydroxides that can react with both acids and alkalis are called **amphoteric**. The overall system is like this:

$$[Al(H_2O)_6]^{3+} \rightleftharpoons [Al(H_2O)_3(OH)_3] \rightleftharpoons [Al(OH)_4]^-$$
$$\text{acidic} \qquad\qquad \text{neutral} \qquad\qquad \text{alkaline}$$

Chromium(III) gives a similar set of reactions.

When sodium hydroxide solution is added to an aqueous solution of a chromium salt, a light green precipitate of chromium(III) hydroxide $[Cr(H_2O)_3(OH)_3]$ is formed initially.

As more sodium hydroxide is added, the precipitate dissolves to give a green solution of chromate(III) ions. In excess alkali, the species is hexahydroxochromate(III) ions $[Cr(OH)_6]^{3-}$.

$$[Cr(H_2O)_6]^{3+} + 3OH^- \rightleftharpoons [Cr(H_2O)_3(OH)_3] + 3H_2O$$
$$\text{light green precipitate}$$

$$[Cr(H_2O)_3(OH)_3] + 3OH^- \rightleftharpoons [Cr(OH)_6]^{3-} + 3H_2O$$
$$\text{green solution}$$

If dilute hydrochloric acid is added to $[Cr(OH)_6]^{3-}$, first there is a light green precipitate of $[Cr(H_2O)_3(OH)_3]$ and then this re-dissolves to give a dark green solution of $[Cr(H_2O)_4Cl_2]^-$.

Chromate(VI)/dichromate(VI) is another example of an equilibrium affected by the addition of acid or alkali:

$$2CrO_4^- \text{ (aq)} + 2H^+\text{(aq)} \rightleftharpoons Cr_2O_7^{2-}\text{(aq)} + H_2O\text{(l)}$$
$$\text{yellow} \qquad\qquad\qquad\qquad \text{orange}$$

Ammonia solution (weak base)

Similar reactions occur if aqueous ammonia is added to metal–aqua ions. This is because ammonia solution is alkaline on account of the equilibrium:

$$NH_3\text{(aq)} + H_2O\text{(l)} \rightleftharpoons NH_4^+\text{(aq)} + OH^-\text{(aq)}$$

The hydroxide ions that are formed in the equilibrium precipitate the metal as a hydroxide. If ammonia is in excess, most transition metal hydroxides react to form soluble ammine complexes. However, aluminium does not form an ammine complex, so it is considered insoluble in aqueous ammonia (although it is very slightly soluble in a large excess of ammonia). The substitution of the water ligand by ammonia in transition metal complex ions also leads to a colour change (see section 15.3).

6 Consider the equilibrium:

$$[Fe(H_2O)_6]^{3+} + H_2O \rightleftharpoons [Fe(H_2O)_5(OH)]^{2+} + H_3O^+$$
$$\text{pale violet} \qquad\qquad\qquad\qquad \text{brown}$$

A series of changes take place on the addition of different solutions to this equilibrium. Explain the changes that take place, using equations in each case.

a Adding dilute hydrochloric acid changes the colour of the above solution from brown to pale violet.

b Adding dilute sodium hydroxide solution changes the colour of the solution from pale violet to brown.

c Adding excess sodium hydroxide solution gives a brown precipitate, which is insoluble in 2 mol dm^{-3} sodium hydroxide solution.

KEY FACTS

■ Adding strong alkalis to transition metal salts gives hydroxide precipitates.

■ Metals ions in water can be removed by adding a chemical that will react to form a precipitate.

■ Metal hydroxides that can react with acids and alkalis are amphoteric.

■ Adding acids or alkalis to solutions containing hydrolysis equilibria shifts the equilibria.

■ Metal(III) ions in excess of strong alkali give soluble complex ions containing the hydroxo ligand.

17.6 Reactions of carbonate with aqua ions

Carbonates, such as sodium carbonate, can also be used to precipitate metal hydroxides from solution. Different reactions are possible, depending on the oxidation state of the metal.

Oxidation state +2

Metal(II)–aqua ions hydrolyse as:

$$[M(H_2O)_6]^{2+} + H_2O \rightleftharpoons [M(H_2O)_5(OH)]^+ + H_3O^+$$

If carbonate ions are present and the concentration of H^+ is sufficient, the following equilibria move to the right and the carbonate ions form carbon dioxide, which is evolved as a gas:

$$CO_3^{2-} + H_3O^+ \rightleftharpoons HCO_3^- + H_2O$$

$$HCO_3^- + H_3O^+ \rightleftharpoons CO_2 + 2H_2O$$

However, metal(II) ions produce only a very weakly acidic solution, so there are very few H_3O^+ ions in the solution. In the presence of the carbonate ions added, there will not be sufficient concentration of H_3O^+ to release carbon dioxide gas. Since metal(II) carbonates are not very soluble in water, they form precipitates. Some important examples are:

$$[Fe(H_2O)_6]^{2+} + CO_3^{2-} \rightleftharpoons FeCO_3 + 6H_2O$$
pale green solution   green precipitate

$$[Co(H_2O)_6]^{2+} + CO_3^{2-} \rightleftharpoons CoCO_3 + 6H_2O$$
pink solution   pink precipitate

$$[Cu(H_2O)_6]^{2+} + CO_3^{2-} \rightleftharpoons CuCO_3 + 6H_2O$$
blue solution   green-blue precipitate

Oxidation state +3

The metal(III) ions hydrolyse in much the same way as the metal(II) ions, but to a much greater extent. An ion of charge 3+ is much more polarising than 2+, so has a greater attraction for the metal oxygen electrons in the co-ordinate bond. This weakens the O–H bond in the water ligand, so more H_3O^+ ions are formed (Fig. 5).

Now there are sufficient H_3O^+ ions to react with the carbonate ions and form carbon dioxide:

$$2H_3O^+ + CO_3^{2-} \rightarrow CO_2 + 3H_2O$$

This removes the H_3O^+ ions and the equilibrium moves to the right. Eventually, the net equilibrium is set up:

$$[M(H_2O)_6]^{3+} + H_2O \rightleftharpoons [M(H_2O)_5(OH)]^{2+} + H_3O^+$$

then:

$$[M(H_2O)_5(OH)]^{2+} + H_2O \rightleftharpoons [M(H_2O)_4(OH)_2]^+ + H_3O^+$$

then:

$$[M(H_2O)_4(OH)_2]^+ + H_2O \rightleftharpoons [M(H_2O)_3(OH)_3] + H_3O^+$$

The metal hydroxide rather than the metal carbonate forms as a precipitate. The overall equation is:

$$2[M(H_2O)_6]^{3+} + 3CO_3^{2-} \rightarrow 2[M(H_2O)_3(OH)_3] + 3CO_2 + 3H_2O$$

Fig. 5 Fe^{3+} is strongly polarising: it weakens the O–H bond, releasing H^+ to form H_3O^+

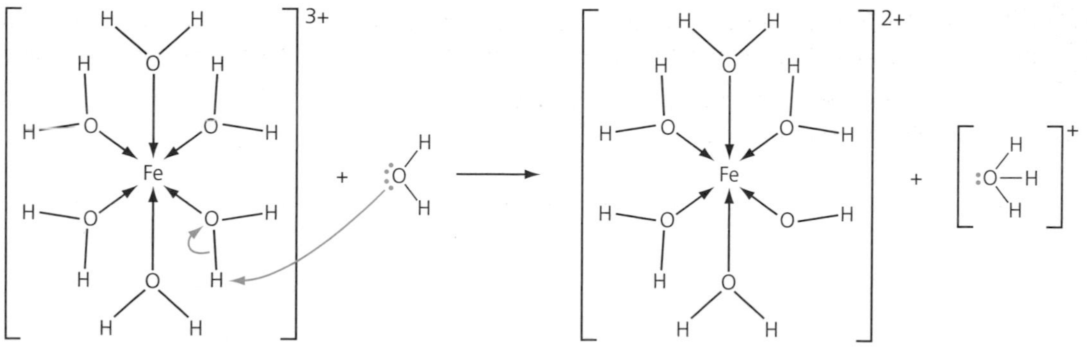

For chromium(III), the following reactions illustrate the above:

$$[Cr(H_2O)_6]^{3+} + H_2O \rightleftharpoons [Cr(H_2O)_5(OH)]^{2+} + H_3O^+$$

then:

$$[Cr(H_2O)_5(OH)]^{2+} + H_2O \rightleftharpoons [Cr(H_2O)_4(OH)_2]^+ + H_3O^+$$

then:

$$[Cr(H_2O)_4(OH)_2]^+ + H_2O \rightleftharpoons [Cr(H_2O)_3(OH)_3] + H_3O^+$$

$$2H_3O^+ + CO_3^{2-} \rightarrow CO_2 + 3H_2O$$

The metal hydroxide precipitates, carbon dioxide is evolved and the overall equation is:

$$2[Cr(H_2O)_6]^{3+} + 3CO_3^{2-} \rightarrow 2[Cr(H_2O)_3(OH)_3] + 3CO_2 + 3H_2O$$
$$\text{green precipitate}$$

No one has ever made a metal(III) carbonate from an aqueous solution. Compounds such as $Cr_2(CO_3)_3$ do not exist. This relative instability can again be explained in terms of polarising power. The 3+ ion is so strongly polarising that it would distort the charge cloud on the carbonate ion and so it forms a metal hydroxide in solution and carbon dioxide. These products have a greater overall stability than the theoretical metal(III) carbonate. The same argument would apply to the relative stabilities of the metal(III) oxide and carbon dioxide compared with the metal(III) carbonate. The carbonate ion is readily polarised but the smaller oxide ion is more difficult to polarise (Tables 2 and 3).

Table 2 Reactions of metal(II) ions with bases					
Aqueous M(II) ion Solution	Base added OH⁻ little	2 M OH⁻ excess	NH₃ little	NH₃ excess	CO₃²⁻
Mg(II) colourless	White precipitate $Mg(OH)_2$	Does not dissolve	White precipitate	Does not dissolve	White precipitate carbonate $MgCO_3$
Mn(II) pale pink $[Mn(H_2O)_6]^{2+}$	White precipitate $Mn(OH)_2$ Rapidly oxidised by air: turns brown	Does not dissolve	White precipitate Rapidly oxidised by air: turns brown		Forms a carbonate $MnCO_3$
Fe(II) green $[Fe(H_2O)_6]^{2+}$	Green precipitate $Fe(OH)_2$	Does not dissolve	Green precipitate Easily oxidised by air: turns brown	Turns brown in air.	Green precipitate carbonate $FeCO_3$
Co(II) pink $[Co(H_2O)_6]^{2+}$	Blue-green ppt. $Co(OH)_2$ Easily oxidised by air: turns brown	Does not dissolve	Blue precipitate Easily oxidised by air: turns brown	Pale straw solution: Turns dark brown . in air	Pink precipitate carbonate $CoCO_3$
Ni(II) green $[Ni(H_2O)_6]^{2+}$	Green precipitate $Ni(OH)_2$	Does not dissolve	Green precipitate	Dissolves to give blue/purple solution $[Ni(NH_3)_6]^{2+}$	Green precipitate carbonate $NiCO_3$
Cu(II) blue $[Cu(H_2O)_6]^{2+}$	Pale blue precipitate $Cu(OH)_2$	Does not dissolve	Pale blue precipitate $Cu(OH)_2$	Deep blue solution $[Cu(NH_3)_4(H_2O)_2]^{2+}$	Green-blue precipitate $CuCO_3$
Zn(II) colourless $[Zn(H_2O)_6]^{2+}$	White precipitate $Zn(OH)_2$	Dissolves Colourless solution $[Zn(H_2O)_2(OH)_4]^{2-}$	White precipitate $Zn(OH)_2$	Does not dissolve	White precipitate $ZnCO_3$
Pb(II) colourless $[Pb(H_2O)_6]^{2+}$	White precipitate $Pb(OH)_2$	Dissolves Colourless solution $[Pb(H_2O)_2(OH)_4]^{2-}$	White precipitate $Pb(OH)_2$	Does not dissolve	White precipitate $PbCO_3$

Table 3 Reactions of metal(III) ions with bases

Aqueous M(III) ion Solution	Base added OH⁻ little	OH⁻ excess	NH₃ little	NH₃ excess	CO₃²⁻
Fe(III) violet $[Fe(H_2O)_6]^{3+}$ Appears brown due to hydrolysis	Brown precipitate $[Fe(H_2O)_3(OH)_3]$	Does not dissolve	Brown precipitate $[Fe(H_2O)_3(OH)_3]$	Does not dissolve	Brown precipitate of hydroxide $[Fe(H_2O)_3(OH)_3]$ and CO_2 evolved
Cr(III) ruby $[Cr(H_2O)_6]^{3+}$	Green precipitate $[Cr(H_2O)_3(OH)_3]$	Dissolves Green solution $[Cr(OH)_6]^{3-}$	Green precipitate $[Cr(H_2O)_3(OH)_3]$	Purple solution $[Cr(NH_3)_6]^{3+}$	Green precipitate of hydroxide $[Cr(H_2O)_3(OH)_3]$ and CO_2 evolved
Al(III) colourless	White precipitate $Al(OH)_3$	Dissolves Colourless solution $[Al(OH)_4]^-$	White precipitate $Al(H_2O)_3(OH)_3$	Slightly soluble	White precipitate of hydroxide $Al(OH)_3$ and CO_2 evolved

7 Write down what would you expect to see, and write the equations for the reactions when a solution of sodium carbonate is added to:

a a green aqueous solution of nickel(II) chloride;

b a yellow aqueous solution iron(III) chloride.

KEY FACTS

■ Metal(III) ions produce greater acidity than metal(II) ions and this determines the products of the reaction.

■ Metal(II) aqua ions give precipitates of carbonates on addition of sodium carbonate solution.

■ Metal(III) aqua ions give precipitates of hydroxides and carbon dioxide gas on addition of sodium carbonate solution.

■ A colour change occurs during the reaction.

17.7 Substitution of ligands in transition metal compounds

As with the acid–base reactions in section 17.5, ammonia molecules can affect the hydrolysis equilibrium in an aqua complex. Ammonia can behave as a base, with the equilibrium:

$$NH_3(aq) + H_2O(l) \rightleftharpoons NH_4^+(aq) + OH^-(aq)$$

The equilibrium for the metal(II) ions exists as before (section 17.4):

$$[M(H_2O)_6]^{2+} + H_2O \rightleftharpoons [M(H_2O)_5(OH)]^+ + H_3O^+$$

In water, the equilibrium between the hydroxide ions and H_3O^+ ions is established:

$$OH^- + H_3O^+ \rightleftharpoons 2H_2O$$

This removes H_3O^+ ions and moves the equilibrium completely to the right. If more OH⁻ is added, the second equilibrium:

$$[M(H_2O)_5(OH)]^+ + H_2O \rightleftharpoons [M(H_2O)_4(OH)_2] + H_3O^+$$

is also displaced to the right by OH⁻ ions reacting with the H_3O^+ formed and $[M(H_2O)_4(OH)_2]$, metal(II) hydroxide, is precipitated. Even though ammonia is a weak base, the concentration of OH⁻ is sufficient to shift the equilibrium to the right to allow the precipitate to form.

We can write the overall equation as:

$$[M(H_2O)_6]2^+ + 2OH^- \rightleftharpoons [M(H_2O)_4(OH)_2] + 2H_2O$$

The addition of OH^- shifts the equilibrium to the right and eventually the precipitate forms.

As well as the acid–base reactions giving the precipitate, most transition metal ions give a further sequence of reactions where the aqua ligands are substituted by ammine ligands. When a water molecule is replaced by any other ligand, it is a **substitution reaction** and, because a lone pair of electrons is being donated, it is an example of **nucleophilic substitution**. Ligand replacement in a complex ion is taking place continually, and the picture we represent is the 'average'.

There are two types of ligand:

- Neutral ligands – molecules that have no overall charge, e.g. H_2O and NH_3
- Anionic – ligands that carry a negative charge e.g. Cl^- and OH^-

These complex ions formed by ligand replacement are soluble, so a precipitate formed by the acid–base reaction will re-dissolve when it forms the complex ion. A series of equilibria are set up and the overall equation for the substitution reaction is:

$$[M(H_2O)_6]^{2+} + 6NH_3 \rightleftharpoons [M(NH_3)_6]^{2+} + 6H_2O$$

This reaction can be written as occurring in six steps (Fig. 6).

 8 What is the general pattern in the reactions in Fig. 6?

In these reactions the ammonia is being written as donating a lone pair of electrons to the central metal ion, so it is a **Lewis base**. In the hydrolysis equilibria, ammonia was being considered as donating a lone pair to a proton, so it is also a Lewis base.

Different complexes in the sequence in Fig. 6 will have different stabilities. For the first substitution:

$$K_c = \frac{[[M(H_2O)_5(NH_3)]^{2+}][H_2O]}{[[M(H_2O)_6]^{2+}][NH_3]}$$

but the overall equilibrium can be defined by the equilibrium constant:

$$K_c = \frac{[[M(NH_3)_6]^{2+}][H_2O]^6}{[[M(H_2O)_6]^{2+}][NH_3]^6}$$

Effects of changing ligands

Since the presence of a ligand affects the split between the energy levels in the d orbitals and hence the frequency of light absorbed, changing the ligands often changes the colour of the complex. The size and charge of the ligand and the strength of the bond formed by donating the lone pairs of electrons into the central metal ion are the most significant factors in changing properties.

The ammonia and water ligands are similar in size, and both form octahedral complexes with transition metal ions. There is no change in shape. Cobalt(II) ions in ammonia solution and copper(II) ions in ammonia solution are examples of this system. Ligand substitution also causes a colour change in the cobalt and copper systems, but no change in co-ordination number.

When concentrated aqueous ammonia is added to a solution containing pink cobalt(II) ions, a green-blue precipitate of cobalt(II) hydroxide forms at first. This is the result of the acidity reaction of the cobalt(II) ions:

Fig. 6 Steps in ammonia ligand substitution for metal(II) complexes

$$[M(H_2O)_6]^{2+} + NH_3 \rightleftharpoons [M(NH_3)(H_2O)_5]^{2+} + H_2O$$

$$[M(NH_3)(H_2O)_5]^{2+} + NH_3 \rightleftharpoons [M(NH_3)_2(H_2O)_4]^{2+} + H_2O$$

$$[M(NH_3)_2(H_2O)_4]^{2+} + NH_3 \rightleftharpoons [M(NH_3)_3(H_2O)_3]^{2+} + H_2O$$

$$[M(NH_3)_3(H_2O)_3]^{2+} + NH_3 \rightleftharpoons [M(NH_3)_4(H_2O)_2]^{2+} + H_2O$$

$$[M(NH_3)_4(H_2O)_2]^{2+} + NH_3 \rightleftharpoons [M(NH_3)_5(H_2O)]^{2+} + H_2O$$

$$[M(NH_3)_5(H_2O)]^{2+} + NH_3 \rightleftharpoons [M(NH_3)_6]^{2+} + H_2O$$

$[Co(H_2O)_6]^{2+}(aq) + 2NH_3(aq) \rightarrow [Co(H_2O)_4(OH)_2](aq) + 2NH_4^+(aq)$
  pink   blue-green precipitate

If excess aqueous ammonia is added, the precipitate dissolves. A pale brown (straw-coloured) solution forms:

$[Co(H_2O)_4(OH)_2](aq) + 6NH_3 \rightarrow [Co(NH_3)_6]^{2+} + 4H_2O + 2OH^-$
  blue/green   pale yellow

We can write the overall reaction as:

$[Co(H_2O)_6]^{2+}(aq) + 6NH_3(aq) \rightarrow [Co(NH_3)_6]^{2+} + 6H_2O$
  pink   pale yellow

When preparing compounds containing the $[Co(NH_3)_6]^{2+}$ ion, it is important to keep the solution of the hexaamminecobalt(II) ion away from air because the hexaamminecobalt(II) ion is easily oxidised in air to form the hexaamminecobalt(III) ion:

$$[Co(NH_3)_6]^{2+} \xrightarrow{O_2} [Co(NH_3)_6]^{3+}$$
pale yellow   brown
octahedral   octahedral

Sometimes, when ligands are replaced, not all of them are exchanged. The ligands are exchanged in a stepwise manner (see Fig. 6) and there is an equilibrium constant for each step. The most common example of partial exchange is the formation of the tetraamminebisaquacopper(II) complex. This complex is formed by simply adding dilute ammonia solution to copper sulphate solution until the precipitate of copper hydroxide re-dissolves, giving a deep blue solution.

$[Cu(H_2O)_6]^{2+} + 2NH_3 \rightleftharpoons [Cu(NH_3)_4(OH)_2] + 2NH_4^+$
blue solution   pale blue precipitate

$[Cu(H_2O)_6]^{2+} + 4NH_3 \rightleftharpoons [Cu(NH_3)_4(H_2O)_2]^{2+} + 4H_2O$
blue solution   deep blue solution

However, under these conditions the equilibrium does not go any further. We can have an idea why this happens if we look at the structure of the complexes. In both complexes, two of the ligands are further away from the copper(II) ion than the other four (see Fig. 7).

You can see from Fig. 7 that the four ligands in the square planar arrangement are replaced by the ammonia molecules. The long bonds are weaker and break most easily, so in solution they are most likely to be replaced by the surrounding ligands. In aqueous solution these are most likely to be aqua ligands.

To substitute all the aqua ligands, the concentration of the ammonia must be increased. One way to achieve this is to use liquid ammonia instead of aqueous ammonia solution.

9 How should increasing the ammonia concentration affect the equilibrium?

Fig. 7 Copper(II) ions in concentrated aqueous ammonia

When concentrated aqueous ammonia is added to copper(II) ions, a blue precipitate of copper(II) hydroxide forms:

$[Cu(H_2O)_6]^{2+} + 2NH_3 \rightleftharpoons [Cu(OH_2)(H_2O)_4] + 2NH_4^+$
  blue   pale blue

When excess aqueous ammonia is added, the precipitate dissolves to give a deep blue solution of tetraamminebisaquacopper(II) ions:

$[Cu(H_2O)_6]^{2+} + 4NH_3 \rightleftharpoons [Cu(NH_3)_4(H_2O)_2]^{2+} + 4H_2O$
  blue   blue-violet
                            tetraamminebisaquacopper(II) ion

Note: when naming a complex with two or more different types of ligand, you must list the ligands in alphabetical order (disregarding prefixes like di-, tri- and so on).

Only four of the six aqua ligands are substituted by ammonia. The structure is:

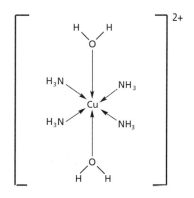

249

17.8 Substitution of H₂O ligands by Cl⁻ ligands

The octahedral structure is clearer in other complex ions. The hydrated cobalt(II) ion $[Co(H_2O)_6]^{2+}$ is pink and is octahedral in shape (Fig. 9). The co-ordination number of cobalt in this complex ion is six. The cobalt ion is bonded to six ligands.

When the chloride ion concentration is high, the blue tetrachlorocobaltate(II) $[CoCl_4]^{2-}$ ion forms. This is tetrahedral.

The chloride anion is much larger than the chlorine atom from which it is formed (see Fig. 8). The chloride ion has the electronic configuration of a noble gas, with four lone pairs of electrons. When the chloride ion acts as a ligand, one of these four pairs is donated to the metal ion.

Concentrated hydrochloric acid with a higher solubility, has a higher concentration of chloride ions than sodium chloride. Adding concentrated hydrochloric acid to a solution containing a metal–aqua ion gives a high enough concentration of Cl⁻ ions to displace the equilibrium:

$$[M(H_2O)_6]^{2+} + 4Cl^- \rightleftharpoons [MCl_4]^{2-} + 6H_2O$$

$$[Co(H_2O)_6]^{2+} + 4Cl^- \rightleftharpoons [CoCl_4]^{2-} + 6H_2O$$

pink	deep blue
octahedral	tetrahedral

Fig. 8 Comparative sizes of some atoms and ions

Adding concentrated HCl changes pink $[Co(H_2O)_6]^{2+}$ ions to blue $[CoCl_4]^{2-}$ ions.

A deep blue solution of ions is formed. If water is added, the solution turns pink again, as the equilibrium is pushed to the left. Adding more concentrated hydrochloric acid turns it deep blue again, pushing the equilibrium to the right (Fig. 9). This time, the colour change is due not only to the change in ligand, but also to the change in co-ordination number.

You can use the different colours of these cobalt complexes to test for the presence of water. Cobalt chloride paper contains high

Fig. 9 Equilibrium between hexaaquacobalt(II) and tetrachlorocobaltate(II)

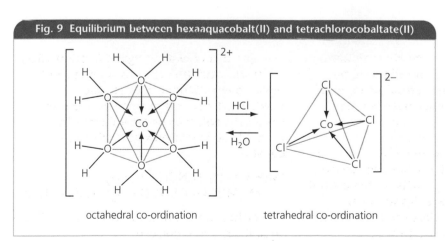

octahedral co-ordination   tetrahedral co-ordination

levels of chloride ions and so the blue $[CoCl_4]^{2-}$ ion is present. If you moisten cobalt chloride paper, it turns pink. Water molecules replace the chloride ligands and the hydrated cobalt(II) ion forms.

10

a For the complex $[CoCl_4]^{2-}$ what is the co-ordination number and oxidation state of cobalt?

b Write an equation for the reaction when concentrated hydrochloric acid is added to aqua cobalt(II) ions. What is the change in co-ordination number?

The co-ordination number changes because the large chloride ion is negatively charged. The ligands around the central metal ion of the complex repel each other. With larger ligands, there comes a point at which a tetrahedral structure is more stable than an octahedral structure, so the co-ordination

number changes. For M^{2+}(aq) ions the general equation for ligand substitution by chloride ions is:

$$[M(H_2O)_6]^{2+} + 4Cl^- \rightleftharpoons [MCl_4]^{2-} + 6H_2O$$

When concentrated hydrochloric acid is added to aqueous copper(II) ions, the colour changes from blue to yellow/green. Each solution contains complex copper(II) ions.

Under completely anhydrous conditions the complex ion $[CuCl_4]^{2-}$ (tetrachloro-cuprate(II)) is yellow, but even with concentrated hydrochloric acid there is a significant concentration of aqua ligands and a yellow/green complex ion forms. Although $[CuCl_4]^{2-}$ is yellow, there will always be some replacement by aqua ligands from any water present. Colour absorption by the ligands splitting the energy levels will give a mixture of yellow and blue, so the solution appears green. It is acceptable to consider the yellow/green complex as $[CuCl_4]^{2-}$ (tetrachlorocuprate(II)).

When this solution is diluted, an equilibrium is set up between the $[CuCl_4]^{2-}$ ion and the hydrated copper(II) ion (Fig. 10):

$$[CuCl_4]^{2-}(aq) + 6H_2O_{(l)} \rightarrow [Cu(H_2O)_6]^{2+}(aq) + 4Cl^-(aq)$$
yellow   blue

The tetrachlorocuprate(II) ion is an anionic (negative) complex ion. The central copper(II) ion is bonded with four chloride ions, using the lone pairs on the chloride ions. The co-ordination number of the copper is four and the ion is tetrahedral in shape.

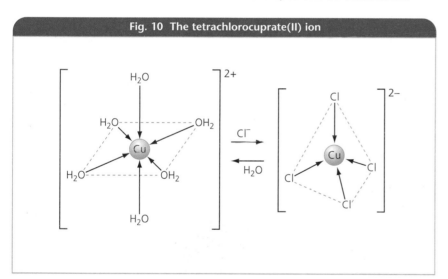

Fig. 10 The tetrachlorocuprate(II) ion

17.9 Type of ligand

A further factor which affects the stability of complexes is the number of links a single ligand can form with the central metal ion. The ligands discussed so far have been **unidentate** (section 15.2), i.e. one pair of electrons is donated from each ion or molecule. A **bidentate** ligand, e.g. ethanedioate $C_2O_4^{2-}$ will donate two electron pairs into the available orbitals on the central metal ion. A **hexadentate** ligand has six donor atoms with lone electron pairs. EDTA^{4-} is a hexadentate ligand and this is an

example of a **chelating** ligand (section 15.2).

Again, equilibria exist for these ligand replacements. Ethane-1,2-diamine, $H_2NCH_2CH_2NH_2$, often written as (en) for simplicity, will replace aqua ligands in a metal(II)–aqua complex according to the equation:

$$[M(H_2O)_6]^{2+} + H_2NCH_2CH_2NH_2$$
$$\rightleftharpoons [M(H_2O)_4(H_2NCH_2CH_2NH_2)]^{2+} + 2H_2O$$

Further substitution can occur until all the aqua ligands are replaced:

$$[M(H_2O)_4(H_2NCH_2CH_2NH_2)]^{2+} + H_2NCH_2CH_2NH_2$$
$$\rightleftharpoons [M(H_2O)_2(H_2NCH_2CH_2NH_2)_2]^{2+} + 2H_2O$$

$$[M(H_2O)_2(H_2NCH_2CH_2NH_2)_2]^{2+} + H_2NCH_2CH_2NH_2$$
$$\rightleftharpoons [M(H_2NCH_2CH_2NH_2)_3]^{2+} + 2H_2O$$

metal(III) trisethane-1,2-diamine complex
(tris means 'three')

The position of equilibrium for these complexes lies well over to the right and the equilibrium constant for the formation of trisethane-1,2-diamine complexes is approximately 10^{20}. This much higher equilibrium constant means that these complexes are very stable when compared with the aqua complexes. Metal(III) trisethane-1,2-diamine complexes can also be formed:

$$[M(H_2O)_6]^{3+} + 3H_2NCH_2CH_2NH_2 \rightleftharpoons M(H_2NCH_2CH_2NH_2)_3]^{3+} + 6H_2O$$

These have equilibrium constants of the order of 10^{30}, so are even more stable compared with the metal(II) complexes.

EDTA^{4-} has the structure as shown in Fig. 11. The lone pairs in EDTA are on the two nitrogen atoms and on four of the oxygen atoms. EDTA^{4-} forms one-to-one complexes with metal(II) atoms, such as [Cu(EDTA)]$^{2-}$ (Fig. 12). The complexes formed by polydentate ligands are very stable, because the central cation is firmly held by many co-ordinate bonds.

The bidentate ligands can form similar structures, e.g. [Cu(en)$_3$]$^{2+}$ (Fig. 13).

There are mixed complexes containing bidentate and unidentate ligands, e.g. [CuCl$_2$(en)$_2$]$^{2+}$ (Fig. 14). These complexes involving multidentate or chelating ligands such as EDTA^{4-} will be more stable and have higher K_c values than those with unidentate ligands, such as NH$_3$.

Fig.12 Displayed diagram of EDTA/Cu

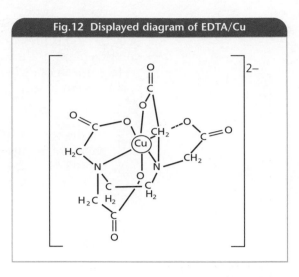

Fig. 13 Chelating effect of ethylenediamine

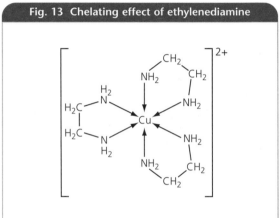

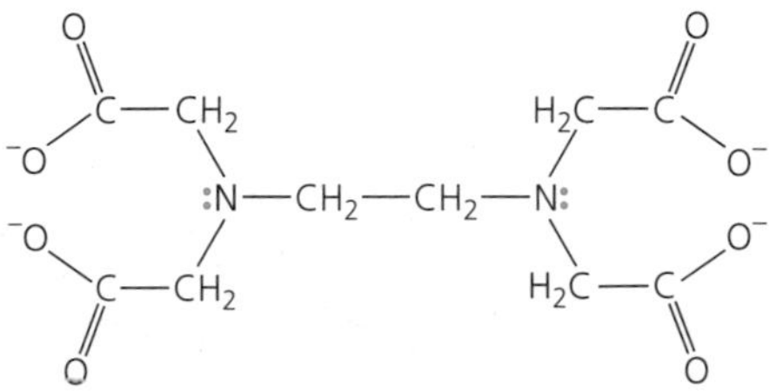

Fig. 14 Mixed complex of ethane-1,2-diamine and chloro ligands

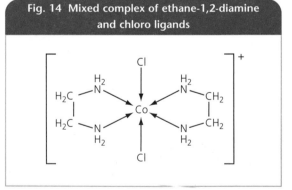

Fig. 11 Displayed formula of EDTA^{4-}

11 Write equations for the exchanges that would take place if the following pairs of chemicals are mixed.

a aqueous [Co(H$_2$O)$_6$]$^{2+}$ with EDTA^{4-};

b [Fe(H$_2$O)$_6$]$^{3+}$ and ethanedioate (C$_2$O$_4$$^{2-}$) ions. Draw this structure.

Fig. 15 Increasing entropy effect of chelating ligands

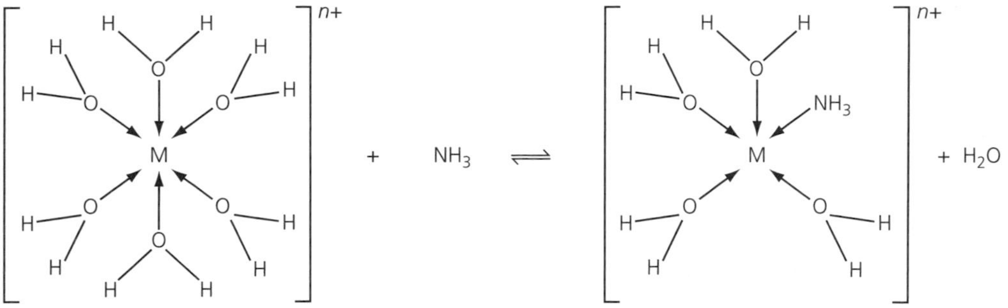

2 particles $\longrightarrow$ 2 particles

1 ammine ligand replaces 1 aqua ligand: minimal entropy change

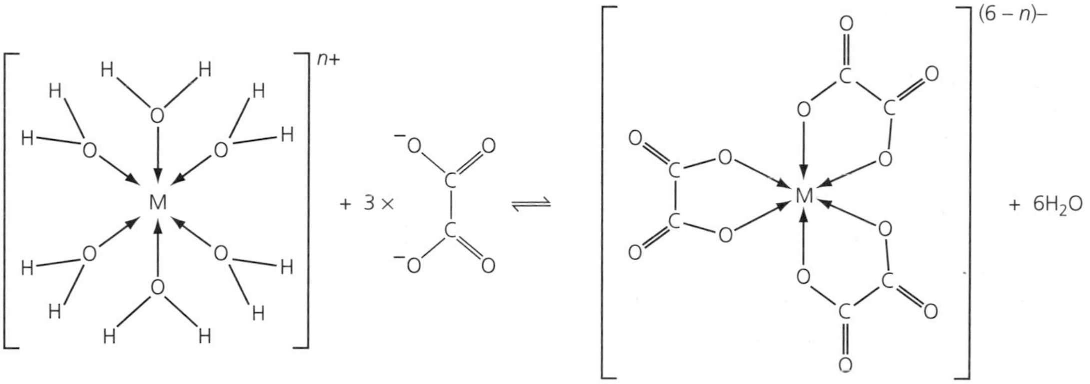

4 particles $\longrightarrow$ 7 particles

3 ethanedioate ligands replace 6 aqua ligands: entropy increases

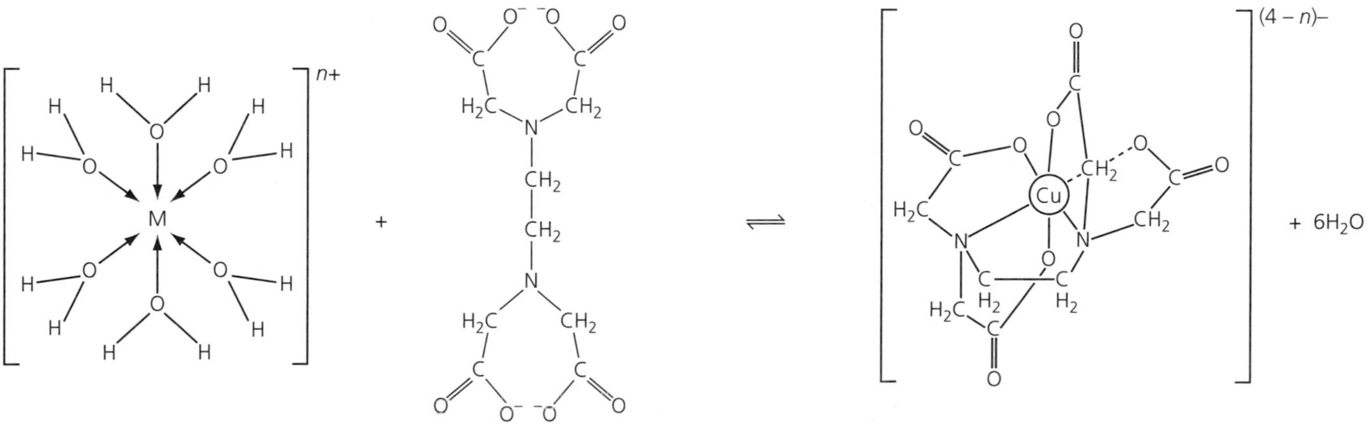

2 particles $\longrightarrow$ 7 particles

1 EDTA^{4-} ligand replaces 6 aqua ligands: larger entropy increase

Ultimately, the value of K_c will depend upon:

- the charge on the ion
- the size and charge of the ligands
- the co-ordination number of the central metal ion
- the number of bonds per ligand
- the chelate effect

Entropy and stability of complexes

The stabilities of complexes with chelating ligands can be explained in terms of the increase in entropy of the system. The overall feasibility of a reaction is determined by the change in free energy (see Chapter 12) calculated using:

$$\Delta G^{\ominus} = \Delta H^{\ominus} - T\Delta S^{\ominus}$$

If $\Delta G^{\ominus}$ has a negative value, the reaction is feasible. ΔH is the enthalpy term and if ΔH becomes more negative, energy is released and the system moves to a lower energy, so the reaction is more favoured. S is the entropy term and this can be thought of as the degree of disorder in the system (Chapter 12). When entropy increases in a reaction, $\Delta S^{\ominus}$ is positive, and $-T\Delta S^{\ominus}$ will have a more negative value. Then, $\Delta G^{\ominus}$ will be more negative and the reaction will be more favoured.

The effect can be illustrated using the ligands H_2O, NH_3 and $EDTA^{4-}$ as examples. If a transition metal ion is dissolved in water, it is surrounded by aqua ligands. When another unidentate ligand replaces a water molecule, it may donate more strongly and this will give a negative $\Delta H^{\ominus}$ value and the exchange is likely to be favoured.

If ammonia solution is added to $[Cu(H_2O)_6]^{2+}$, then some ligands are replaced:

$$[Cu(H_2O)_6]^{2+} + 4NH_3 \rightleftharpoons [Cu(NH_3)_4(H_2O)_2]^{2+} + 4H_2O$$
$$\text{5 particles} \qquad\qquad \text{5 particles}$$

In this example, although the $\Delta H^{\ominus}$ value has changed, the entropy of the system will be very similar, $\Delta S^{\ominus}$ will be small and $T\Delta S^{\ominus}$ insignificant. This is because one particle moving randomly in the aqueous solution has been replaced by one water molecule, which also will move about randomly in the solution. So the total number of particles remains constant and there is minimal change in the disorder.

If a multidentate ligand such as $EDTA^{4-}$ replaces the unidentate ligands, one $EDTA^{4-}$ which is a hexadentate ligand replaces six water molecules. This means that previously there was one ($EDTA^{4-}$) particle moving randomly and now there are six (water molecules) moving randomly. The system now has a much higher entropy value, so the $-T\Delta S^{\ominus}$ term for the reaction is large.

$$[Cu(H_2O)_6]^{2+} + EDTA^{4-} \rightarrow [Cu(EDTA)]^{2-} + 6H_2O$$
$$\text{2 particles} \qquad \rightarrow \qquad \text{7 particles}$$

More particles gives a greater number of possible arrangements, this gives a greater disorder, this increases the entropy.

The enthalpy change in ligand substitution reactions is usually small, so the large value of $-T\Delta S^{\ominus}$ means that $\Delta G^{\ominus}$ is negative and the reaction is feasible. Ligand substitution occurs and the reaction proceeds essentially to completion (Fig. 15).

12 Why does a more negative $\Delta H^{\ominus}$ value mean a more stable complex ion?

KEY FACTS

- Ammonia and water molecules can act as neutral ligands.

- Ammonia will replace water in aqua complexes, with no change of co-ordination number.

- The reactions of transition metal complexes often result in a colour change, because of a change of ligand, and/or a change of co-ordination number, and/or a change of oxidation state.

- Ligand substitution reactions of ammonia or water by chloride ions involve a change of co-ordination number from 6 to 4.

- Ligands are commonly unidentate, bidentate or hexadentate.

- Chelating ligands form very stable complexes.

- Forming complexes with chelating ligands increases the entropy of the system.

APPLICATION

Soil contamination

British Chrome Chemicals Ltd has manufactured chrome for nearly 150 years and has accumulated vast amounts of waste materials. Also, in mining for heavy metals such as copper, silver and lead, soil contamination and water contamination built up in these areas. Nowadays, we are much more environmentally aware, and dispose of waste products more carefully. However, it is estimated that there could be up to 100 000 seriously contaminated sites in Britain. Water was an essential resource for mining and other metal manufacturing processes, and the land and water courses have been contaminated by leaching.

If a metal ion is present in water, a hydrolysis reaction can occur.

1 Write equations for $Cu^{2+}(aq)$ hydrolysis and $Cr^{3+}(aq)$ hydrolysis in water. Why are they called hydrolysis reactions?

2 Explain why hydrolysis is more significant with Cr^{3+} than Cu^{2+}.

3 State how the hydrolysis will affect the pH of the water.

At low pH, nearly all metals are more soluble than in neutral water. Oxides and hydroxides have high lattice energies, so many of them are insoluble. If a complex ion can be formed, more hydration can take place, releasing more energy to break down the lattice.

4 Generally, soft water contains more aluminium ions (dissolved from soil and rocks) than hard water. Explain why there are higher amounts of dissolved copper in soft water than in hard water.

In water, the following equilibrium can exist for complex ions of chromium(III):

$$[Cr(H_2O)_6]^{3+} + Cl^- \rightleftharpoons [Cr(H_2O)_5Cl]^{2+} + H_2O$$
$$\text{violet} \qquad\qquad\qquad\qquad \text{green}$$

5 Describe and explain the changes that will take place when more chloride ions are added.

6 Write equations for the reactions which occur when $Cr^{3+}(aq)$ ions are treated with dilute sodium hydroxide solution until present in excess.

7 If $Cu^{2+}(aq)$ reacts with Cl^- the aqua complex changes its co-ordination number. Write an equilibrium for this reaction.

8 $Cu^{2+}(aq)$ can be removed from contaminated water using $EDTA^{4-}$.
a Explain why EDTA is described as a multidentate complex.
b Explain why this is such a stable complex.

Copper, often found in high concentrations in mining areas, is toxic to plants and micro-organisms especially at low pH. This copper can leach into water courses. The formation of the $[Cu(EDTA)]^{2-}$ complex can be used to determine amounts of copper in water because the complex is so stable.

9 If 100 cm³ of a water sample was titrated with $EDTA^{4-}$ and 4.80 cm³ of 0.05 mol dm⁻³ $EDTA^{4-}$ were needed to reach the endpoint:
a calculate the number of moles of $EDTA^{4-}$ used in the titration.
b calculate the number of moles of copper complexed.
c calculate the concentration of copper as mg dm⁻³ in the original sample.

1 Read the passage below. Identify each of **A**, **B**, **C**, **D**, **E**, **F**, **G**, **H** and **I** and write equations for all the reactions occurring.

A is a black solid which dissolves in water to form a blue solution which contains a cation **B** and an anion **C**.

The addition of aqueous ammonia to the blue solution gives initially a blue precipitate **D** which dissolves when an excess of aqueous ammonia is added giving a deep blue solution containing species **E**.

The addition of concentrated hydrochloric acid to the blue solution of **A** gives a yellow–green solution containing species **F**.

The addition of aqueous silver nitrate to the blue solution of **A** gives a cream precipitate **G**. Precipitate **G** is insoluble in dilute aqueous ammonia but dissolves forming a colourless solution containing species **H** when concentrated aqueous ammonia is added. Precipitate **G** also dissolves when an excess of an aqueous solution of sodium thiosulphate is added giving a colourless solution containing species **I**. (15)

AQA June 2000 CH05 Q6

2
a i) Distinguish between the terms *Lewis base* and *reducing agent*.
ii) By means of an equation, in each case, show how a bromide ion can behave as a Lewis base in one reaction and as a reducing agent in another reaction. (4)
b Describe by stating essential reagents and conditions how, starting from potassium dichromate(VI), you would obtain a solution containing each of the following ions as the only chromium species: $Cr^{3+}(aq)$ and $Cr^{2+}(aq)$. Give an equation for the reaction in each case. (9)
c Predict what you would observe, and give the formula of the chromium-containing product obtained in each case, when solid sodium carbonate is added to aqueous solutions of each of the following ions: $Cr^{3+}(aq)$ and $Cr^{2+}(aq)$. (5)
AQA June 2000 CH05 Q3

3
a When aqueous cobalt(II) chloride is treated with aqueous ammonia, a precipitate forms.
i) Give the formula of this precipitate and write an equation, or equations, to show how it is formed.
ii) This precipitate dissolves when an excess of aqueous ammonia is added and a pale brown solution is formed. Give the formula of the cobalt species present in the pale brown solution and write an equation to show how it is formed from the

precipitate.
iii) State what is observed when this pale brown solution is allowed to stand in air and give the formula of the new cobalt species formed. (8)
b In order to determine the concentration of a solution cobalt(II) chloride, a 25.0 cm³ sample was titrated with a 0.0168 M solution of EDTA⁴⁻; 36.2 cm³ were required to reach the end-point. The reaction occurring in the titration is:

$$[Co(H_2O)_6]^{2+} + EDTA^{4-} \rightarrow [Co(EDTA)]^{2-} + 6H_2O$$

i) What type of ligand is EDTA⁴⁻?
ii) Calculate the molar concentration of the cobalt(II) chloride solution.
iii) Suggest an alternative analytical method for determining the concentration of a solution which contains only cobalt(II) chloride. (15)
AQA June 2000 CH05 Q2

4
a i) Describe what you would observe when anhydrous aluminium chloride is added to an excess of water. Write an equation for the reaction.
ii) Write an equation to show why an aqueous solution containing aluminium ions is acidic.
iii) Describe what you would observe when solid sodium carbonate is added to a solution containing ions. Write an equation (or equations) for the reaction which occurs. (8)
b i) describe what is observed when dilute ammonia solution is added, dropwise until in excess, to a solution containing aluminium ions. Give the formula of the final aluminium-containing species.
ii) How would your observations differ if dilute sodium hydroxide solution was used instead of dilute ammonia solution?
Give the formula of any different aluminium-containing species formed. (4)
NEAB June 1997 CH05 Q6

5
a Using 'M' to represent the metal, write an equation to show that metal(III)–aqua ions can behave as Brønsted–Lowry acids in aqueous solution.
b Explain why metal(III)–aqua ions are stronger acids than metal(II)–aqua ions.
c By giving observations and the formula of iron-containing products in each case, show how sodium carbonate solution could be used to distinguish between aqueous solutions of iron(II) sulphate and iron(III) sulphate.
NEAB February 1997 CH05 Q4

Synoptic questions

1 Copy the diagram below which shows the distribution curve for the energies of the molecules of a gas in a given system, E_a is the energy of activation for a particular reaction.

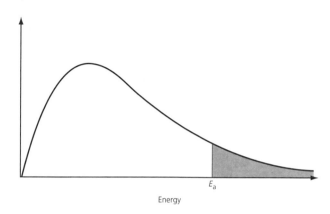

Energy

a State the meaning of the term *activation energy* for a chemical reaction.

b i) On your copy of the diagram, label the vertical axis on the diagram.

ii) Why does the curve start at the origin?

iii) Why does the curve not intersect the horizontal axis at higher energies?

iv) What does the shaded area represent?

c Name one physical property of the molecules which increases as their energy increases.

d Draw a curve on the graph representing the distribution of energies for the same molecules at a higher temperature.

e By reference to the curve you have drawn in **d**, and referring to the value of the activation energy in each case, explain the effect on the rate of a chemical reaction of:

i) the higher temperature,

ii) using a catalyst,

iii) increasing concentration.

f A reaction sequence which may contribute to the formation of acid rain is given below.

$NO(g) + \frac{1}{2}O_2(g) \rightarrow NO_2(g)$

$NO_2(g) + SO_2(g) \rightarrow NO(g) + SO_3(g)$

Discuss the role of the NO in this reaction sequence.

g The following data were obtained from studies of the reaction between NO and O_2 in a vessel at constant temperature.

Experiment	1	2	3
Initial total pressure of NO and O_2/atm	1.00	1.60	2.00
Initial partial pressure of NO/atm	0.40	0.40	0.80
Initial rate of reaction/atm s^{-1}	1.08	2.16	8.64

Use the information given to determine the order of the reaction, the rate equation and the value of the rate constant at that temperature. Show how you derive your answers.

2 a Discuss (without experimental details) how you would determine the order of reaction with respect to the

reactant **A** and to the reactant **B** for the overall reaction:

$nA + mB \rightarrow pC + qD$

by using measurements in the early stages of reaction only.

b Discuss the mechanism of the photo-chlorination of methane when methane is present in a large excess. Give the chemical equations and the name for each stage in the reaction, and classify any products formed as major or minor.

c When chlorine is exposed to ultra-violet light in a vessel which also contains a mixture of the gaseous compounds X and Y, the products XCl and YCl are formed as shown in the reactions below. The order with respect to each reactant is one.

$Cl + X \rightarrow XCl \quad k_1$

$Cl + Y \rightarrow YCl \quad k_2$

i) State the rate equations for the formation of XCl and YCl.

ii) When the concentration of $X = 3.0 \times 10^{-3}$ mol dm^{-3} and the concentration of $Y = 9.0 \times 10^{-3}$ mol dm^{-3}, the rates of formation of XCl and YCl are 6.0×10^{-5} and 1.5×10^{-5} mol s^{-1} respectively. Calculate the ratio of the rate constants (k_1/k_2) for the reactions.

3 a Define pH and state how you would measure the pH of a solution.

b Explain why the pH values of equimolar solutions of hydrochloric acid and ethanoic acid are different.

c Give an approximate pH value (or a pH range) for each of the following solutions. In each case explain your answer.

i) aqueous 0.1 mol dm^{-3} $FeCl_3$

ii) aqueous 0.1 mol dm^{-3} Na_2SO_4

iii) aqueous 0.1 mol dm^{-3} Na_2CO_3

d i) Write an equation for the reaction between sodium hydroxide and ethanoic acid.

ii) Use the following data show how the pH changes during a titration when aqueous 0.100 mol dm^{-3} NaOH is added to 10 cm^3 of aqueous ethanoic acid.

Volume 0.100 mol dm^{-3}

NaOH(aq)/cm^3	0.0	1.0	2.0	4.0	6.0	7.0	8.0	8.5	10.0	14.0
pH		2.9	4.0	4.3	4.7	5.2	5.5	6.4	11.2	12.0	12.4

i) Use the data to plot a titration curve on graph paper.

ii) State the pH range for the equivalence point and explain why the pH changes rapidly in the region of the equivalence point.

iii) Calculate the initial concentration of the ethanoic acid.

iv) Select a suitable indicator for the titration and explain why your indicator is suitable.

v) Using any suitable pH value, calculate K_a for ethanoic acid. Show your working.

e Sodium ethanoate and ethanoic acid can be mixed together to make a buffer solution.

i) What is meant by the term *buffer*?

ii) Explain how a solution of ethanoic acid/sodium ethanoate acts as a buffer when a small amount of hydrochloric acid is added.

257

f A buffer solution was made by mixing 10.5 cm³ of 0.800 mol dm⁻³ of sodium hydroxide with 25.0 cm³ of 0.920 mol dm⁻³ ethanoic acid. Calculate the concentrations of ethanoate ions (ignore the small number of ethanoate ions formed by dissociation of the excess acid), the concentration of acid remaining and hence the pH of the solution.

4 a What do you understand by the Lewis theory and the Brønsted–Lowry theory of acids and bases? Illustrate your answer with examples with each type.

b What is the difference between a strong acid and a weak acid?

c You are given a 0.1 mol dm⁻³ solution of an unknown monobasic acid in water. Describe an experiment you would carry out in order to decide if it is a strong acid or a weak acid, and show how the results of this experiment would be interpreted.

d Explain why the equivalence point of the titration between sodium hydroxide and ethanoic acid is not pH 7.

e What is a buffer solution? Give **one** example of its use.

f Buffers can be *acidic* or *basic*. Explain the mode of action of an acid buffer.

g How could you make an acidic buffer if you were given 200 cm³ 0.5 mol dm⁻³ of the following solutions and no other chemicals? You may not need to use all of the solutions.

nitric acid	propanoic acid
ammonia	potassium hydroxide

Give the approximate volumes of solutions which you would use to make as much as possible of the buffer.

5 a Define pH. Describe briefly how you would measure the pH of a solution.

b Sketch titration curves showing the change in pH which results when an excess of a 1.00 mol dm⁻³ sodium hydroxide solution is added gradually to 25 cm³ of:

  i) 1.00 mol dm⁻³ hydrochloric acid
  ii) 1.00 mol dm⁻³ ethanedioic acid
  iii) 1.00 mol dm⁻³ ethanoic acid
  ($K_a = 1.8 \times 10^{-5}$ mol dm⁻³)

On your sketches, indicate the volumes of the sodium hydroxide used at the various equivalence points

c Indicator solutions can be considered as weak acids in the following equilibrium.

  HIn $\rightleftharpoons$ H⁺ + In⁻
  colour 1   colour 2

  i) How does this equilibrium explain the action of acid–base indicators in acid–base titrations?
  ii) Write an expression for the equilibrium constant K_{In} at the end-point for the indicator.
  iii) Write an expression for the pH at the end-point for the indicator.
  iv) What factors would you consider in order to select a suitable indicator for a particular titration?

6 Esters **Q**, **R** and **S** all have branched-chain structures. Each contains 58.8% carbon and 9.8% hydrogen. Each produces a mass spectrum with $m/z = 102$ as the highest value.

Q has only two peaks in its low resolution proton n.m.r. spectrum, but **R** and **S** each have three peaks.

Hydrolysis of **R** in acid conditions forms compounds **T** and **U**. The low-resolution proton n.m.r. spectrum of **T** has three peaks and that of **U** has two peaks. Infra-red spectra of **T** and **U** are shown below.

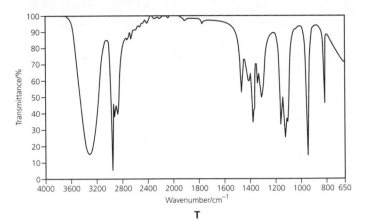

T

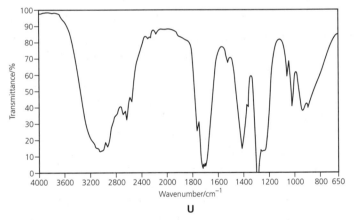

U

a Deduce the structures of compounds **Q**, **R**, **S**, **T** and **U** and explain how the mass spectrometry, n.m.r. and infrared data support your conclusions. Name each structure. Refer to the infra-red data in the Data section to help you answer this question.

b Calculate the mass of carbon dioxide produced and the volume of oxygen consumed (measured at 25 °C and 95 kPa) when 1.20 g of **Q** is burned in excess oxygen. [$R = 8.31$ J K⁻¹ mole⁻¹]

c Outline a four-stage synthesis of compound **S** starting from propene, methanol and standard inorganic reagents.

7 In the presence of $AlCl_3$, ethanoyl chloride reacts with the dialkylbenzene **A**, (C_8H_{10}), to form a single compound **B**, $C_{10}H_{12}O$. Compound **B** is converted into **C**, $C_{10}H_{14}O$, on treatment with $NaBH_4$. Compound **D** is produced when compound **C** is warmed with concentrated H_2SO_4.

Compound **B** has a strong absorption band in the infra-red at

1685 cm^{-1}, compound **C** has a broad absorption at 3340 cm^{-1} and compound **D** has an absorption band close to 1630 cm^{-1}.

a i) Show how the information provided in the question and infra-red data from the Data section can be used to deduce structures for compounds **A**, **B**, **C** and **D**. Name compound **A**.

   ii) Explain the significance of the fact that **B** is obtained as a single compound.

   iii) With the aid of appropriate diagrams, explain briefly why compound **C** can exist in stereoisomeric forms.

c Showing the repeat unit, draw the structure of the polymer which could be derived from molecule **D** and state the type of polymerisation involved.

d The conversion of **C** to **D** can also occur in the gas phase in the form of an equilibrium process.

   i) Write an equation for this process and expressions for the corresponding K_c and K_p values. State their units.

   ii) State how, and explain why, the applied pressure should be changed to increase the yield of **D** at equilibrium.

   iii) At constant temperature and a total pressure of 80 kPa, 13.8 g of **C** produces an equilibrium mixture containing 5.40 g of **D**. Calculate the value of K_p.

e Describe the procedures and results for chemical tests which could be used to distinguish samples of **B**, **C** and **D**.

8 Acid-catalysed dehydration of an optically active compound **A**, $C_4H_{10}O$, yields two isomeric products **B** and **C**, C_4H_8. Compound **B** exists in stereoisomeric forms. The reaction between **C** and hydrogen bromide produces compound **D**, C_4H_9Br, which yields **A** on hydrolysis. Oxidation of **A** gives compound **E**, C_4H_8O.

Compound **A** has a broad absorption at 3350 cm^{-1} in the infra-red, compound **E** has a strong absorption band at 1715 cm^{-1}, and compounds **B** and **C** each have significant absorption bands close to 1650 cm^{-1}.

a Use the infra-red data from the Data section and the information provided in the question to deduce structures for compounds **A**, **B**, **C**, **D** and **E**, explaining the stereoisomerism shown by **B**. Name each of these compounds.

b Predict the main characteristics of the low resolution n.m.r. spectra for compounds **A**, **B**, **C**, **D** and **E**, explaining the cause of each.

c Use bond energy values from the Data section to estimate the enthalpy change for the conversion of one mole of **A** to one mole of **C**.

d Draw and name the chain and positional isomers of molecule **A**, and discuss whether they could, or could not, be distinguished using mass spectrometry.

e **A** can be converted to **C** in a gas-phase equilibrium reaction. Write an equation for this process and explain, in terms of ΔG, why the production of **C** is favoured by the use of a higher temperature. State one other advantage and one disadvantage of using a higher temperature to produce **C**.

9 **a** What do you understand by the terms *enthalpy*, *entropy* and *free energy*?

Sulphur dioxide reacts with oxygen to form sulphur trioxide according to the equation:

$2SO_2(g) + O_2(g) \rightleftharpoons 2SO_3(g)$

Data for this reaction are shown in the table below.

	$\Delta H_f^{\ominus}/kJ\ mol^{-1}$	$S^{\ominus}/J\ K^{-1}\ mol^{-1}$
$SO_3(g)$	−396	+257
$2SO_2(g)$	−297	+248
$O_2(g)$		+204

b Determine the standard enthalpy, the standard entropy and standard free energy changes at 298 K for this reaction.

c The reaction is said to be feasible. In terms of free energy change, explain the meaning of the term *feasible*. Calculate the temperature at which the reaction between sulphur dioxide and oxygen ceases to be feasible.

d At a fixed temperature and a total pressure of 540 kPa, a vessel contains an equilibrium mixture of $SO_3(g)$ (0.050 mol), $O_2(g)$ (0.080 mol) and $2SO_2(g)$ (0.070 mol). Determine the equilibrium partial pressure of each of the three gases in this mixture, and hence the value of the equilibrium constant, K_p, for the reaction, stating its units.

e Many chemical reactions are exothermic and feasible but some are endothermic and still feasible. Give an example of one such endothermic chemical reaction and explain why it is feasible.

f Sulphuric acid which is manufactured from SO_3 mixes with water and the process is highly exothermic. Ammonium nitrate dissolves in water and the reaction is endothermic to the extent that any moisture on the outside of the beaker will freeze to ice. What factors are contributing to the enthalpy changes in these two examples? Explain why they are both feasible.

10 **a** Describe an experiment to determine accurately the enthalpy of combustion of a flammable substance.

b The enthalpy of formation of methane cannot be determined directly. Show how it may be determined indirectly from the following enthalpies of combustion: C(s) −393.5 kJ mol^{-1}; H$_2$(g) −285.9 kJ mol^{-1}; CH$_4$(g) −890.3 kJ mol^{-1}.

c The enthalpies of hydrogenation of cyclohexene and benzene are −120 and −208 kJ mol^{-1} respectively. Explain why the value for benzene is not three times the value for cyclohexene and estimate, with reasons, the enthalpies of hydrogenation of cyclohexa-1,4-diene and cyclohexa-1,3-diene.

cyclohexa-1,4-diene   cyclohexa-1,3-diene

11 **a** Explain the variations in atomic radius and in first ionisation energy across an eight-membered row of the Periodic Table.

b Show how the properties of **i)** melting point and

ii) conductivity for the elements in Period 3 (Na to Ar) depend on structure and bonding.

c The pH values obtained when 0.1 mol of each of the following oxides of the Period 3 elements is added to a litre of water are as shown.

Oxide	Na_2O	MgO	Al_2O_3	SiO_2	P_4O_{10}	SO_3
pH	13	9	7	7	3	1

Explain this trend and write equations for the reactions which occur.

12 a Define the term *electronegativity*.

b Use the electronegativity values given below to discuss the bonding in and the melting points of each of the compounds formed between chlorine and each of the elements Na to Si.

Element	Na	Mg	Al	Si	Cl
Electronegativity value	0.9	1.2	1.5	1.8	3.0

c How do each of the compounds in **b** react with water? State what you would observe in each case and discuss the trend in the pH of the resulting solutions.

d Use the electronegativity values and first ionisation enthalpies given below for Group I elements to discuss the predicted trend in melting point of the elements, and discuss how the reactivity with water will change from Li to Cs. (First ionisation enthalpies in kJ mol^{-1})

Element	Li	Na	K	Rb	Cs
Electronegativity value	1.0	0.9	0.8	0.8	0.7
First ionisation enthalpy	520	496	419	403	376
Melting point/K	454	371	336	312	302

13 An aromatic hydrocarbon **D**, C_7H_8, is converted into compound **E**, $C_9H_{10}O$, on treatment with ethanoyl chloride in the presence of $AlCl_3$. When **E** is treated with $NaBH_4$, it is converted into **F**, $C_9H_{12}O$. Compound **G** is formed when **F** is warmed with concentrated H_2SO_4. Note that compound **E** is formed as a mixture of three isomers.

Compound **E** has a strong absorption band in the infra-red at 1685 cm^{-1}, compound **F** has a broad absorption at 3340 cm^{-1} and compound **G** has an absorption band close to 1630 cm^{-1}.

a Show how the information provided in the question and infra-red data from the Data section can be used to deduce the structures, including isomers, of compounds **D**, **E**, **F** and **G** respectively. Choose one of the isomers of compound **E** to show the formation of compounds **F** and **G**.

b Predict the structures of molecules **H** and **J** formed by reacting molecule **G** with HBr followed by excess ammonia in ethanol. Explain why your answer for **H** is preferred over an alternative positional isomer.

Molecule **J** is a weak base. Its K_c (K_a) is 7.14×10^{-8} mol dm^{-3} at 25 °C. Calculate the pH of a solution of **J** of concentration 0.005 mol dm^{-3}.

d When reacted with acidified potassium nitrate(III) [potassium nitrite] at 25 °C, molecule **J** releases nitrogen gas and is converted to the corresponding alcohol. The following data show the rate of release of nitrogen for different concentrations of **J**, with the concentrations of

acid and nitrate(III) constant.

[J]/mol dm^{-3}	0.001	0.002	0.003	0.004
Rate/cm^3 N_2 s^{-1}	4.2	8.3	12.5	16.6

Use these data to deduce the order of reaction with respect to **J**. Write the rate equation and deduce the value and units of the rate constant.

14 When warmed with excess aqueous silver nitrate solution, 2.40 g of impure 1-iodobutane produced 2.45 g of silver iodide and an organic compound, **X**.

a Write balanced equations for the reactions occurring during these changes and name the organic product, **X**.

b Calculate the percentage purity of the sample of 1-iodobutane.

c By stating each essential reaction condition and writing corresponding balanced equations, describe a possible synthesis of pentanoic acid from **X**.

d A solution of pentanoic acid of concentration 0.050 mol dm^{-3} is found to have a pH of 3.10 at 25 °C. Calculate the value of K_a for pentanoic acid.

e 0.100 moles of pentanoic acid is allowed to equilibrate with 0.200 moles of ethanol at 25 °C. At equilibrium, the reaction mixture is found to contain 0.040 moles of unreacted pentanoic acid.
   i) Write an equation for the reaction between pentanoic acid and ethanol, naming the organic product.
   ii) Write an expression for K_c for this reaction.
   iii) Use the data given to calculate a value for K_c.

15 The following table shows the results of three experiments to investigate the rate of the reaction between compounds **J** and **K**. All three experiments were carried out at the same temperature.

	Experiment		
	1	**2**	**3**
Initial concentration of **J**/mol dm^{-3}	0.50×10^{-3}	1.00×10^{-3}	1.50×10^{-3}
Initial concentration of **K**/mol dm^{-3}	1.00×10^{-3}	1.00×10^{-3}	1.50×10^{-3}
Initial rate /mol dm^{-3} s^{-1}	0.15×10^{-3}	0.60×10^{-3}	1.35×10^{-3}

a Use the data in the table to deduce the order of reaction with respect to **J** and the order of reaction with respect to **K**. Hence write an overall rate equation for the reaction.

b In a reaction between compounds **L** and **M**, the order of reaction with respect to **L** is two and the order of reaction with respect to **M** is one. Given that the initial rate of reaction is 4.00×10^{-4} mol dm^{-3} s^{-1} when the initial concentration of **L** is 2.00×10^{-2} mol dm^{-3} and the initial concentration of **M** is 5.00×10^{-2} mol dm^{-3}, calculate the value of the rate constant for this reaction at this temperature and deduce its units.

c The rate equation for the decomposition of a compound **N** has a rate constant with the unit s^{-1}. The rate constant is 4.31×10^3 s^{-1} at 700 K and 1.78×10^4 s^{-1} at a temperature *T*. Use this information to deduce the overall order of reaction and whether temperature T is greater or smaller than 700 K.

d The decomposition of ethanal vapour, shown below, is catalysed by iodine vapour.

$$CH_3CHO(g) \rightarrow CH_4(g) + CO(g)$$

   i) State the qualitative effect, if any, of increasing the temperature, and of removing the catalyst, on the activation energy and on the rate of the above reaction.

   ii) In the absence of the catalyst, the mechanism is thought to be a chain reaction involving the following steps:

   Step 1 $CH_3CHO \rightarrow {}^\bullet CH_3 + {}^\bullet CHO$
   Step 2 $CH_3CHO + {}^\bullet CH_3 \rightarrow CH_4 + CH_3{}^\bullet CO$
   Step 3 $CH_3{}^\bullet CO \rightarrow {}^\bullet CH_3 + CO$

   Name the two different types of reaction step, Step 1 and Step 2, suggest a possible by-product of the reaction, and name the type of step in which it is formed.

e Protonation of alcohol **A** and subsequent loss of water, produces the intermediate **B**.

$$CH_3CH_2{-}\underset{\underset{OH}{|}}{\overset{\overset{CH_3}{|}}{C}}{-}CH_2CH_3 \qquad CH_3CH_2{-}\underset{+}{\overset{\overset{CH_3}{|}}{C}}{-}CH_2CH_3$$

      A   **B**

   i) Name alcohol **A**, and name the type of species of which intermediate **B** is an example. Draw the structures of the two alkenes which can be formed from species **B** by removal of a proton. Label as **C** the alkene which shows geometrical isomerism.

   ii) Outline a mechanism showing the nucleophilic attack by water on the intermediate **B** and the subsequent formation of alcohol **A**.

   iii) State what final colour you would see if alcohol **A** were warmed with acidified potassium dichromate(VI) and explain your answer. Draw two structural isomers of alcohol **A** which form branched-chain ketones when heated with acidified potassium dichromate(VI), but which could not form alkene **C** on dehydration.

16 Hydrogen peroxide decomposes rapidly at room temperature in the presence of **some** metal oxides according to the following equation: $H_2O_2(aq) \rightarrow H_2O(l) + O_2(g)$

a A student proposes that copper(II) oxide and cobalt(II) oxides will catalyse the reaction but zinc(II) oxide will not. Plan a series of experiments that this group of students could carry out in order to verify this idea. You should give details of all the apparatus needed, all procedures and how the results would be used for verification.

b Why is it likely that copper(II) oxide and cobalt(II) oxides will be effective catalysts but that zinc(II) oxide will not.

c **i)** Why would these oxides be described as examples of *heterogeneous* catalysts?

   ii) Give an example of *homogeneous* catalysis.

Hydrogen peroxide is produced in the body as a by-product of respiration. Hydrogen peroxide is toxic and the enzymes in liver catalyse its decomposition. If the effectiveness of liver as a catalyst is measured at different pH values the results show a curve as below.

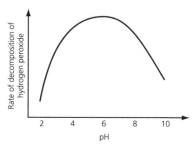

These enzymes are proteins. How does the chemical composition of proteins indicate that the effectiveness of these catalysts will be pH dependent? Plan an experiment or series of experiments that could be used to verify the curve shown in the graph.

17 a Explain the meaning of the terms
   i) *ligand*,
   ii) *complex ion*.

b Using the ligands H_2O and Cl^-, give examples of the complexes formed by
   i) two transition metals and
   ii) two non-transition metals.
   In each case, give the coordination number of the metal, state the shape of the complex and give any appropriate equations.

c The oxidation–reduction behaviour of transition metal compounds is influenced by
   i) a change of ligand and
   ii) a change of pH.
   Give one example in each case to illustrate these effects and write the chemical equations for the change.

18 a Describe an experiment to determine the dissociation constant of a weak monobasic acid. Give practical details and outline the underlying principles.

b Boron trifluoride and methoxymethane form a complex which dissociates endothermically on heating to give the following equilibrium mixture:

$$(CH_3)_2O{\rightarrow}BF_3(g) \rightleftharpoons (CH_3)_2O(g) + BF_3(g)$$

When 3.41 g of the complex were heated in a vessel of volume 7.2×10^{-3} m^3, the final equilibrium pressure was 2.21×10^4 Pa at 100 °C. Calculate:
   i) the degree of dissociation of the complex at 100 °C,
   ii) the value of the equilibrium constant K_p at 100 °C.

c Discuss qualitatively the effect on the position of the equilibrium in **b** of
   i) raising the temperature,
   ii) carrying out the dissociation in the presence of a quantity of methoxymethane.

261

d Discuss qualitatively how this reaction can be endothermic and yet still feasible.

19 a Explain why metal ions become hydrated in water.

b For hydrated transition metal ions, explain the term hydrolysis. Give the factors which determine the extent of hydrolysis, and write equations for
  i) a substitution reaction,
  ii) a hydrolysis (acidity) reaction.

c In each of the following reactions, state what you would observe. Give the type(s) of reaction occurring, and give equations.
  i) Aqueous cobalt(II) chloride is treated with concentrated aqueous ammonia, added dropwise until in excess, and the mixture is allowed to stand in air.
  ii) Aqueous cobalt(II) chloride is treated with concentrated hydrochloric acid, added dropwise until in excess.
  iii) Aqueous chromium(III) sulphate is treated with aqueous sodium carbonate, added dropwise until in excess.

d Explain why reactions such as the above involve colour changes.

20 a Describe, with the aid of a labelled diagram, how you would measure the standard electrode potential of a metal. State the standard conditions needed.

b Disproportion occurs when an element in one oxidation state is simultaneously oxidised and reduced. Use the following information to illustrate and to explain the process of disproportionation.

$$MnO_4^-(aq) + e^- \rightarrow MnO_4^{2-}(aq) \qquad E^\ominus = +0.60V$$
$$4H^+(aq) + MnO_4^{2-}(aq) + 2e^- \rightarrow MnO_2(s) \qquad E^\ominus = +1.55V$$

c Discuss, with one example in each case, two factors which influence the stability with respect to oxidation or reduction of a transition metal ion in aqueous solution.

d Write an equation for the reaction which occurs when chlorine gas is bubbled into water, and explain why this reaction can be considered as disproportionation.

21 a The Fe^{2+} ion was detected in a sample of river water taken near a chemical works by a group of students as part of an environmental project. The sample of water was concentrated 10 000 times and the concentration of Fe^{2+} ion present determined by titration using acidified potassium manganate(VII) solution.
The following results were obtained: $25.00 \ cm^3$ of the concentrated river water required $12.50 \ cm^3$ of a $0.020 \ 00 \ mol \ dm^{-3}$ potassium manganate(VII) solution.
  i) Suggest a chemical test to identify the Fe^{2+} ion in the concentrated river water.
  ii) Outline carefully, giving full practical details, how you would carry out the titration to obtain the above results.
  iii) Calculate the concentration of the Fe^{2+} ion in $mol \ dm^{-3}$ and determine the original concentration of the Fe^{2+} ion in the river.
  iv) Suggest one laboratory method of obtaining a pure sample of water from river water.

b The alums are a group of compounds having the general formula $Q^+R^{3+}(SO_4^{2-})_2.12H_2O$. When dissolved in water, alums behave like mixtures of their constituent ions. Some combinations of Q^+ and R^{3+} are shown in the table below.

Alum	Q^+	R^{3+}
A	K^+	Fe^{3+}
B	Na^+	Al^{3+}
C	NH_4^+	V^{3+}

The pH values of separate aqueous solutions of A and B are within the range 2-4. When aqueous sodium carbonate solution was added to separate aqueous solutions of A and B, effervescence occurred in both cases together, with a rust-coloured precipitate in A and a white precipitate in B. Interpret these observations with the help of equations, and predict what you might expect to observe when aqueous sodium carbonate solution is added to an aqueous solution of the vanadium-containing alum C.

22 a Indicate how the concept of oxidation as electron transfer is related to the definition in terms of the addition of oxygen. Give appropriate examples.

b Explain how a metal produces a potential difference when a piece of metal is placed a solution of its ions. Define standard electrode potential and describe how it can be measured.

c A piece of cobalt metal in a $1 \ mol \ dm^{-3}$ solution of $Co^{2+}(aq)$ is connected to another half-cell consisting of platinum in a $1 \ mol \ dm^{-3}$ solution of $Cl^-(aq)$ through which chlorine at a pressure of 1 atm is bubbled. Cobalt acts as the negative electrode. Using the data below, write down the spontaneous cell reaction and identify which species is oxidised and which reduced. Deduce the standard electrode potential of the Co^{2+}/Co couple. Describe what would be the effect on the cell voltage of:
  i) increasing the pressure of chlorine,
  ii) decreasing the concentration of $Co^{2+}(aq)$.
  [e.m.f. of cell: 1.63 V; $\frac{1}{2}Cl_2(g) + e^- \rightarrow Cl^-(aq)$
  $$E^\ominus = +1.36 \ V]$$
  Explain your answers in terms of Le Chatelier's principle.

23 a Describe, with the aid of a labelled diagram, an experiment to measure the standard electrode potential of silver, and write an equation representing the cell reaction.

b Construct a cycle of the Born–Haber type for the formation of silver ions in aqueous solution from solid silver. Name the enthalpy change in each step.

c By reference to the following data, discuss if it is possible by reactions using these species to prepare fluorine and chlorine.

$$\frac{1}{2}F_2(g) + e^- \rightarrow F(aq) \qquad E^\ominus = +2.87 \ V$$
$$\frac{1}{2}Cl_2(g) + e^- \rightarrow Cl^-(aq) \qquad E^\ominus = +1.36 \ V$$
$$MnO_4^-(aq) + 8H^+(aq) + 5e^- \rightarrow Mn^{2+}(aq) + 4H_2O(l) \qquad E^\ominus = +1.51 \ V$$
$$MnO_2(s) + 4H^+(aq) + 2e^- \rightarrow Mn^{2+}(aq) + 2H_2O(l) \qquad E^\ominus = +1.23 \ V$$

Periodic table

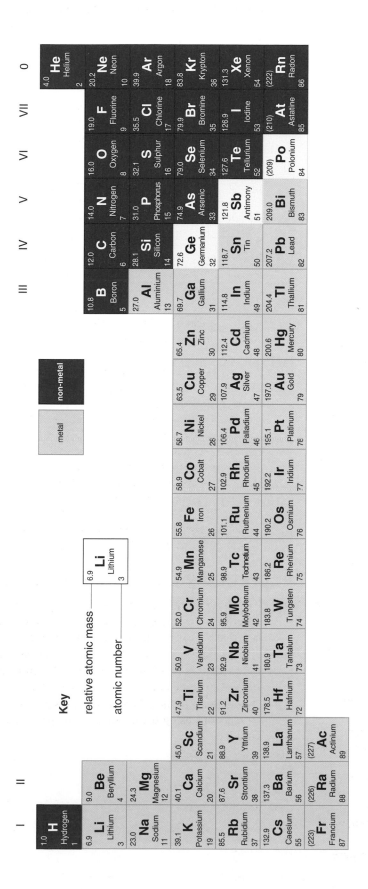

	Key	
relative atomic mass —	6.9 **Li**	
atomic number —	Lithium 3	

non-metal

metal

58–71 Lanthanides

90–103 Actinides

Units

Chemists usually use the International System of Units (Système International, or SI). The base SI units that are most often used in chemistry are shown below.

Base SI units

Quantity	Unit name	Symbol
length	metre	m
mass	kilogram	kg
time	second	s
electric current	ampere	A
temperature	kelvin	K
amount of substance	mole	mol

For convenience, any of the prefixes below may be used with any unit: for example, the kilometre (1 km = 10^3 m) and the milliampere (1 mA = 10^{-3} A) are often useful.

Prefixes for units

Prefix	Symbol	Meaning
tera	T	10^{12}
giga	G	10^9
mega	M	10^6
kilo	k	10^3
deci	d	10^{-1}
centi	c	10^{-2}
milli	m	10^{-3}
micro	μ	10^{-6}
nano	n	10^{-9}
pico	p	10^{-12}

Other units can be derived from the base units. For example, energy is normally measured in joules (symbol J), or multiples of joules (kJ, MJ), defined in terms of base units as $kg\ m^2\ s^{-2}$. Some non-SI units can be converted to SI units as shown below.

Units conversions

Unit	Symbol	SI equivalent
length	metre	m
atomic mass unit	U	1.661×10^{-27} kg
atmosphere	atm	101 325 Pa
degree Celsius	°C	1 K
litre	dm^3	$10^{-3}\ m^3$
tonne	t	10^3 kg

Formulae

Ideal gas equation

$pV = nRT$

Amount of substance

$$\text{number of moles} = \frac{\text{mass}}{M_r}$$

Equilibrium law

For the reaction

$mA + nB \rightleftharpoons pC + qD$ the equilibrium constant K_c is given by

$$K_c = \frac{[C]^p[D]^q}{[A]^m[B]^n}$$

where values in square brackets are equilibrium concentrations.

pH

$pH = -\log_{10}[H^+]$

Ionic product of water

$K_w = [H^+][OH^-]$
$\quad\ = 1 \times 10^{-14}\ mol^2\ dm^{-6}$ at s.t.p.

Order of reaction

For a zero-order reaction: $\quad \dfrac{-d[A]}{dt} = k_0$

For a first-order reaction: $\quad \dfrac{-d[A]}{dt} = k_1[A]$

For a second-order reaction: $\quad \dfrac{-d[A]}{dt} = k_2[A]^2$

Important values, constants and standards

Constant	Symbol	Value
molar gas constant	R	$8.31\ J\ K^{-1}\ mol^{-1}$
Faraday constant	F	$9.65 \times 10^4\ C\ mol^{-1}$
Avogadro constant	L	$6.02 \times 10^{23}\ mol^{-1}$
Planck constant	h	$6.63 \times 10^{-34}\ J\ Hz^{-1}$
speed of light in a vacuum	c	$3.00 \times 10^{-8}\ m\ s^{-1}$
mass of proton	m_p	1.67×10^{-27} kg
mass of neutron	m_n	1.67×10^{-27} kg
mass of electron	m_e	9.11×10^{-31} kg
electronic charge	e	1.60×10^{-19} C
molar volume of gas	V_m	$22.4\ dm^3\ mol^{-1}$ (at s.t.p.)
specific heat capacity of water		$4.18\ kJ\ kg^{-1}\ K^{-1}$

s.t.p. is approximately 101 kPa and 273 K (0 °C)

Bond angles

Compound	Sequence	Angle	Bond	Bond length/nm
CCl_4	Cl—C—Cl	109.5	Cl—C	0.177
CH_4	H—C—H	109.5	H—C	0.109
C_2H_4	H—C—H	117.3	H—C	0.109
C_6H_6 (benzene)	C—C—C	120.0	C—C	0.1397
H_2O	H—O—H	104.5	H—O	0.096
NH_3	H—N—H	107.0	H—N	0.101
PCl_5	Cl—P—Cl	120. 0	Cl—P	0.204
SF_6	F—S—F	90.0	F—S	0.156

Bond lengths and bond energies

Bond	in	Bond length /nm	Bond energy /kJ mol⁻¹	Bond	in	Bond length /nm	Bond energy /kJ mol⁻¹
Br—Br	Br_2	0.228	193	O—Si	$SiO_2(s)$	0.161	466
Br—H	HBr	0.141	366	O=Si	$SiO_2(g)$	–	638
Cl—Cl	Cl	0.199	243	O=Si	SiO	–	805
Cl—H	HCl	0.127	432	P—P	P_4	0.221	198
F—F	F_2	0.142	158	P=P	P_2	0.189	485
F—H	HF	0.092	568	C—C	average	0.154	347
I—I	I_2	0.267	151	C=C	average	0.134	612
H—I	HI	0.161	298	C≡C	average	0.120	838
H—H	H_2	0.074	435	C—H	average	0.108	413
H—Si	SiH_4	0.148	318	C—H	CH_4	0.109	435
H—Ge	GeH_4	0.153	285	C—F	average	0.138	467
H—N	NH_3	0.101	391	C—F	CH_3F	0.139	452
H—P	PH_3	0.144	321	C—F	CF_4	0.132	485
H—As	AsH_3	0.152	297	C—Cl	average	0.177	346
H—O	H_2O	0.096	464	C—Cl	CCl_4	0.177	327
H—S	H_2S	0.134	364	C—Cl	C_6H_5Cl	0.170	–
H—Se	H_2Se	0.146	313	C—Br	average	0.194	290
Na—Na	Na_2	0.308	72	C—Br	CBr_4	0.194	285
K—K	K_2	0.392	49	C—I	average	0.214	228
N—N	N_2H_4	0.145	158	C—I	CH_3I	0.214	234
N=N	$C_6H_{14}N_2$	0.120	410	C—N	average	0.147	286
N≡N	N_2	0.110	945	C=N	average	0.130	615
N—O	HNO_2	0.120	214	C≡N	average	0.116	887
N=O	NOF, NOCl	0.114	587	C—N	phenylamine	0.135	–
N=P	PN	0.149	582	C—O	average	0.143	358
O—O	H_2O_2	0.148	144	C—O	CH_3OH	0.143	336
O—O	O_3	0.128	302	C=O	CO_2	0.116	805
O=O	O_2	0.121	498	C=O	HCHO	0.121	695
S—S	S_8	0.205	266	C=O	aldehydes	0.122	736
S=S	S_2	0.189	429	C=O	ketones	0.122	749
O—S	SO_3	0.143	469	C=O	CO	0.113	1077
Si—Si	Si(s)	0.235	226	C—Si	$(CH_3)_4Si$, SiC(s)	0.187	307

Infrared spectroscopy: characteristic absorption bands

Compound	Wavelength/nm
C—H stretching vibrations	
alkane	3376–3505
alkene	3231–3322
arene	3300
aldehyde	3448–3546
	3603–3703
C—H bending vibrations	
alkane	6734–7326
arene	11 364–14 300
O—H stretching vibrations	
alcohols	
(not hydrogen-bonded)	2740–2786
(hydrogen-bonded	2667–3125
carboxylic acids	
(hydrogen-bonded)	3030–4000
Carbon–halogen stretching vibrations	
C—F	7142–10 000
C—Cl	12 500–16 667
C—Br	16 667–20 000
C—I	about 20 000
C=C stretching vibrations	
alkene	5991–6079
arene	6250–6897
C=O stretching vibrations	
aldehydes, saturated alkyl group	5747–5814
ketones	5882–5952
carboxylic acids	
saturated alkyl	5797–5882
aryl (from arene)	5882–5952
esters (saturated)	5714–5763

Infrared wavenumber data

Bond		Wavenumber/cm^{-1}
C—H		2850–3300
C—C		750–1100
C=C		1620–1680
C=O		1680–1750
C—O		1000–1300
O—H	(alcohols)	3230–3550
O—H	(acids)	2500–3000

Selected data for some elements

Element	Symbol	Atomic number	Stable mass number (% abundance)	Molar mass /g mol^{-1}	Melting point/°C	Boiling point/°C	Electro-negativity[1]	Atomic radius[2] /nm	Ionic radius[3] /nm	Ionisation energies /kJ mol^{-1} 1st	2nd	3rd	4th
Aluminium	Al	13	27(100)	27.0	660	2467	1.5	m0.143	+30.053	578	1817	2745	11578
Argon	Ar	18	40(99.6), 36(0.34), 38(0.063)	39.9	−189	−186	–	v0.190	–	1521	2666	3931	5771
Barium	Ba	56	138(71.7), 137(11.32), 136(7.81), 135(6.59)	137.3	725	1640	0.9	m0.224	+20.136	503	965	–	–
Beryllium	Be	4	9(100)	9.01	1278	2970	1.5	m0.112	+20.027	900	1757	14849	21007
Boron	B	5	11(80.3), 10(19.7)	10.8	2300	2550	2.0	m0.098	+30.012	801	2427	3660	25026
Bromine	Br	35	79(50.5), 81(49.5)	79.9	−7	59	2.8	c0.114	−0.195	1140	2100	3500	4560
Caesium	Cs	55	133(100)	132.9	29	669	0.7	m0.272	+10.170	376	2420	3300	–
Calcium	Ca	20	40(96.97), 44(2.06), 42(0.64)	40.1	839	1484	1.0	m0.197	+20.100	590	1145	4912	6474
Carbon	C	6	12(98.9), 13(1.1)	12.0	3652	4827	2.5	c0.077	–	1086	2353	4621	6223
Chlorine	Cl	17	35(75.5), 37(24.5)	35.5	−101	−35	3.0	c0.099	−0.180	1251	2297	3822	5158

Element	Symbol	Atomic number	Stable mass number (% abundance)	Molar mass /g mol⁻¹	Melting point/°C	Boiling point/°C	Electro-negativity[1]	Atomic radius[2] /nm	Ionic radius[3] /nm	Ionisation energies /kJ mol⁻¹ 1st	2nd	3rd	4th	Symbol
Chromium	Cr	24	52(83.8), 53(9.55), 50(4.31)	52.0	1857	2670	1.6	m0.129	+³0.062	653	1592	2987	4740	Cr
Cobalt	Co	27	59(100)	58.9	1495	2870	1.8	m0.125	+²0.065	758	1646	3232	4950	Co
Copper	Cu	29	63(69.1), 65(30.9)	63.5	1083	2567	1.9	m0.128	+²0.073	746	1958	3554	5330	Cu
Fluorine	F	9	19(100)	19.0	-220	-188	4.0	c0.071	-0.133	1681	3374	6051	8408	F
Helium	He	2	4(100)	4.0	-270	-269	–	v0.180	–	2372	5251	–	–	He
Hydrogen	H	1	1(99.98), 2(0.015)	1.0	-259	-253	2.1	c0.037	-0.208	1312	–	–	–	H
Iodine	I	53	127(100)	126.9	114	184	2.5	c0.133	-0.215	1008	1846	3200	–	I
Iron	Fe	26	56(91.7), 54(5.8), 57(2.2)	55.8	1535	2750	1.8	m0.126	+³0.055	759	1561	2958	5290	Fe
Krypton	Kr	36	84(56.9), 86(17.4), 82(11.5), 83(11.5)	83.8	-157	-152	–	v0.200	–	1351	2368	3565	5070	Kr
Lithium	Li	3	7(92.6), 6(7.4)	6.9	171	1342	1.0	m0.157	+¹0.074	520	7298	11815	–	Li
Magnesium	Mg	12	24(78.6), 25(10.1), 26(11.3)	24.3	649	1107	1.2	m0.160	+²0.072	738	1451	7733	10541	Mg
Manganese	Mn	25	55(100)	54.9	1244	1962	1.5	m0.137	+²0.067	717	1509	3249	4940	Mn
Neon	Ne	10	20(90.9), 22(8.8)	20.2	-248	-246	–	v0.160	–	2081	3952	6122	9370	Ne
Nickel	Ni	28	58(67.8), 60(26.2), 62(3.7)	58.7	1455	2730	1.8	m0.125	+²0.070	737	1753	3394	5300	Ni
Nitrogen	N	7	14(99.6), 15(0.4)	14.0	-210	-196	3.0	c0.075	-³0.171	1402	2856	4578	7475	N
Oxygen	O	8	16(99.8), 18(0.2)	16.0	-218	-183	3.5	c0.073	-²0.140	1314	3388	5301	7469	O
Phosphorus	P	15	31(100)	31.0	44 (white)	280 (white)	2.1	c0.110	-³0.190	1012	1903	2912	4957	P
Potassium	K	19	39(93.2), 41(6.8)	39.1	63	760	0.8	m0.235	+¹0.138	419	3051	4412	5877	K
Rubidium	Rb	37	85(72.15), 87(27.85)	85.5	39	686	0.8	m0.250	+¹0.149	403	2632	3900	5080	Rb
Scandium	Sc	21	45(100)	45.0	1541	2831	1.3	m0.164	+³0.075	631	1235	2389	7089	Sc
Silicon	Si	14	28(92.2), 29(4.7), 30(3.1)	28.1	1410	2355	1.8	c0.118	+⁴0.040	789	1577	3232	4356	Si
Sodium	Na	11	23(100)	23.0	98	883	0.9	m0.191	+¹0.102	496	4563	6913	9544	Na
Strontium	Sr	38	88(82.6), 86(9.9), 87(7.0)	87.6	769	1384	1.0	m0.215	+²0.113	550	1064	4210	5500	Sr
Sulphur	S	16	32(95), 34(4.2), 33(0.8)	32.1	119	445	2.5	c0.102	-²0.185	1000	2251	3361	4564	S
Titanium	Ti	22	48(74), 46(8.0), 47(7.3), 49(5.5)	47.9	1660	3287	1.5	m0.147	+⁴0.061	658	1310	2653	4175	Ti
Vanadium	V	23	51 (99.7), 50 (0.3)	50.9	1890	3380	1.6	m0.135	+³0.064	650	1414	2828	4507	V
Xenon	Xe	54	Many	131.3	-112	-107	–	v0.220	–	1170	2047	3100	–	Xe
Zinc	Zn	30	Several	65.4	420	907	1.6	m0.137	+²0.075	906	1733	3833	5730	Zn

[1] Pauling electronegativity index   [2] m = metallic radius; v = van der Waals radius; c = covalent radius   [3] Superscript shows the charge on the ion

Selected data for some inorganic compounds

Compound	Formula	State	Molar mass/g mol^{-1}	T_m/K	T_b/K	ΔH/kJ mol^{-1}
Aluminium fluoride	AlF_3	s	84.0	1564 (sub)	–	–1504
Aluminium chloride	$AlCl_3$	s	133.3	463	451 (sub)	–704
Aluminium oxide	Al_2O_3	s	102.0	2345	3253	–1676
Caesium fluoride	CsF	s	151.9	955	1524	–553
Caesium chloride	CsCl	s	168.4	918	1563	–443
Caesium oxide	Cs_2O	s	281.8	763 (in N_2)	673 (dec)	–346
Carbon monoxide	CO	g	28.0	74	82	–110
Carbon dioxide	CO_2	g	44.0	217 (at 5.2atm)	195	–393
Hydrogen fluoride	HF	g	20.0	190	293	–271
Hydrogen chloride	HCl	g	36.5	158	188	–92.3
Hydrogen bromide	HBr	g	80.9	185	206	–36.4
Hydrogen iodide	HI	g	127.9	222	238	26.5
Water	H_2O	l	18.0	273	373	–286
Hydrogen sulphide	H_2S	g	34.1	188	212	–20.6
Lithium fluoride	LiF	s	25.9	1118	1949	–616
Lithium chloride	LiCl	s	42.4	878	1613	–408.6
Lithium oxide	Li_2O	s	29.9	>1973	–	–598
Magnesium chloride	$MgCl_2$	s	95.2	987	1685	–641
Magnesium oxide	MgO	s	40.3	3125	3873	–602
Hydrazine	N_2H_4	l	32.0	275	387	50.6
Ammonia	NH_3	g	17.0	195	240	–46.1
Nitrogen chloride	NCl_3	l	120.4	<233	<344	230.1
Phosphorus(III) choride	PCl_3	l	137.3	161	349	–320
Phosphorus(V) choride	PCl_5	s	208.2	435 (sub)	440 (dec)	–443
Silicon(IV) chloride	$SiCl_4$	l	169.9	203	331	–687
Silicon dioxide	SiO_2	s	60.1	1883	2503	–911
Sodium fluoride	NaF	s	42.0	1266	1968	–574
Sodium chloride	NaCl	s	58.4	1074	1686	–411
Sodium bromide	NaBr	s	102.9	1020	1663	–361
Sodium oxide	Na_2O	s	62.0	1548 (sub)	–	–414
Sulphur(II) chloride	SCl_2	g	103.0	195	332 (dec)	–20
Sulphur(IV) chloride	SCl_4	l	173.9	243	258 (dec)	–56
Sulphur(IV) oxide	SO_2	g	64.1	200	263	–297
Sulphur(VI) oxide	SO_3	l	80.1	290	318	–441

s = solid, l = liquid, g = gas

T_m melting point; sub = sublimes; dec = decomposes

T_b boiling point at 1 atmosphere

ΔH_f Standard molar enthalpy change of formation (i.e. at 298 K and 1 atmosphere)

Selected data for some organic compounds

Compound	Formula	State	Molar mass/g mol^{-1}	T_m/K	T_b/K	ΔH_c/kJ mol^{-1}	ΔH_f/kJ mol^{-1}
Alkanes							
Methane	CH_4	g	16.0	91.1	109.1	−890	−75
Ethane	CH_3CH_3	g	30.1	89.8	184.5	−1560	−85
Propane	$CH_3CH_2CH_3$	g	44.1	83.4	231.0	−2219	−104
Butane	$CH_3(CH_2)_2CH_3$	g	58.1	134.7	272.6	−2876	−126
Pentane	$CH_3(CH_2)_3CH_3$	l	72.2	143.1	309.2	−3509	−173
Hexane	$CH_3(CH_2)_4CH_3$	l	86.2	178.1	342.1	−4163	−199
Alkenes							
Ethene	$CH_2{=}CH_2$	g	28.1	104.1	169.4	−1411	+52
Propene	$CH_2{=}CHCH_3$	g	56.1	87.8	266.8	−2717	−0.4
trans-But-2-ene	$CH_3CH{=}CHCH_3$	g	56.1	167.6	274.0	−2705	−12
cis-But-2-ene	$CH_3CH{=}CHCH_3$	g	56.1	134.2	276.8	−2709	−8
Arenes							
Benzene	C_6H_6	l	78.1	278.6	353.2	−3267	+49
Haloalkanes							
Fluoromethane	CH_3F	g	34.0	131.3	194.7	−	−247
Chloromethane	CH_3Cl	g	50.5	176.0	248.9	−764.0	−82
Bromomethane	CH_3Br	g	94.9	179.5	276.7	−770	−37
Iodomethane	CH_3I	l	141.9	206.7	315.5	−815	−15
Dichloromethane	CH_2Cl_2	l	84.9	178.0	313.1	−606	−124
Trichloromethane	$CHCl_3$	l	119.4	209.6	334.8	−474	−135
Tetrachloromethane	CCl_4	l	153.8	250.1	349.6	−360	−130
Alcohols							
Methanol	CH_3OH	l	32.0	179.2	338.1	−726	−239
Ethanol	CH_3CH_2OH	l	46.1	155.8	351.6	−1367	−277
Propan-1-ol	$CH_3CH_2CH_2OH$	l	60.1	146.6	370.5	−2021	−303
Propan-2-ol	$CH_3CHOHCH_3$	l	60.1	183.6	355.5	−2006	−318
Butan-1-ol	$CH_3(CH_2)CH_2OH$	l	74.1	183.6	390.3	−2676	−327
Pentan-1-ol	$CH_3(CH_2)CH_2OH$	l	88.2	194.1	411.1	−3329	−354
Hexan-1-ol	$CH_3(CH_2)_4CH_2OH$	l	102.2	226.4	431.1	−3984	−379
Aldehydes							
Methanal	HCHO	g	30.0	181.1	252.1	−571	−109
Ethanal	CH_3CHO	g	44.1	152.1	293.9	−1167	−191
Propanal	CH_3CH_2CHO	l	58.1	192.1	321.9	−1821	−217
Ketones							
Propanone	CH_3COCH_3	l	58.1	177.8	329.3	−1816	−248
Butanone	$CH_3CH_2COCH_3$	l	72.1	186.8	352.7	−2441	−276
Carboxylic acids							
Methanoic	HCOOH	l	46.0	281.5	373.7	−254	−425
Ethanoic	CH_3COOH	l	60.1	289.7	391.0	−874	−484
Propanoic	CH_3CH_2COOH	l	74.1	252.3	414.1	−1527	−511
Butanoic	$CH_3CH_2CH_2COOH$	l	88.1	268.6	438.6	−2183	−534

s = solid, l = liquid, g = gas

T_m melting point

T_b boiling point at 1 atmosphere

ΔH_c Standard molar enthalpy change of combustion

ΔH_f Standard molar enthalpy change of formation (i.e. at 298 K and 1 atmosphere)

Chapter 1

1   a   The curve shows the number of particles with particular energies, and the peak shows that more particles have this energy value than any other energy value.
    b   The value for both axes is zero at the origin, so no particles in the gas have zero energy.
    c   The curve will not touch the axis after zero energy because there is not a limit to the energy that a particular particle may receive.
2   a   Reactions occur at the surface of a solid, so the greater the surface area, the faster will be the rate. A 2 cm cube has a surface area of $6 \times 4 = 24$ cm^2. Eight 1 cm cubes have a total area of $8 \times 6 = 48$ cm^2, so the rate increases by a factor of 2.
    b   Only those particles with an energy greater than or equal to the activation energy can produce collisions that result in a reaction. Increasing the temperature increases the number of particles with the activation energy, so more particles can react. A 10 °C rise in temperature will approximately double the number of particles having energy greater than or equal to activation energy, so a 10 °C rise in temperature will approximately double the rate of reaction.
3   a   The reactant is used up fastest at the start of the reaction when its concentration is greatest. At this point the slope of the curve is at its steepest.
    b   For the same reasons, the rate at which the product is formed is fastest at the start: product formation rate decreases as the reaction proceeds.
4   a   order = the power to which the concentration terms are raised. In the rate equation, the power of NO is 2, so the reaction is second order with respect to NO.
    b Cl_2 is to the power 1, so the reaction is first order w.r.t. Cl_2.
    c   Overall order = the sum of the powers to which the reactant concentration terms are raised. In the rate equation, the powers are $2 + 1 = 3$, so the reaction is third order.
    d   Units for k = (units of rate)/(units of concentration) to their respective powers: (mol dm^{-3} s^{-1})/[(mol dm^{-3})$^2 \times$ mol dm^{-3}] = mol^{-2} dm^6 s^{-1}.
5   a   Graph A   b   Graph B
    c   As the temperature is increased the value of the rate constant increases: k = rate/concentration terms; the rate increases with temperature and the concentrations have not changed.

Chapter 2

1   a   Assuming that H_2 and I_2 are mixed at the start of the reaction, curve A represents their concentration and curve B represents the concentration of HI. At the start, the concentration of the reactants is high and decreases as they react. Conversely, HI is not present at the start of the reaction and its concentration increases as it is formed.
    b   Increasing the temperature will increase the rate of the forward and the reverse reactions (but not equally).
2   a $K_c = [CO][NO_2]/[CO_2][NO]$. K_c has no units.
    b $K_c = [C_2H_4][H_2]/[C_2H_6]$. Units of K_c = mol dm^{-3}
    c $K_c = [NO_2]^2/[NO]^2[O_2]$. Units of K_c = mol^{-1} dm^3
3   a $K_c = [CuA][H^+]^2/[Cu^{2+}][H_2A]$
    b   The equilibrium lies well over to the right, so the value of K_c will be quite large.
    c   At equilibrium, in mol dm^{-3}, $[Cu^{2+}] = 0.0001$, $[H_2A] = 0.0001$, $[H^+] = 0.0049^2$
    d $K_c = 11.76$ mol dm^{-3}
    e   It shifts the equilibrium from right to left (Le Chatelier).
    f   The value of K_c is unchanged by changes in concentration.
4   If the total pressure is increased, the equilibrium shifts to the side with fewer molecules (Le Chatelier).
    a   Left 3 molecules, right 2 molecules: equilibrium shifts from left to right.
    b   Left 1 molecule, right 2 molecules: equilibrium shifts from right to left.
    c   Left 2 molecules, right 2 molecules: equilibrium unchanged.
    d   Left 2 molecules, right 1 molecule: equilibrium shifts from left to right.
5   Assuming no ethanol has formed:
    number of moles of ethene = 1000/28 = 35.7 mol
    number of moles of water = 1000/18 = 55.6 mol
    total number of moles = 91.3 mol
    mole fraction of ethene = 35.7/91.3 = 0.39
    mole fraction of water = 55.6/91.3 = 0.61
    partial pressure of ethene $0.39 \times 5 \times 10^6 = 1.95 \times 10^6$ Pa
    partial pressure of water $0.61 \times 5 \times 10^6 = 3.05 \times 10^6$ Pa
6   a

	C_2H_4	$+$	H_2O	$\rightleftharpoons$	C_2H_5OH
start of reaction / mol 0	1		1		
equilibrium / mol	1 − 0.812 0.188		1 − 0.812 0.188		0.812 0.812

    b $p(C_2H_4) = \dfrac{0.188 \times 2.02 \times 10^5}{1.188} = 3.20 \times 10^4$ Pa
        $p(H_2O) = \dfrac{0.188 \times 2.02 \times 10^5}{1.188} = 3.20 \times 10^4$ Pa
        $p(C_2H_5OH) = \dfrac{0.812 \times 2.02 \times 10^5}{1.188} = 1.39 \times 10^5$ Pa
    c $K_p = \dfrac{p(C_2H_5OH)}{p(H_2O)\, p(C_2H_4)} = \dfrac{1.39 \times 10^5}{(3.20 \times 10^4)(3.20 \times 10^4)}$
        $= 1.36 \times 10^{-4}$
    d   Doubling the total pressure does not affect the value of K_p but it does increase the yield of ethanol (moves to the side with the fewer gas molecules).
7   a   Only a small amount of product is formed because K_c is very small, so the reaction can be considered as unlikely to occur.
    b   At the higher temperature the value of K_c will be higher and the yield of NO will increase
    c   Endothermic because K_c has increased with increase in temperature, $K_c = [NO]^2/[N_2][O_2]$ so the yield of NO (right hand side of equation) increases. Le Chatelier's principle predicts that if temperature is increased a reaction will shift in the endothermic direction.

8   At midnight (the start of the graph), the concentration of NO is quite low. As traffic starts to build up the concentration of NO increases. Following the production of NO then it is converted to NO_2 by oxidation from the oxygen in the atmosphere. When the concentration of NO is sufficiently high the rate of formation of NO_2 will be significant and the NO is being removed as it is formed so this reaches a maximum value. Also as the light intensity increases the concentration of NO_2 decreases as it undergoes a series of photochemical reactions and one of the products is 'ground level ozone'. As the level of traffic decreases the concentration of NO_2 decreases. There is a further slight increase in the level of NO from increased traffic but this is being removed as it is formed by the 'ground level ozone' present in the atmosphere from the previous reactions. From then the levels of NO and NO_2 will decrease.
9   a   The reaction to produce ethanol is exothermic so the equilibrium yield of ethanol will be decreased by an increase in temperature.
    b   The increase in temperature will decrease the equilibrium yield but the rate of reaction will increase so ethanol is produced faster. So over a given time more ethanol is obtained.

Application Haemoglobin and gas exchange
1   a $K_c = [HbO_2]/[Hb][O_2]$ b   mol^{-1} dm^3.
    c   Exercise uses the O_2 from HbO_2 in the muscles and this shifts the equilibrium from left to right to replace the O_2 in the HbO_2.
2   a $K_c = [HbCO_2][O_2]/[HbO_2][CO_2]$ b K_c has no units.
3   a $K_c = [H^+][HCO_3^-]/[CO_2][H_2O]$
    b   Because water is present in such a large excess compared to the other species in the equilibrium, its concentration changes negligibly so it can be incorporated into a new constant - the acid dissociation constant.
    c 4.48×10^{-7} mol dm^{-3}
4   a   moles $O_2 = 0.2/32 = 6.25 \times 10^{-3}$
        moles $N_2 = 0.8/28 = 2.86 \times 10^{-2}$
        total moles = $6.25 \times 10^{-3} + 2.86 \times 10^{-2} = 3.4821 \times 10^{-2}$
        mole fraction $O_2 = 6.25 \times 10^{-3} / 3.48 \times 10^{-2} = 1.79 \times 10^{-1}$
        mole fraction $N_2 = 2.8571 \times 10^{-2} / 3.4821 \times 10^{-2} = 8.21 \times 10^{-1}$
        partial pressure O_2 at sea level = $1.7949 \times 10^{-1} \times 1.01 \times 10^5 = 1.81 \times 10^4$
        partial pressure N_2 at sea level = $8.2051 \times 10^{-1} \times 1.01 \times 10^5 = 8.29 \times 10^4$
        partial pressure O_2 at 3000m = $1.7949 \times 10^{-1} \times 7.0 \times 10^4 = 1.26 \times 10^4$
        partial pressure N_2 at 3000m = $8.2051 \times 10^{-1} \times 7.0 \times 10^4 = 5.74 \times 10^4$
    b   At higher altitudes the air is more rarefied so that the concentration (partial pressure) of oxygen is lower. If the partial pressure of oxygen in the lungs is lower, less oxygen dissolves in the blood. Therefore equilibrium (1) moves from right to left. This means that there less HbO_2 is available for respiration.
5   If the temperature is increased the equilibrium shifts from right to left reducing the HbO_2 available to the muscles.

Chapter 3

1   a $HCl(aq) + KOH(aq) \rightarrow KCl(aq) + H_2O(l)$
        $H^+(aq) + OH^-(aq) \rightarrow H_2O(l)$
    b $H_2SO_4(aq) + 2NaOH(aq) \rightarrow 2NaCl(aq) + 2H_2O(l)$
        $2H^+(aq) + OH^-(aq) \rightarrow 2H_2O(l)$
2   a   The hydrogens in methane are held by covalent bonds between elements with similar electronegativities, so there is no polarisation of the C–H bond. In CH_3COOH one of the hydrogens is attached to an oxygen atom, an element with a high electronegativity. The difference in electronegativity means there is a polarisation of the O–H bond and it dissociates more easily, releasing H^+ (H_3O^+) ions.
    b $HCOOH(aq) + H_2O(l) \rightleftharpoons HCOO^- + H_3O^+(aq)$
3   a $CuO(s) + H_2SO_4(aq) \rightleftharpoons CuSO_4(aq) + H_2O(l)$
        base   acid
    b $NH_4^+(aq) + OH^-(aq) \rightleftharpoons NH_3(aq) + H_2O(l)$
        acid   base
    c $CH_3COO^-(aq) + H_3O^+(aq) \rightleftharpoons CH_3COOH(aq) + H_2O(l)$
        base   acid
4   $K_a = \dfrac{[H^+] \times [CH_3(CH_2)_2COO^-]}{[CH_3(CH_2)_2COOH]}$
5   The strength of the acid depends upon the value of K_a, stronger acids have higher K_a values. K_a for reaction1: sulphuric(IV) acid → hydrogensulphate is 1.5×10^{-2} mol dm^{-3} which is higher than 6.2×10^{-8} mol dm^{-3} for reaction 2: hydrogensulphate → sulphate. Therefore reaction 1 is the stronger acid.
6   pH $= -\log_{10}[H^+]$,
    a $-\log_{10}0.2 = 0.70$ b $-\log_{10} 1.2 \times 10^{-2} = 1.92$ c $-\log_{10} 6.7 \times 10^{-5} = 4.17$
7   pH $= -\log_{10}[H^+]$, $= -\log_{10} 2.0 = -0.30$
8   a $[H^+] = 10^{-pH} = 10^{-5.49} = 3.24 \times 10^{-6}$ mol dm^{-3}
    b $[H^+] = 10^{-2.75} = 1.78 \times 10^{-3}$ mol dm^{-3}.
    c $[H^+] = 10^{+0.5} = 3.16$ mol dm^{-3}.
9   $[H^+] = \sqrt{\text{acid concentration} \times K_a} = \sqrt{0.01 \times 1.34 \times 10^{-5}} = 3.66 \times 10^{-4}$,
    pH $= -\log_{10}[H^+] = -\log_{10}3.66 \times 10^{-4} = 3.44$.
10 $K_w = 10^{-14} = [H^+][OH^-]$, $[OH^-] = 10^{-7}$ mol dm^{-3}
11 a $[H^+] = K_w/[OH^-] = 10^{-14}/0.01 = 10^{-12}$, pH $= -\log_{10}[H^+] = -\log_{10}10^{-12} = 12$
    b $[H^+] = K_w/[OH^-] = 10^{-14}/3.0 = 3.33 \times 10^{-15}$ pH $= -\log_{10}[H^+] = -\log_{10}3.33 \times 10^{-15} = 14.5$
12 $[H^+]^2 = [CH_3CH_2COOH] \times 10^{-pKa} = 0.45 \times 10^{-4.87} = 6.07 \times 10^{-6}$,
    $[H^+] = \sqrt{6.07 \times 10^{-6}} = 2.46 \times 10^{-3}$, pH $= -\log_{10}2. \times 10^{-3} = 2.61$
13 a $K_a = [H^+]^2/[HA] = (7.94 \times 10^{-4})^2/0.01 = 6.30 \times 10^{-5}$
    b   This value for K_a is a small value, so benzoic acid is a weak acid.
    c $pK_a = -\log_{10}6.30 \times 10^{-5} = 4.2$
14 a $K_a = 10^{-pKa} = 10^{-3.075} = 8.41 \times 10^{-4}$.
    b $[H^+] = \sqrt{Ka \times [HA]} = \sqrt{8.41 \times 10^{-4} \times 0.01} = \sqrt{8.41 \times 10^{-6}} = 2.90 \times 10^{-3}$
    c   pH $= -\log_{10}2.90 \times 10^{-3} = 2.54$

15 A weak acid

16 a

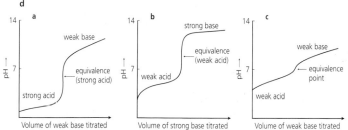

b Equivalence point pH 8.5

17 a The equation for this reaction is:
$Na_2CO_3(aq) + H_2SO_4(aq) \rightleftharpoons 2NaCl(aq) + H_2O + CO_2$
1 mol of sodium carbonate reacts with 1 mol of sulphuric acid. Using the equation:
no. of moles = molarity of soln. × volume of soln
in a solution   (in mol dm⁻³)   (in dm³)
For the sodium carbonate solution:
no. of moles of Na_2CO_3 = $0.1 \times 0.025 = 2.5 \times 10^{-3}$ mol
i.e. 2.5×10^{-3} mol of sulphuric acid is needed to neutralise the sodium carbonate solution, so that: $2.5 \times 10^{-3} = 0.125 \times$ volume of acid
Therefore, volume of sulphuric acid
$2.5 \times 10^{-3}/0.125 = 2.0 \times 10^{-2}$ dm³ = 20.0 cm³

b Using the equation:
no. of moles = molarity of soln. × volume of soln
in a solution   (in mol dm⁻³)   (in dm³)
For the sodium carbonate solution:
no. of moles of Na_2CO_3 = $0.1 \times 0.025 = 0.0025$ mol
From the balanced chemical equation for the reaction between sodium carbonate and hydrochloric acid we can see that for every mole of sodium carbonate neutralised we need twice the number of moles of hydrochloric acid, i.e. 2×0.0025. Therefore, for the hydrochloric acid solution, the number of moles of HCl = 0.005 = molarity of soln × 0.035
So the molarity of the HCl solution = 0.143 mol dm⁻³.

18 a methyl orange   b phenolphthalein   c none suitable
d

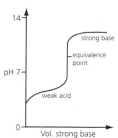

19 a NH_3 is a weak base:
$NH_3(aq) + H_2O(l) \rightleftharpoons NH_4^+(aq) + OH^-(aq)$
The salt NH_4Cl provides a supply of NH_4^+, ions. If H^+ ions are added to the buffer solution the ions will react with NH_3 to produce NH_4^+ ions, thereby removing H^+ ions from the solution:
$NH_3(aq) + H^+(aq) \rightleftharpoons NH_4^+(aq)$
If OH⁻ ions are added to the buffer they will react with NH_4^+ ions to produce NH_3, thereby removing OH⁻ ions from solution:
$NH_4^+(aq) + OH^-(aq) \rightleftharpoons NH_3(aq) + H_2O(l)$

b $[H^+] = \dfrac{K_a \times [CH_3CH_2COOH]}{[CH_3CH_2COO^-]}$ = $1.3 \times 10^{-5} \times 0.05/0.02 = 3.25 \times 10^{-5}$
pH = $-\log_{10} 3.25 \times 10^{-5} = 4.49$

20 a $[H^+] = \dfrac{K_a \times [CH_3CH_2COOH]}{[CH_3CH_2COO^-]}$
no of moles of CH_3CH_2COOH = $25.0 \times 2/1000 = 5 \times 10^{-2}$
no of moles of $CH_3CH_2COO^-$ = $20.0 \times 1.5/1000 = 3 \times 10^{-2}$
$[H^+] = 1.34 \times 10^{-5} \times (5 \times 10^{-2})/(3 \times 10^{-2}) = 2.23 \times 10^{-5}$ pH = $-\log_{10}[H^+] = 4.65$

b The addition of acid shifts the equilibrium to the left
$CH_3CH_2COOH(aq) \rightleftharpoons CH_3CH_2COO^-(aq) + H^+(aq)$
$H^+(aq)$ reacts with $CH_3CH_2COO^-(aq)$ to give $CH_3CH_2COOH(aq)$ so decrease the number of moles of $CH_3CH_2COO^-$ by the number of moles of HNO_3 added and increase the number of moles of CH_3CH_2COOH by the same amount
Moles of HNO_3 = $10.0 \times 0.05/1000 = 5 \times 10^{-4}$ mol
Moles of $CH_3CH_2COO^-(aq)$ after adding HNO_3 = $3 \times 10^{-2} - 5 \times 10^{-4} = 2.95 \times 10^{-2}$
Moles of $CH_3CH_2COOH(aq)$ after adding HNO_3 = $5 \times 10^{-2} + 5 \times 10^{-4} = 5.05 \times 10^{-2}$
$[H^+] = \dfrac{K_a \times [CH_3CH_2COOH]}{[CH_3CH_2COO^-]}$ = $1.34 \times 10^{-5} \times 5.05 \times 10^{-2}/2.95 \times 10^{-2}$ = 2.294×10^{-5}, pH = $-\log_{10}[H^+] = 4.64$
the pH of the solution has changed from 4.65 to 4.64 i.e. by 0.01 pH units, so the addition of a significant amount of acid has had little effect on the pH of the solution. The solution has acted as a buffer.

Application

1 Strong base: one which is fully ionised in solution
Alkali: a base which is soluble in water

2 $CH_3(CH_2)_{16}COOH + H_2O \rightleftharpoons CH_3(CH_2)_{16}COO^- + H_3O^+$
For a weak acid this equilibrium lies well to the left, so the H_3O^+ concentration is low.
$CH_3(CH_2)_{16}COOH + NaOH \rightarrow CH_3(CH_2)_{16}COONa + H_2O$

3 $K_a = [CH_3(CH_2)_{16}COO^-] [H^+]/[CH_3(CH_2)_{16}COOH]$
$K_a = [A^-] [H^+]/[HA] \approx [H^+]^2/[HA] [H^+]^2 = [HA] \times K_a$
$[H^+] = \sqrt{(\text{acid concentration} \times K_a)} = \sqrt{(0.5 \times 3.98 \times 10^{-6})} = 3.66 \times 10^{-4}$,
pH = $-\log_{10}[H^+] = -\log_{10} 3.66 \times 10^{-4} = 3.44$.

4 Making soap: stearic acid, a weak acid, is neutralised with sodium hydroxide, a strong base, so the titration curve is that for a weak acid/strong base. The equivalence point (mid point on vertical part of curve) gives pH = 9 approx, so a solution of sodium stearate will be slightly alkaline.

5 A stronger acid is more dissociated, so the equilibrium HA $\rightleftharpoons$ H⁺ + A⁻ lies further to the right and K_a will have a higher value.

6 a Reaction equation: $C_{12}H_{25}OSO_2OH + NaOH \rightarrow C_{12}H_{25}OSO_2ONa + H_2O$
no. of moles in a solution
= molar conc. of soln. (mol dm⁻³) × volume of soln. (dm³)
no. of moles of acid = $(0.270 \times 25.0)/1000 = 6.75 \times 10^{-3}$.
no of moles of alkali needed (1:1 stoichiometry) = 6.75×10^{-3}
vol. NaOH = 1000 × no. of moles/molar conc. of soln.
= $(1000 \times 6.75 \times 10^{-3})/0.15 = 45.0$ cm³

b For a strong acid reacting with a strong base, the equivalence point is at pH 7 and any of the usual indicators would be suitable.

7 A buffer is a solution which resists a change in pH. There are acidic buffers (weak acid with its strong base salt) and basic buffers (weak base with its strong acid salt).

8 The buffering action works because the weak acid or base maintains a store of associated ions which are released when required. The buffering action on the skin is due to a mixture of lactic acid (LaH) and lactate (La⁻).
$LaH(aq) + H_2O(l) \rightleftharpoons La^-(aq) + H_3O^+(aq)$ $NaLa(aq) \rightleftharpoons Na^+(aq) + La^-(aq)$
The presence of lactate ions pushes the lactic acid equilibrium towards undissociated acid.
$LaH(aq) + H_2O(l) \rightleftharpoons La^-(aq) + H_3O^+(aq)$
$La^-(aq)$ present in large amounts from salt drives equilibrium to the left.
If an acid shampoo is added to the buffer the equilibrium adjusts; H_3O^+ combines with some of the large amount of La⁻ to give more undissociated acid:
$LaH(aq) + H_2O(l) \rightleftharpoons La^-(aq) + H_3O^+(aq)$ (equilibrium moves to the left).
If alkaline shampoo is used the equilibrium adjusts by moving to the right:
$LaH(aq) + H_2O(l) \rightleftharpoons La^-(aq) + H_3O^+(aq)$. $OH^-(aq)$ reacts with $H_3O^+(aq)$ and removes it from the mixture. Some of the large amount of LaH present dissociates to give more H_3O^+. The buffer maintains the equilibrium reaction and the pH remains steady, except in the presence of large quantities of acid or alkali. K_a remains constant throughout. K_a can be used to show that $[H^+]$ remain fairly constant.

$K_a = \dfrac{[La^-(aq)][H_3O^+(aq)]}{[LaH(aq)]}$ $[H_3O^+(aq)] = \dfrac{K_a[LaH(aq)]}{[La^-(aq)]}$

If both $[La^-(aq)]$ and $[LaH(aq)]$ are large, then small changes in their concentrations will not affect the overall ratio significantly, so $[H_3O^+(aq)]$ remains fairly constant.

9 a $K_a = [La^-] [H^+]/[LaH]$
b $K_a = [La^-] [H^+]/[LaH]$ $[H^+] = [LaH] \times K_a/[La^-]$
$[H^+] = (0.012 \times 10^{-3} \times 8.4 \times 10^{-4})/(3.2 \times 10^{-3}) = 3.15 \times 10^{-6}$,
pH = $-\log_{10}[H^+] = -\log_{10} 3.15 \times 10^{-6} = 5.5$

10 Moles of H^+ in 10.0 cm³ of 0.2 millimol dm⁻³ shampoo
= $(10.0 \times 0.2 \times 10^{-3})/1000 = 2 \times 10^{-6}$ mol
Moles of $La^-(aq)$ after adding shampoo = $(3.2 \times 10^{-3}) - (2 \times 10^{-6})$
= 3.198×10^{-3}
Moles of $LaH(aq)$ after adding shampoo = $(0.012 \times 10^{-3}) + (2 \times 10^{-6})$
= 1.4×10^{-5}

$[H^+] = \dfrac{K_a[LaH]}{[La]}$ = $(8.4 \times 10^{-4} \times 1.4 \times 10^{-5})/(3.198 \times 10^{-3}) = 3.677 \times 10^{-6}$

pH = $-\log_{10}[H^+] = -\log_{10} 3.667 \times 10^{-6} = 5.44$
The pH of the solution has changed from 5.5 to 5.44, i.e. by 0.06 pH units, so the addition of a the shampoo has had little effect on the pH of the skin. The skin has acted as a buffer.

Chapter 4

1   heptane, octane and decane

2   a 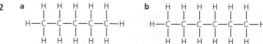   b

3   2-methylpropane   4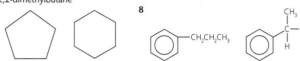

5   hexane
CH₃CH(CH₃)CH₂CH₂CH₃ 2-methylpentane
CH₃CH(CH₃)CH(CH₃)CH₃ or CH₃C(CH₃)₂CH₂CH₃ 2,3-dimethylbutane or
2,2-dimethylbutane

7   (cyclopentane)   (cyclohexane)   8 (propylbenzene)   (isopropylbenzene)

9

i bromoethane
CH₃CH₂Br

ii propanoic acid
CH₃CH₂—C(=O)OH

iii propanal
CH₃CH₂—C(=O)H

iv butan-2-one
CH₃—C(=O)—CH₂CH₃

v methylethanoate
CH₃—C(=O)—O—CH₃

vi 2-chloropropane
CH₃—CH(Cl)—CH₃

vii 4-methyl-1-nitrobenzene

viii pent-2-ene
CH₃—CH=CH—CH₂CH₃

ix methanoic acid
H—C(=O)OH

x methylpropene
CH₃—C(CH₃)=CH₂

xi butan-1-ol
CH₃CH₂CH₂CH₂OH

xii 2-methylpropan-2-ol
CH₃—C(CH₃)(OH)—CH₃

xiii but-1,3-diene
CH₂=CH—CH=CH₂

xiv 2-amino-2-phenylethanoic acid

xv N-methylaminomethane
CH₃—NH—CH₃

xvi methanoic ethanoic anhydride
H—C(=O)—O—C(=O)—CH₃

xvii butanoyl chloride
CH₃CH₂CH₂—C(=O)Cl

xviii ethyl methanoate
H—C(=O)—O—CH₂CH₃

10 i   bromomethane
    iii   butan-2-ol
    v   hexanal
    vii   propanone
    ix   1,1-dibromoethane
    xi   phenylmethanoic (benzoic) acid
    xiii butanoyl chloride
    xv   butyl ethanoate
    xvii 2-methylhex-3-ene
    xix   2-hydroxybutanenitrile
    ii   chlorobenzene
    iv   ethanoic acid
    vi   aminoethane
    viii 1,2-dibromoethane
    x   but-1-ene
    xii   ethylpropanoate
    xiv   butanoic anhydride
    xvi   hexan-3-one
    xviii propanenitrile

11 CH₃CH(CH₃)CH₂CH₂CH₃ 2-methylpentane
    CH₃CH₂CH(CH₃)CH₂CH₃ 3-methylpentane
    CH₃C(CH₃)₂CH₂CH₃ 2,2-dimethylbutane
    CH₃CH(CH₃)CH(CH₃)CH₃ 2,3-dimethylbutane

12 a CH₃CH₂CH₂CHO butanal   CH₃CH(CH₃)CHO 2-methylpropanal
       CH₃COCH₂CH₃ butan-2-one
    b CH₃CH₂CH₂OH propan-1-ol   CH₃CH(OH)CH₃ propan-2-ol
       CH₃OCH₂CH₃ methoxyethane
    c CH₃CH₂CH₂COOH butanoic acid   CH₃CH(CH₃)COOH 2-methylpropanoic acid
       CH₃CH₂COOCH₃ methyl propanoate   CH₃COOCH₂CH₃ ethyl ethanoate
       HCOOCH₂CH₂CH₃ 1-propyl methanoate
       HCOOCH(CH₃)CH₃ 2-propyl methanoate

13

b cis-pent-2-ene   trans-pent-2-ene   e cis-cyclohexan-1,4-diol   trans-cyclohexan-1,4-diol

14 b (structures)   c (structures)
    d (structures)

Application A

Drawn structural formulae where applicable.
1   CH₃CH₂CH₂CH₂CH₂CH₃ and CH₃C(CH₃)₂CH₂CH(CH₃)CH₃
2   No: they have different total numbers of carbons atoms.
3,4 1   2-methylhexane   CH₃CH(CH₃)CH₂CH₂CH₃
    2   3-methylhexane   CH₃CH₂C*H(CH₃)CH₂CH₃
    3   2,3-dimethylpentane   CH₃CH(CH₃)C*H(CH₃)CH₂CH₃
    4   2,4-dimethylpentane   CH₃CH(CH₃)CH₂CH(CH₃)CH₃
    5   2,2-dimethylpentane   CH₃C(CH₃)₂CH₂CH₂CH₃
    6   3,3-dimethylpentane   CH₃CH₂C(CH₃)₂CH₂CH₃
    7   2,2,3-trimethylbutane CH₃C(CH₃)₂CH(CH₃)CH₃
5   1   1-bromoheptane   CH₃CH₂CH₂CH₂CH₂CH₂Br
    2   2-bromoheptane   CH₃CH₂CH₂CH₂CH₂C*HBrCH₃
    3   3-bromoheptane   CH₃CH₂CH₂CH₂C*HBrCH₂CH₃
    4   4-bromoheptane   CH₃CH₂CH₂CHBrCH₂CH₂CH₃
6   hept-1-ene   CH₃CH₂CH₂CH₂CH₂CH=CH₂
    hept-2-ene   CH₃CH₂CH₂CH₂CH=CHCH₃
7   Hept-1-ene has no geometrical isomers.

cis-hept-2-ene   trans-hept-2-ene

Application B

Drawn structural formulae where applicable.
1   CH₃COCH₃   2 CH₃CH₂CHO   propanal
3   CH₃CH₂CH₂OH propan-1-ol   CH₃CH₂COOH propanoic acid
4   CH₃CH₂COOCH₂CH₃ ethyl propanoate
    CH₃COOCH₂CH₂CH₃ propyl ethanoate. Positional isomers
5   a propan-2-ol   b 2-bromopropane   c 2-aminopropane
    d 2-methylpropanenitrile   e 1-amino-2-methylpropane

Chapter 5

1   a methanal   methanoic acid   b propanal   propanoic acid
    H—C(=O)H   H—C(=O)OH   CH₃CH₂—C(=O)H   CH₃CH₂—C(=O)OH

2   Cr(+6) to (+3)
3
CH₃CH₂CH₂CH₂OH + [O] ⟶ CH₃CH₂CH₂—C(=O)H + H₂O
                              butanal
then:
CH₃CH₂CH₂—C(=O)H + [O] ⟶ CH₃CH₂CH₂—C(=O)H
                              butanoic acid

4   a CH₃CH₂CH₂CH(OH)CH₃   pentan-2-ol
    b (C₆H₅)—CH(OH)CH₃   1-phenyl ethanol

5   (cyclohexanol) —OH + [O] ⟶ (cyclohexanone) + H₂O
    cyclohexanone

6   a
CH₃CH₂—C(=O)H + HCN ⟶ CH₃CH₂—C(H)(OH)(CN)
2-hydroxybutanenitrile

    b CH₃—C(=O)—CH₂CH₃ + HCN ⟶ CH₃—C(OH)(CN)—CH₂CH₃
2-hydroxy-2-methyl butanenitrile

7 a (structures) **b** (structures)

8 a Warm separate samples with dilute, acidified dichromate solution. Tertiary alcohol will remain orange; others will change from orange to green.
b Distil the organic product from the other samples from (a). Warm each with Fehling's solution. Primary alcohol product (aldehyde) will cause a change from blue to red-brown ppt; secondary alcohol will remain blue.

9 a $CH_3CH_2\text{—CHO} + 2[H] \longrightarrow CH_3CH_2CH_2OH$ propan-1-ol
b $CH_3CH_2CH_2\text{—CO—}CH_3 + 2[H] \longrightarrow CH_3CH_2CH_2\text{—CH(OH)—}CH_3$ pentan-2-ol
c $CH_3CH_2CH_2\text{—COOH} + 2[H] \longrightarrow CH_3CH_2CH_2\text{—CHO} + H_2O$ butanal
then $CH_3CH_2CH_2\text{—CHO} + 2[H] \longrightarrow CH_3CH_2CH_2CH_2OH$ butan-1-ol

10 methanoic acid < ethanoic acid < benzoic acid

11 $CH_3CH_2CH_2\text{—COOH} + CH_3CH_2CH_2OH \longrightarrow CH_3CH_2CH_2\text{—COO—}CH_2CH_2CH_3 + H_2O$ propyl butanoate

12 a–c (structures)

13 a $CH_3\text{—CO—}OCH_2CH_3 + H_2O \longrightarrow CH_3\text{—COOH} + CH_3CH_2OH$ ethanoic acid ethanol
b $CH_3CH_2\text{—CO—}OCH_3 + NaOH \longrightarrow CH_3CH_2\text{—COO}^-Na^+ + CH_3OH$ sodium propanoate methanol

Application A
1 CH_3CH_2OH
2 Primary. OH group is positioned at the end of a carbon chain.
3 $CH_3CH_2OH + [O] \rightarrow CH_3CHO + H_2O$; ethanal
$CH_3CHO + [O] \rightarrow CH_3COOH$; ethanoic acid
a $Cr_2O_7^{2-}$ and Cr^{3+} **b** Reduction
4 Ketones are not oxidised by $Cr_2O_7^{2-}$

5 $CH_3CH_2CH_2CH_2OH$; $CH_3CH(OH)CH_2CH_3$; $CH_3C(CH_3)(OH)CH_3$; CH_3COOH; $CH_3CH_2CH_2COOH$
a Butan-1-ol and butan-2-ol because primary and secondary alcohols are oxidised by $Cr_2O_7^{2-}$.
b 2-methylpropan-2-ol, ethanoic acid and butanoic acid because tertiary alcohols and carboxylic acids are not oxidised by $Cr_2O_7^{2-}$.
c 1-butylbutanoate; $CH_3CH_2CH_2CO.OCH_2CH_2CH_2CH_3$
6 $CH_3CH_2CH_2CO.OCH_2CH_2CH_2CH_3 + H_2O \longrightarrow CH_3CH_2CH_2COOH + CH_3CH_2CH_2CH_2OH$

Application B
1 $C_6H_{12}O_6$ for both. They are isomers.
2 Glucose – circles around C1; squares around C2, C3, C4 and C5
Fructose – circles around C1 and C6; squares around C2, C3, and C4
3 Glucose is an aldehyde; fructose is a ketone.
4 When heated with Fehling's solutions, only glucose should produce a brick red precipitate because only the glucose structure contains a reducing aldehyde group
5 By transfer of [H], the ketone group is reduced to a secondary alcohol group and the primary alcohol group is oxidised to an aldehyde group.
6 Aldehyde group, –CHO, changed to carboxylic acid group, –COOH
7 Ketone group, C=O, changed to HO–C–CN
8 Nucleophilic addition
(mechanism)
9 Nucleophilic addition – C5OH adds across C=O at C1
10 Same as answer to Question 6
11 Change any one (or more) of the –OH groups in fructose to –O–COCH₃. Catalyst.

Chapter 6
1 The 1,2-isomer has a double bond between the carbons that have the bromine atoms attached; the 1,6-isomer has a single bond.
2 The 1,5-isomer would be identical to the 1,3-isomer.
3 (structures NO₂ → NH₂) $+ 6[H] \longrightarrow + 2H_2O$
4 an electrophilic substitution reaction
5 (structure) $\text{—H} + O_2 \longrightarrow$ phenol $+ CH_3\text{—CO—}CH_3$

Application A
1 (structures) 2-nitromethylbenzene 3-nitromethylbenzene 4-nitromethylbenzene
2 (structures) 2,4-dinitromethylbenzene 2,6-dinitromethylbenzene
3 electrophilic substitution by nitronium ion (NO_2^+)
4 **a** $HNO_3 + 2H_2SO_4 \longrightarrow NO_2^+ + H_3O^+ + 2HSO_4^-$
b (mechanism)
c (mechanism) $\longrightarrow + H^+$

273

5 increase
6 decrease
7

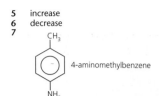

4-aminomethylbenzene

8 An azo dye

Application B
1 a lone pair acceptor
2

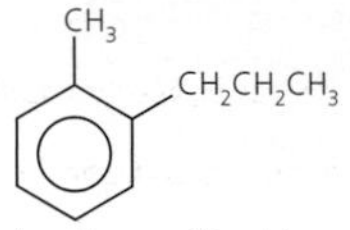

CH$_3$

CH$_2$CH$_2$CH$_3$

(or other positional isomers)

3

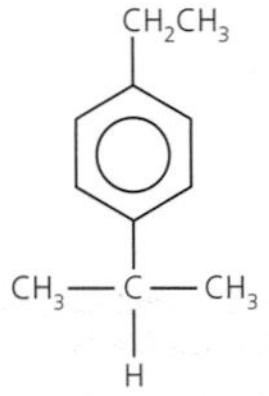

CH$_2$CH$_3$

CH$_3$—C—CH$_3$
|
H

(or other positional isomers)

4

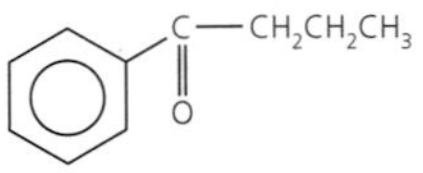

—C—CH$_2$CH$_2$CH$_3$
‖
O

5 React benzene with chloromethane (CH$_3$Cl) or chloroethane (CH$_3$CH$_2$Cl) in presence of AlCl$_3$ Friedel–Crafts catalyst.

6
a CH$_3$Cl + AlCl$_3$ ⟶ CH$_3^+$ + AlCl$_4^-$

b + $^+$CH$_3$ ⟶ (+) CH$_3$, H

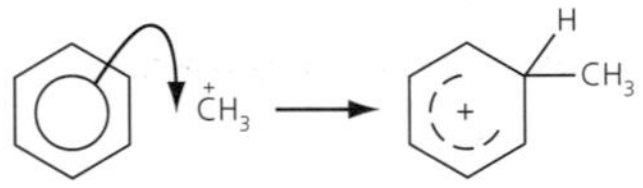

c (+) CH$_3$, H ⟶ CH$_3$ + H$^+$

d H$^+$ + AlCl$_4^-$ ⟶ HCl + AlCl$_3$

7 a + CH$_2$=CH$_2$ ⟶ CH$_2$CH$_3$

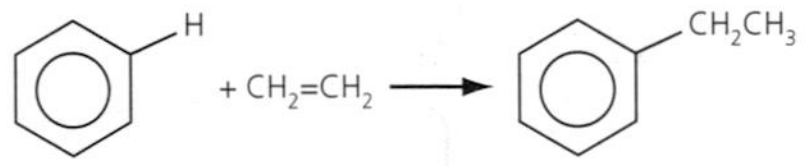

b cracking of petroleum oil

8 a CH$_2$CH$_3$ ⟶ CH=CH$_2$ + H$_2$

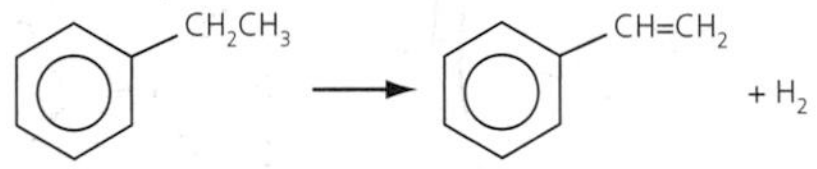

b poly(phenylethene)
c

—C—C—C—C— etc
(with H and phenyl substituents)

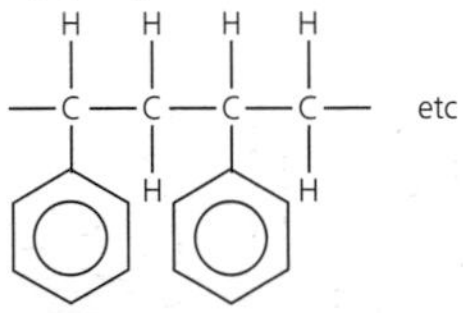

d packing and thermal insulation

Chapter 7

1 a CH$_3$CH$_2$—N—CH$_2$CH$_2$CH$_2$CH$_3$ (with H on N)
b CH$_3$CH$_2$—N—CH$_2$CH$_2$CH$_3$ (with CH$_3$ on N)

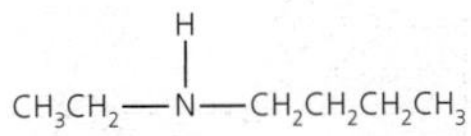

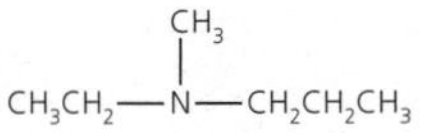

c secondary; tertiary
2 a bromoethane   **b** 2-bromopropane
c 1,6-dibromohexane
3 diethylamine, triethylamine and tetramethylammonium ion
4 a 2-bromobutane + KCN → 2-methylbutanenitrile
2-methylbutanenitrile + LiAlH$_4$ → 2-methylbutylamine
b No, because the required amine has the NH$_2$ group bonded midway along the carbon chain, whereas the nitrile reduction introduces the group at the end of the chain.

5

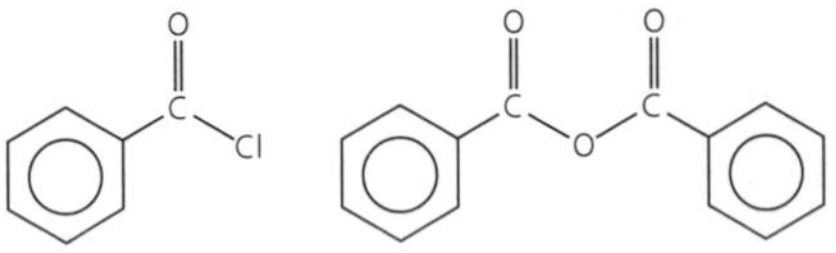

(benzoyl chloride, benzoic anhydride structures)

6
a 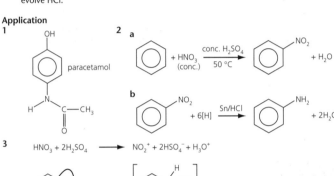 + H$_2$O ⟶ 2 benzoic acid (—COOH)

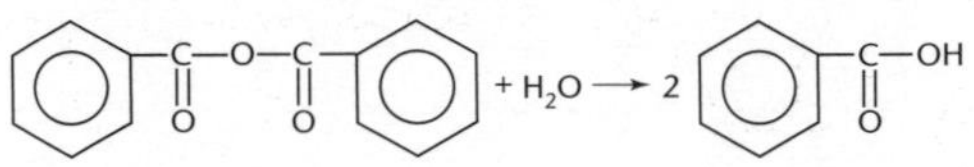

benzoic acid

b CH$_3$—CH$_2$—C—Cl + NH$_3$ ⟶ CH$_3$—CH$_2$—C—NH$_2$ + HCl
(C=O)

propanamide

c CH$_3$CH$_2$CH$_2$—C—O—C—CH$_2$CH$_2$CH$_3$ + CH$_3$CH$_2$NH$_2$

⟶ CH$_3$CH$_2$CH$_2$—C—N—CH$_2$CH$_3$ + CH$_3$CH$_2$CH$_2$—C—OH
(O N) (O)

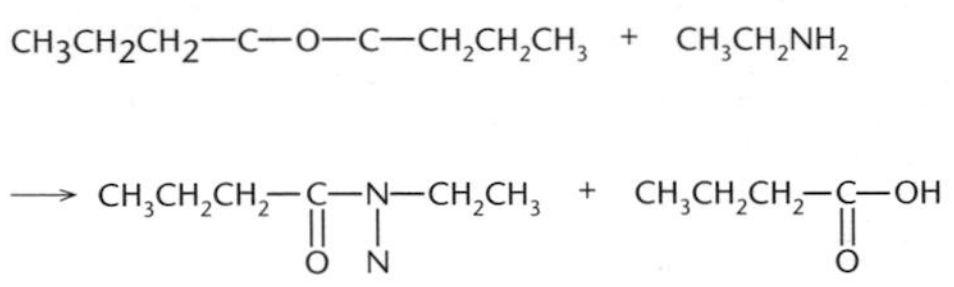

N - ethyl butanamide   butanoic acid

7 Ethanoic anhydride is used because: it is cheaper; reacts more slowly; does not evolve HCl.

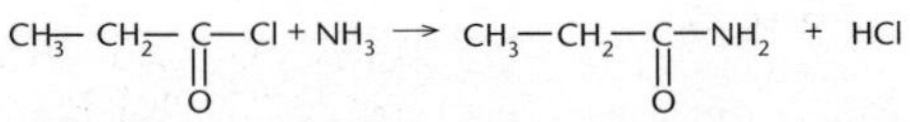

Application
1

OH

N—C—CH$_3$ (with H on N, C=O)

paracetamol

2 a + HNO$_3$ (conc.) →[conc. H$_2$SO$_4$][50 °C] NO$_2$ + H$_2$O

b NO$_2$ + 6[H] →[Sn/HCl] NH$_2$ + 2H$_2$O

3 HNO$_3$ + 2H$_2$SO$_4$ ⟶ NO$_2^+$ + 2HSO$_4^-$ + H$_3$O$^+$

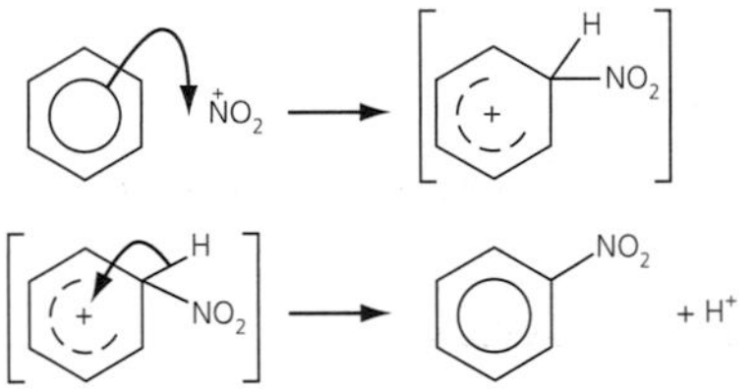

+ $^+$NO$_2$ ⟶ [(+) NO$_2$, H]

[(+) NO$_2$, H] ⟶ NO$_2$ + H$^+$

H$^+$ + HSO$_4^-$ ⟶ H$_2$SO$_4$

4

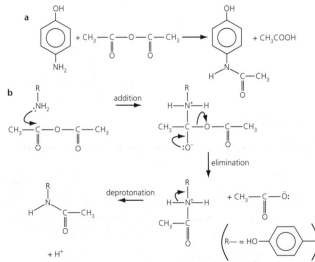

a OH, NH$_2$ + CH$_3$—C—O—C—CH$_3$ ⟶ OH, N—C—CH$_3$ + CH$_3$COOH

b (mechanism: addition, elimination, deprotonation steps)

R = HO—⟨benzene⟩—

c No HCl is produced; it is cheaper.

5

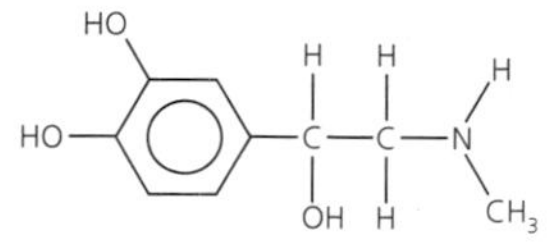

Add excess dilute sulphuric acid.

6 secondary: contains

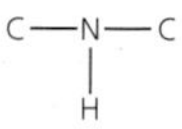

$$C-N-C$$
$$\quad\ \ |$$
$$\quad\ \ H$$

7

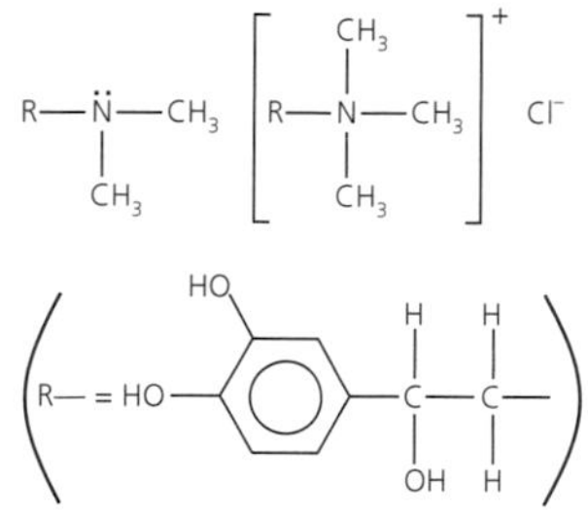

The quaternary amine will be favoured as the proportion of chloromethane is increased.

$$\left(R-=HO \left\langle\text{benzene ring}\right\rangle -C-C-\right)$$

8

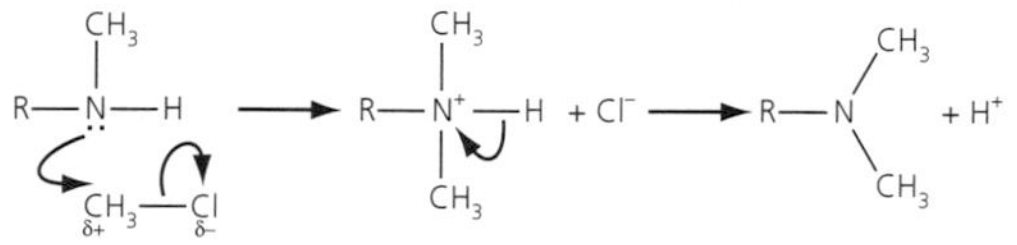

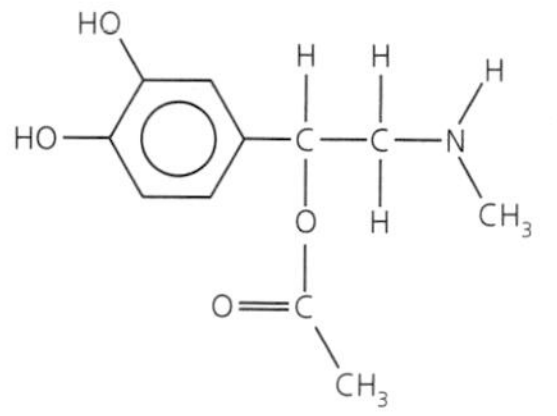

acylation, esterification or condensation

9 a

b HO– bonded to benzene ring
NH in secondary amine group

5

$$HO-C-N-C-NH_2 + H_2O \rightarrow HO-C-NH_2 + HO-C-C-NH_2$$

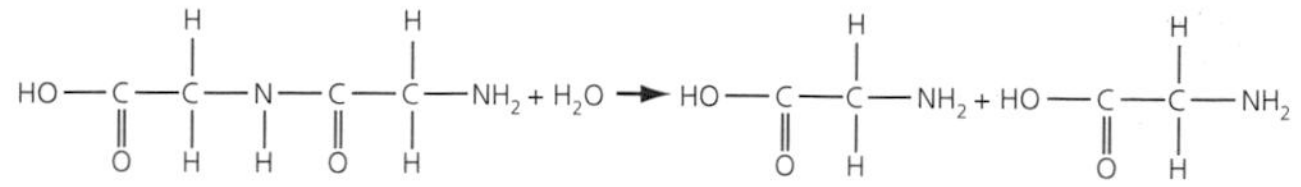

6 Structures for:
alanine HOOCCH(CH$_3$)NH$_2$
aspartic acid HOOCCH(CH$_2$COOH)NH$_2$
lysine HOOCCH(CH$_2$CH$_2$CH$_2$CH$_2$NH$_2$)NH$_2$

7

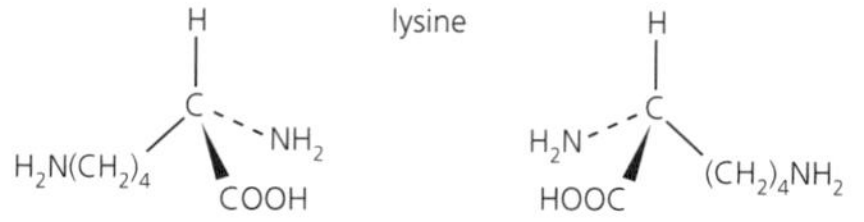

lysine

They are chemically and physically the same except they cause plane polarised light to rotate in opposite directions.

8 Lysine has an extra basic NH$_2$ group, causing it to be most alkaline. Aspartic acid has an extra acid COOH group causing it to be most acidic.

9

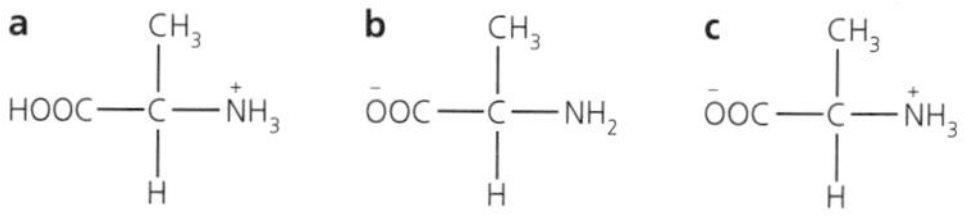

a
$$HOOC-C-\overset{+}{N}H_3$$

b
$$^-OOC-C-NH_2$$

c
$$^-OOC-C-\overset{+}{N}H_3$$

Chapter 9

1 Both involve Van der Waals intermolecular forces, but LDPE molecules are branched, which does not allow them to get close together, so the intermolecular forces are less effective.

2 All bonds in poly(ethene) are single bonds; the molecules are saturated hydrocarbons.

3 HOOC(CH$_2$)$_8$COOH and HO[CO(CH$_2$)$_8$–NH–(CH$_2$)$_6$–NH]$_n$H

4 a

$$CH_3-C-O-CH_2CH_3 + H_2O \rightarrow CH_3-C-OH + CH_3CH_2OH$$

$$CH_3-C-N-CH_2CH_3 + H_2O \rightarrow CH_3-C-OH + CH_3CH_2NH_2$$

b Hydrolysis

c Long-chain polyester and polyamide plastics will be broken down by repeated hydrolysis reactions at the ester or amide links.

Application A

1

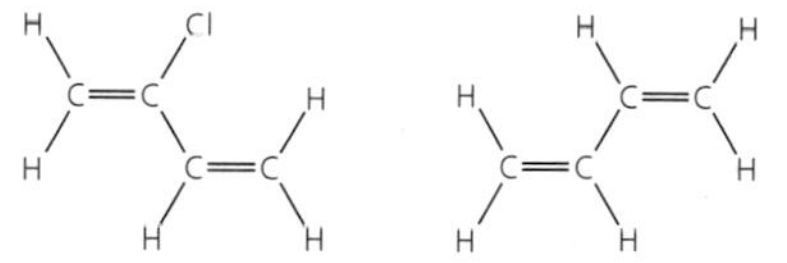

2 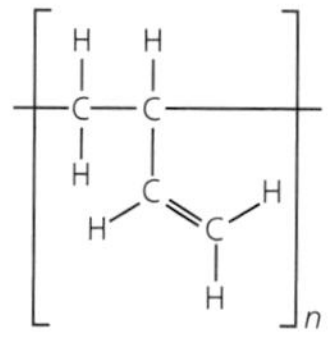 **3** 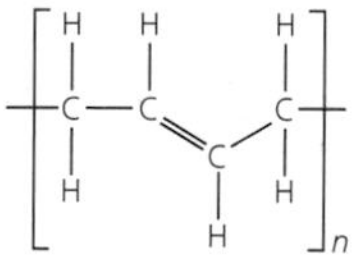 **4**

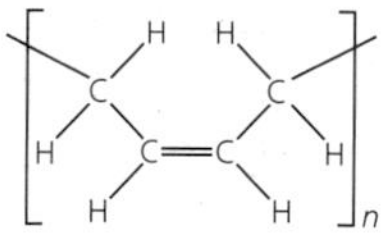

5

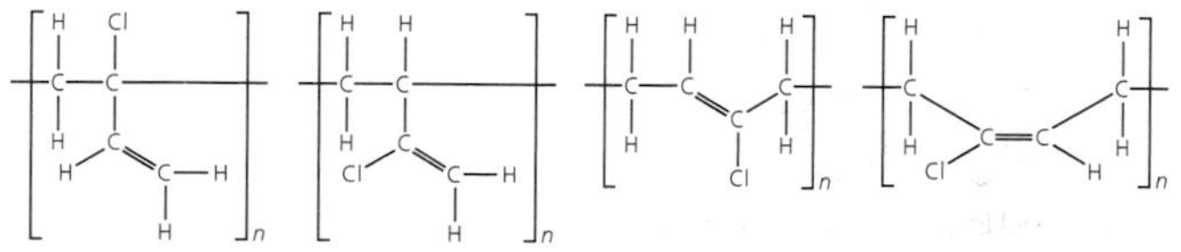

Chapter 8

1 a **b**

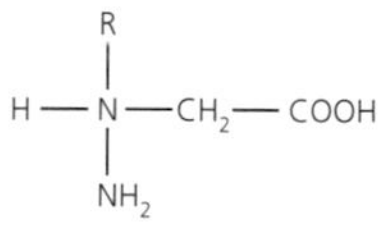

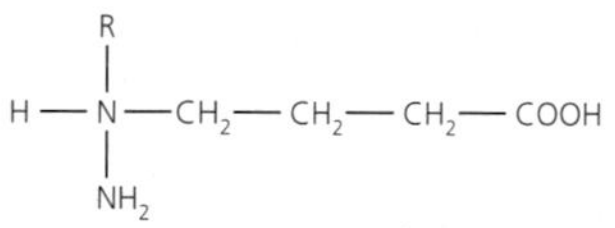

2 No carbon atom with four different groups bonded to it.
3 a negative **b** positive
4 No movement
5

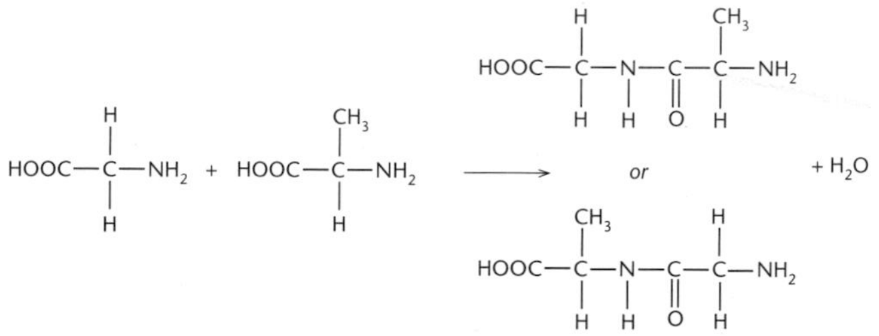

$$HOOC-C-NH_2 + HOOC-C-NH_2 \longrightarrow \quad or \quad + H_2O$$

6 Catalyst, providing H$^+$ ions.

Application

1 Number of amino acid residues shows that **a** and **b** = dipeptides, **c** = tripeptide

2
a **c**

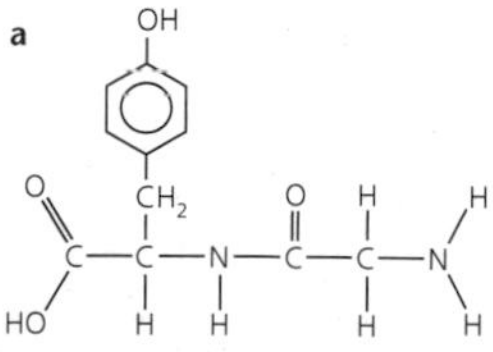

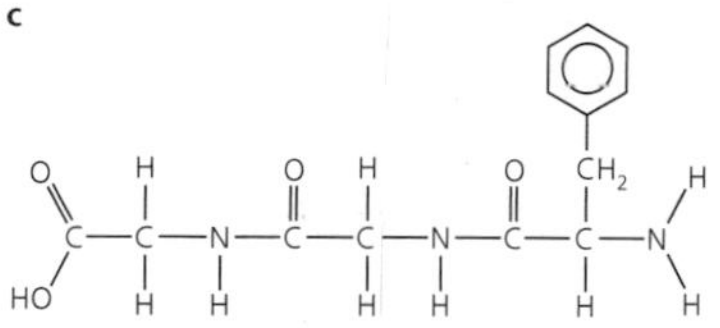

3 HOOC–Tyr–Gly–Gly–Phe–Leu–NH$_2$
$\qquad$ (3) (1) (2) (3)
Hydrolysis at link (1) forms A
Hydrolysis at link (2) forms B
Hydrolysis at link (3) forms C

4 Glycine – no carbon atom is bonded to four different groups.

Application B

1
a **b**

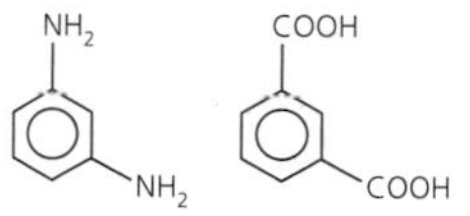

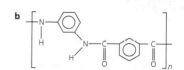

2
a 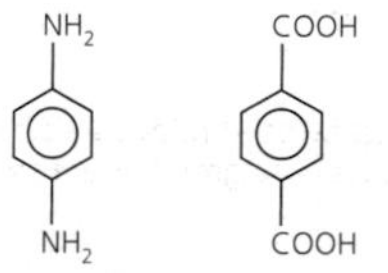 **b**

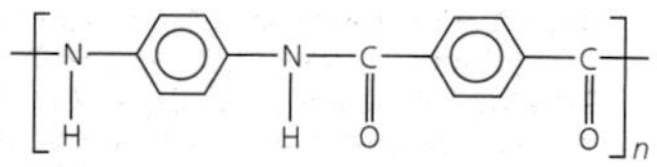

3 Hydrogen bonds

4

Chain 1 ~~~~~ C
hydrogen bond ~~~ O
H
N
Chain 2 ~~~~~ N

5 Kevlar because the '1,4' monomers cause the polymer chain to be less zig-zag shaped, so the molecules lie close together and the intermolecular forces can be most effective.

Chapter 10

1 Less oxygen/air is needed to sustain combustion. **2** Br^+
3 1,2-dibromobutane. **4** $RX + OH^- \rightarrow ROH + X^-$
5 It is partially ionised in solution. **6** $-OH$
7 Water contains OH groups and would cause interference.
8 $Cr^{3+}(aq)$ **9** Cu_2O **10** Ag
11 Apply a test to distinguish an aldehyde from a ketone.
12 Addition–elimination; condensation.
13 Lone pair on N atom of NH_2.
14 Amino acids are buffers allowing a slower rise in pH.
15 i pH (<7).
  ii Dichromate oxidation (green) and Fehling's test (negative).
  iii Boil with NaOH(aq); acidify (HNO_3); add silver nitrate (cream precipitate).
  iv As (ii) but Fehling's test positive.
  v As (ii) and pH (<7).
  vi Add water; add $AgNO_3$ (white precipitate).
  vii Add bromine solution (colourless).
  viii Sodium test (H_2 evolved) and pH (7) and dichromate oxidation (negative).
  ix pH (>7) in water.
16 a Heat with acidified dichromate.
  b Heat with NaOH(aq); acidify (HNO_3); add $AgNO_3$(aq).
  c Add aqueous bromine solution.
  d Fehling's test or silver mirror test.
  e Add sodium metal.
17 $100 \times (1/2 \times 2/5 \times 3/4) = 15\%$.
18 a Excess hot acidified dichromate solution.

$$CH_3CH_2CH(OH)CH_3 + [O] \longrightarrow CH_3CH_2-\underset{\underset{O}{\|}}{C}-CH_3 + H_2O$$

  b CH_3COOH and concentrated sulphuric acid; catalyst warm.

$$CH_3CH_2CH_2OH + CH_3COOH \longrightarrow CH_3CH_2CH_2-O-\underset{\underset{O}{\|}}{C}-CH_3 + H_2O$$

  c Concentrated H_2SO_4; 170°C.

$$CH_3CH_2CH_2CH_2OH \longrightarrow CH_3CH_2CH=CH_2 + H_2O$$

  d $LiAlH_4$ in ethoxyethane.

$$CH_3CH_2CH_2COOH + 4[H] \longrightarrow CH_3CH_2CH_2CH_2OH + H_2O$$

  e Excess ammonia in ethanol; heat and pressure.

$$CH_3CH_2Br + 2NH_3 \longrightarrow CH_3CH_2NH_2 + NH_4Br$$

19 a $CH_3CH_2Cl + NaOH(aq) \xrightarrow{\text{Reflux}} CH_3CH_2OH + NaCl$

$CH_3CH_2OH + 2[O] \xrightarrow[\text{excess. Reflux}]{Cr_2O_7{}^{2-}/H^+ \text{ in}} CH_3COOH + H_2O$

  b $CH_2{=}CH_2 + HBr \longrightarrow CH_3CH_2Br$

$CH_3CH_2Br + KCN \xrightarrow[\text{ethanol soln.}]{\text{Reflux in}} CH_3CH_2CN$

$CH_3CH_2CN + 2H_2O \xrightarrow[H_2SO_4]{\text{Reflux in}} CH_3CH_2COOH + NH_3$

  c $CH_3CH_2OH + 2[O] \xrightarrow[\text{excess. Reflux}]{Cr_2O_7{}^{2-}/H^+ \text{ in}} CH_3CH_2COOH + H_2O$

$CH_3COOH + CH_3CH_2OH \xrightarrow[H_2SO_4 \text{ catalyst}]{\text{Reflux. Conc.}} CH_3CO.OCH_2CH_3 + H_2O$

d

$+ HNO_3(\text{conc.}) \xrightarrow[50°C]{\text{Conc. } H_2SO_4}$ (nitrobenzene) $+ H_2O$

(nitrobenzene) $+ 6[H] \xrightarrow{Sn/HCl}$ (aniline) $+ 2H_2O$

Application A
1 Molecule/functional group(s): muscalure/alkene; thyroxine/halogeno, hydroxy, ether, primary amine, carboxyl; benzedrine/primary amine; grandisol/primary alcohol, alkene; aspirin/ester, carboxyl
2 Test 1: Add indicator. Only aspirin will show a significant acidic pH and only benzedrine will be significantly alkaline. Test 2: Boil remaining three with acidified $K_2Cr_2O_7$. Only grandisol will show an orange-to-green colour change. Test 3: Add aqueous bromine to remaining two. Only muscalure will decolorise brown bromine.

Application B
1

2 CH_3 group must be introduced before NO_2 group to ensure 1,4-isomer can be made. Reduction of NO_2 to NH_2 must occur before oxidation of CH_3 to COOH because COOH can also be reduced.

Step

(1) (benzene) $+ CH_3Cl \xrightarrow{AlCl_3}$ (toluene) $+ HCl$

(2) (toluene) $+ HNO_3 \xrightarrow{H_2SO_4}$ (4-nitrotoluene) $+ H_2O$

(3) (4-nitrotoluene) $+ 4[H] \xrightarrow{Sn/HCl}$ (4-methylaniline) $+ 2H_2O$

(4) (4-methylaniline) $+ 3[O] \xrightarrow[\text{in } OH^-]{KMnO_4}$ (4-aminobenzoic acid) $+ H_2O$

(5) (4-aminobenzoic acid) $+ CH_3CH_2OH \xrightarrow[H_2SO_4]{\text{conc.}}$ (ethyl 4-aminobenzoate) $+ H_2O$

Chapter 11

1 C_2H_4O **2** $C_4H_8O_2$
3 Acid (–COOH), not ester, ketol or aldol $CH_3CH_2CH_2COOH$ or $CH_3CH(CH_3)COOH$
4 1.67×10^{-5} to 2.5×10^{-6} m **5** 1.8×10^{13} to 1.2×10^{14} s^{-1}
6 7.18 to 47.9 kJ mol^{-1}
7 The heavier Br atom causes a lower frequency vibration.
8 Absorptions caused by the solvent molecules need to be eliminated and such absorptions may obscure absorptions caused by the sample molecule.
9 –C–H at $\approx$ 2950 cm^{-1}, =C–H at 3084 cm^{-1}, ≡C–H at 3312 cm^{-1}.
Higher wavenumbers correspond to stronger bonds, so the C–H bonds must be influenced by the adjacent C–C, C=C or C≡C bond.
10 1, C–H 2, C–H 3, C–H 4, O–H 5, C–H 6, C–H
7, C–H 8, C–H 9, C–H 10, C=O 11, C–H

11 a

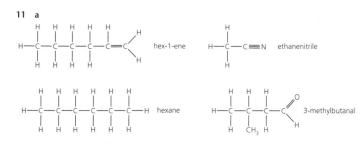

hex-1-ene ethanenitrile

hexane 3-methylbutanal

b hex-1-ene: alkane C–H, alkene C–H, C–C, C=C
ethanenitrile: alkane C–H, C–C, C≡N
3-methylbutanal: alkane C–H, aldehyde C–H, C–C, C=O
c 1: hex-1-ene 2: 3-methylbutanal
3: hexane 4: ethanenitrile

12 They will collide with the electrode, pick up electrons and revert to being neutral particles

13 a X b Z c Y

14 They contain different numbers of neutrons in their nuclei; ^{12}C has six neutrons, ^{13}C has seven.

15 $A_r(Zn) = 65.45$ (4 sf) **16** 9:6:1

17 Pairs of peaks with a mass difference of 2 in a ratio of 3:1 (containing ^{35}Cl and ^{37}Cl).

18 29 and 43

19 The presence of small numbers of molecules containing heavier isotopes (particularly ^{13}C).

20 The spectrum will be much more simpler because of the symmetry of the methoxymethane molecule. It will contain significant signals at 46 ($CH_3OCH_3^+$), 31 (CH_3O^+) and 15 (CH_3^+) only.

21 $C_8H_6O_2$, $C_6H_{14}O_3$, $C_5H_{10}O_4$, $C_4H_6O_5$

22 Propan-1-ol would produce signals at 29 ($C_2H_5^+$) and 43 ($C_3H_7^+$) which propan-2-ol would not.

23 The 10^6 factor scales the chemical shift values so most fall between 1 and 10.

24 The electronegative oxygen atom near to the aldehyde hydrogen atom causes the hydrogen nucleus to resonate at a lower magnetic field.

25 Each hydrogen atom is bonded to the same C atom.

26 i 1 ii 2 iii 3 iv 1 v 2

27 i 3:1 ii 3:1 iii 3:2:1

28 A quartet in a 1:3:3:1 ratio.

29 A quartet (from the CH_3 group) of doublets (from the OH group).

Application

Substance X

1 C_3H_7Cl

2 X contains a saturated C–Cl group.

3 $[3(78) + 80]/4 = 78.5$

4 Empirical formula mass
$= 36 + 7 + 35.5 = 78.5$

5 $CH_3CH_2CH_2Cl$ (1-chloropropane);
$CH_3CHClCH_3$ (2-chloropropane)

6 C–C, C–H (alkane) and C–Cl

7 Common to both isomers.

8 Comparing whole spectrum with spectra of authentic samples of the two isomers, especially in the fingerprint region.

9 $CH_3CH_2CH_2Cl$ produces 3 n.m.r. absorptions, ratio 3:2:2
$CH_3CHClCH_3$ produces 2 n.m.r. absorptions, ratio 6:1, so X = 2-chloropropane.

10 CH_3 groups have one neighbouring H (C–H), therefore doublet.
CH group has six neighbouring H ($2CH_3$), therefore heptet.

Substance Y

1 $C_5H_{10}O$ **2** Y contains C–H and C=O bonds: not aldehyde. **3** 86

4 Empirical formula mass = 60 + 10 + 16 = 86.
So molecular formula = empirical formula = $C_5H_{10}O$

5 $[CH_3]^+$, $[CH_3CO]^+$, or $[C_3H_7]^+$. $CH_3\overset{+}{C}HCH_3$ not $CH_3CH_2CH_2^+$ because no $CH_3CH_2^+$

6

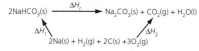

3-methyl butan-2-one

Chapter 12

1 Energy must be supplied to overcome the attractive forces between the positively charged nucleus and the negatively charged electrons.

2 For element **a** there is a large increase in ionisation enthalpy after the 3rd electron has been removed. This indicates that the atom is in Group 3. For similar reasons, element **b** is in Group 4 and element **c** is in Group 6.

3 The energy has to be supplied to break the bonds between atoms.

4

$$2NaHCO_3(s) \xrightarrow{\Delta H_r} Na_2CO_3(s) + CO_2(g) + H_2O(l)$$

$\Delta H_1 \searrow \qquad \nearrow \Delta H_2$

$2Na(s) + H_2(g) + 2C(s) + 3O_2(g)$

clockwise = anticlockwise: $\Delta H_1 + \Delta H_r = \Delta H_2$
$\Delta H_r = \Delta H_2 - \Delta H_1 = -1810.0 - (-1901.6) = +91.6$ kJ mol^{-1}.
The reaction is endothermic.

5 The enthalpy change for the formation of hydrogen chloride:
Breaking the bonds in $Cl_2(g)$ and $H_2(g)$ to form $Cl(g) + 3H(g)$
$\frac{1}{2}Cl_2(g) = \frac{1}{2} \times 234.4 \quad = + 121.7$
$\frac{1}{2}H_2(g) = \frac{1}{2} \times 435.9 \quad = + \underline{217.95}$
total $\quad = + 339.65$
Forming the bonds in HCl(g) = −432.0
Total enthalpy change for bond breaking and bond forming:
$= +339.65 + (−432) = −92.35$ kJ mol^{-1}. The value is quite close to the experimental value, so mean bond enthalpy values are satisfactory here.
Experimental value for $\Delta H_f = −92.3$ kJ mol^{-1}.

6 The steps you need to make are to calculate are: i the standard enthalpy change; ii the standard entropy change; iii the Gibb's free energy change at 298 K. Above what temperature might the reaction proceed spontaneously?
enthalpy change = −1767 −(−1902) = +135 kJ mol^{-1}
Remember the units for entropy include J (not kJ) i.e. J K^{-1} mol^{-1}
Entropy change for $\quad NaHCO_3(s) \rightarrow Na_2CO_3(s) + H_2O(g) + CO_2(g)$
$\qquad\qquad\qquad 2 \times 102 \rightarrow \qquad 135 \quad + \quad 189 \quad + \quad 214$
entropy increase = 538 − 204 = 334 J mol^{-1}
At 298K the combined entropy term $−T\Delta S^{\ominus} = 298 \times 334/1000 = 99.5$ kJ mol^{-1}
At 298K $\Delta G^{\ominus} = \Delta H^{\ominus} − T\Delta S^{\ominus} = 135 -99.5 = 35.5$ kJ mol^{-1}
$\Delta G = \Delta H − T\Delta S$ is positive; the reaction is not thermodynamically feasible at this temperature. The temperature at which feasibility occurs is when $\Delta G^{\ominus} = 0$.
$\Delta G^{\ominus} = \Delta H^{\ominus} − T\Delta S^{\ominus}$. $T\Delta S^{\ominus} = \Delta H^{\ominus}$
$T = \Delta H^{\ominus}/\Delta S^{\ominus} = 135/0.334 = 404$ K
This means that the $NaHCO_3$ must be heated to at least 404 K (131 °C). Note this is only the temperature at which the reaction becomes feasible and it gives no indication of the rate.

Application

1

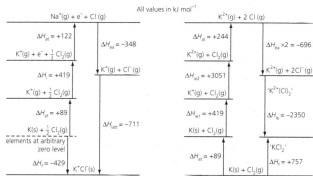

All values in kJ mol^{-1}

Generally an energy input is needed to overcome attractive forces, and energy is released when particles are attracted and move towards each other.
Atomisation is endothermic. An energy input is needed. In metals the lattice of positive ions is held together by the sea of mobile, negative electrons. Covalent molecules – an energy input is needed to form the separate atoms by breaking covalent bonds.
Ionisation enthalpy is endothermic. An energy input is needed. The negative electrons are attracted to the positive nucleus.
Electron affinity is exothermic for the first electron and endothermic for any further electrons. Electrons are attracted towards the positive nucleus so energy is released. If more electrons are introduced the electron-electron repulsion means that an energy input is needed.
Lattice enthalpy is exothermic. There are attractive forces between the positive and negative ions, so energy is released.
ΔH_f for KCl is −429 kJ mol^{-1}, so there is a net release of energy. ΔH_f for KCl_2 is +757 kJ mol^{-1}, the process is endothermic, so KCl is the more favoured energetically.

2

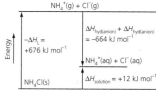

3 For a reaction to be spontaneous Gibb's free energy must be zero or negative and $\Delta G = \Delta H − T\Delta S$. Process can proceed spontaneously because there is an increase in entropy as the ordered crystal arrangement is changing to a disordered system in solution. Even though ΔH is positive the increase in entropy and the combined $−T\Delta S$ is sufficiently negative to make ΔG negative.

4 $\Delta H = mc\Delta T = 200$ g $\times 4.17$ J g^{-1} $K^{-1} \times 16.5$ K $= 13760$ J $= 13.76$ kJ

5 The molar enthalpy change for ammonium nitrate is +26.5kJ mol^{-1} so the number of moles needed is $13.76/26.5 = 0.519$ mol
number of moles = mass $\times M_r = 0.519 \times 80.0 = 41.5$g

Chapter 13

1   –1 , +7

2   a   Nuclear charge is increasing across Period 3, which increases the attraction for the electrons, which decreases the atomic radius.

   b   The metal lattice is held together by the forces of attraction between the positive nucleus and the negative outer (delocalised) electrons. The nuclear charge of magnesium is greater than that for sodium so the attraction for the delocalised electrons is greater, so more energy is needed to overcome these forces.

   c   The nuclear charge for sodium is the least of all the Period 3 elements, so the attraction for the outer electrons is the lowest, so the energy to remove one electron is the least. Across Period 3 nuclear charge is increasing, the inner electron screening remains constant, so the first ionisation energy increases from left to right.

3   The melting point is affected by the force of attraction between the positive and the negative ions. In the fluorides, magnesium 2+ has a greater ionic charge than sodium 1+, so there is a greater force of attraction between ions, so more energy is needed to separate them.

4   The melting point of silicon dioxide is relatively high a for a covalent compound because the structure is macromolecular and is held it together by a covalent bonds. The oxides of sulphur have low melting points. Sulphur(VI) oxide is covalent and unimolecular, sulphur(VI) oxide is a trimer at room temperature, it melts or sublimes very easily.

5   Dissolving in non-polar solvents and not conducting electricity when molten are properties typical of covalent compounds. Although aluminium is a metal, its high ionic charge means it is very polarising and there is considerable distortion of the chloride ion, which produces significant covalent character – hence its covalent properties.

6   Because the aluminium ion is highly polarising it is readily hydrolysed in water producing a high enough concentration of hydroxonium ions to produce a reaction with sodium carbonate solution. Hence bubbles of carbon dioxide are slowly formed. Hydrolysis also produces a white precipitate of aluminium hydroxide. Aluminium carbonate is not formed because of the highly polarising aluminium ion which would distort the carbonate ion, decomposing it.

Application

1   They both have a density less than that for water.

2   $2Na + 2H_2O \rightarrow 2NaOH + H_2$, $2K + 2H_2O \rightarrow 2KOH + H_2$

3   The first ionisation energy for potassium is lower than that for sodium. The value for the melting point for potassium are lower than for sodium.

4   For a substance to dissolve, ΔH_{sol} needs to be negative or only slightly positive. ΔH_{sol} is the difference between the lattice dissociation enthalpy and the hydration enthalpy. For sodium oxide, ΔH_{le} is much less negative than ΔH_{hyd} so ΔH_{sol} is negative and it dissolves. For both magnesium oxide and aluminium oxide, ΔH_{le} is much more negative than ΔH_{hyd} so ΔH_{sol} is positive and it does not dissolve. The increase in entropy on dissolving will not compensate for the large positive ΔH_{sol} (see Chapter 12).

5   $Na_2O + H_2O \rightarrow 2NaOH$

6   $Al_2O_3 + 6HCl \rightarrow Al_2Cl_6 + 3H_2O$ or $Al_2O_3 + 6HCl + 3H_2O \rightarrow 2[Al(H_2O)_6]^{3+} + 6Cl^-$
   $Al_2O_3 + 2NaOH + 3H_2O \rightarrow 2[Al(OH)_4]^- + 2Na^+$.

7   This behaviour is described as amphoteric.The aluminium ion has an overall charge of 3+ which means it has a strong attractive force for the lone pairs in the H_2O molecule, so in alkaline conditions the H_2O molecule will lose H^+ and form the aluminate ion, but in acidic conditions the H^+ ion will be retained and it forms the hexaaquaaluminium ion.

8   $P_4O_{10} + 6H_2O \rightarrow 4H_3PO_4$, $SO_2 + H_2O \rightarrow H_2SO_3$. Both these compounds will then release H^+ ions forming an acidic solution. Phosphorus and sulphur both have high electronegativities so the bonds P–O and S–O are quite strong and the O–H bond is weakened, therefore H^+ ions are released. For sodium and magnesium the electronegativities are lower, so they stay as Na^+ and Mg^{2+} ions thus releasing OH^- ions and the solutions are basic.

9   A low melting and boiling point indicates covalent bonding with weak forces between molecules.

10 Phosphorus has a high electronegativity value so the bonding in PCl_5 is covalent. H_2O molecules will donate into the empty orbitals of the P atom displacing Cl^- ions. As the P–H_2O bonds form the high electronegativity of the P atom means that the O–H bond is weakened and H^+ ions are released.

11 They both would have low melting and boiling points, because they are predominantly covalent compounds. Silicon chloride is more covalent than aluminium chloride, so will have lower melting and boiling points. They both react violently with water forming acidic solutions. Silicon is more electronegative than aluminium so will be more violently hydrolysed on contact with water. Both form acidic solutions and give fumes of HCl(g). $Al_2Cl_6 + 12H_2O \rightarrow 2Al(H_2O)_3(OH)_3 + 6HCl$, $SiCl_4 + 3H_2O \rightarrow H_2SiO_3 + 4HCl$

Chapter 14

1   a = +3, b = +6, c = +6, d = +1.

2   a $Mg^{2+}(aq) + 2e^- \rightleftharpoons Mg(s)$

   b $Fe^{3+}(aq) + e^- \rightleftharpoons Fe^{2+}(aq)$ c $^1/_2Cl_2(g) + e^- \rightleftharpoons Cl^-(aq)$

3   a $Zn(s)|Zn^{2+}(aq) \| Pb^{2+}(aq)|Pb(s)$ b $Zn(s)|Zn^{2+}(aq) \| H^+(aq)|H_2(g)|Pt(s)$
   c $Cu(s)|Cu^{2+}(aq) \| Fe^{3+}(aq),Fe^{2+}(aq)|Pt(s)$

4   If the half-reaction for the electrode has a more negative potential than the calomel electrode, subtract 0.27 V. If the half-reaction for the electrode has a more positive potential than the calomel, add 0.27 V.

5   You would choose two electrodes which gave you as large an e.m.f. as possible, i.e. $Ag(s)|Ag(s)$ cell with $Li^+|Li(s)$ cell. In reality you would need to consider many other factors as well, such as mass (lower mass is preferred), cost (the cheaper the materials and construction costs, the better), and chemical hazards (e.g. reactivity of electrode chemicals with air and water, their toxicity, etc.).

6   The equilibrium will move from left to right towards zinc to reduce the number of zinc ions in solution.

7   a
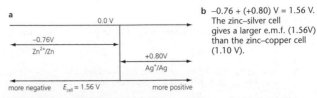

   b   –0.76 + (+0.80) V = 1.56 V. The zinc–silver cell gives a larger e.m.f. (1.56V) than the zinc–copper cell (1.10 V).

   c   A cell of electrodes Mg^{2+}/Mg and Zn^{2+}/Zn produces an e.m.f. of 1.61 V (difference between –2.37 and –0.76). A cell of electrodes Zn^{2+}/Zn and Ag^+/Ag produces an e.m.f. of 1.56 V (difference between –0.76 and +0.80). A cell of electrodes Mg^{2+}/Mg and Ag^+/Ag produces an e.m.f. of 3.17 V (difference between –2.37 and +0.80).

9   For iodine to displace bromine from solution the following two half-reactions must occur:
   $I_2(s) + 2e^- \rightleftharpoons 2I^-(aq)$ $E^{\ominus} = +0.54 V$
   $2Br^-(aq) \rightleftharpoons Br_2(l) + 2e^-$ $E^{\ominus} = -1.07 V$
   The sum of these two half-reactions is:
   $I_2(s) + 2Br^-(aq) \rightleftharpoons 2I^-(aq) + Br_2(l)$ $E^{\ominus}_{cell} = -0.53 V$
   For this reaction to be spontaneous $E^{\ominus}_{cell}$ must be positive. Therefore, iodine does not displace bromine from solution.

10 a   For a spontaneous cell reaction to occur $E^{\ominus}_{cell}$ must be positive, so for the Li/MnO_2 cell: $E^{\ominus}_{cell} = +1.23 -(-3.03) = +4.26 V$

   b   To work out the reaction for the cell, write the more negative half-reaction as an oxidation, and the more positive half-reaction as a reduction. Then balance the number of electrons in each half- reaction and add the two together:
   $$2Li \rightleftharpoons 2Li^+ + 2e^-$$
   $$MnO_2 + 4H^+ + 2e^- \rightleftharpoons Mn^{2+} + 2H_2O$$
   $$2Li + MnO_2 + 4H^+ \rightleftharpoons 2Li^+ + Mn^{2+} + 2H_2O$$

11 a   Zinc, because Zn^{2+}/Zn has a more negative electrode potential than H^+/H_2, so that electrons flow from the zinc half-reaction to the hydrogen ions in the acid, producing hydrogen gas molecules.

   b   Yes. Fe^{3+}/Fe^{2+} has a more negative electrode potential than Br_2/Br^- so electrons flow towards Br_2/Br^-. This oxidises iron(II) ions to iron(III).

   c   No. I_2/I^- has a more negative electrode potential than Fe^{2+}/Fe so electrons flow towards Fe^{2+}/Fe This reduces iron(III) ions to iron(II).

   d   Yes. Fe^{3+}/Fe^{2+} has a more negative electrode potential than $MnO_4^-,H^+/Mn^{2+}$.

Application

1   a $Zn^{2+}(aq) + 2e^- \rightleftharpoons Zn(s)$ b $Mn^{4+}(aq) + e^- \rightleftharpoons Mn^{2+}(aq)$

   c $^1/_2O_2(g) + H^+ + 2e^- \rightleftharpoons OH^-(aq)$

   d   Reduction is more likely because the equilibrium goes from left to right. Therefore, the cell potential will be less negative.

2   a $Zn(s)|Zn^{2+}(aq) \|$ b $Li^+(aq)|Li(s) \|$ c $Pt(s)|H_2(g)|H^+(aq) \|$

   d   Under standard conditions it is the Standard Hydrogen Electrode. It is assigned a value zero and all other standard electrode potentials are measured relative to this.

   e   Hydrogen pressure = 1 bar, temperature = 298 K, $[H^+]$ = 1 mol dm^{-3}.

3   a $Zn(s)|Zn^{2+}(aq) \| Ag^+(aq)|Ag(s)$ b $Pt(s)|H_2(g)|H^+(aq) \| Zn^{2+}(aq)|Zn(s)$

4   a   The zinc/silver cell gives a larger e.m.f. (1.56 V) than the zinc/copper (1.10 V).

   b
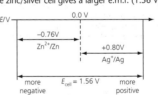

   c   –0.76 + (+0.80) V = 1.56 V

   d   It has a higher cell potential. The cell reactants have a lower density.

5   i   Silver will be displaced. For zinc to displace silver from aqueous silver solution the following two half-reactions must occur:
   $Zn(s) \rightleftharpoons Zn^{2+}(aq) + 2e^-$ $E^{\ominus} = -(-0.76) V$
   $Ag^+(aq) + e^- \rightleftharpoons Ag(s)$ $E^{\ominus} = +0.80 V$
   The sum of these two half-reactions is:
   $Zn(s) + Ag^+(aq) \rightleftharpoons Zn^{2+}(aq) + Ag(s)$ $E^{\ominus}_{cell} = +1.56 V$
   $I_2(s) + 2Br^-(aq) \rightleftharpoons 2I^-(aq) + Br_2(l)$ $E^{\ominus}_{cell} = -0.53 V$
   For this reaction to be spontaneous, $E^{\ominus}_{cell}$ must be positive. Therefore, zinc does displace silver from solution.

   ii   No reaction. This is the reverse of the reaction in part i. $E^{\ominus}_{cell}$ is negative. For this reaction to be spontaneous, $E^{\ominus}_{cell}$ must be positive, so no reaction occurs.

   iii No reaction. For iodine to displace bromine from bromide solution the following two half-reactions must occur:
   $I_2(s) + 2e^- \rightleftharpoons 2I^-(aq)$ $E^{\ominus} = +0.54 V$
   $2Br^-(aq) \rightleftharpoons Br_2(l) + 2e^-$ $E^{\ominus} = -1.07 V$
   The sum of these two half-reactions is:
   $I_2(s) + 2Br^-(aq) \rightleftharpoons 2I^-(aq) + Br_2(l)$ $E^{\ominus}_{cell} = -0.53 V$
   For this reaction to be spontaneous $E^{\ominus}_{cell}$ must be positive. Therefore, iodine does not displace bromine from solution.

   iv   This is the reverse of the reaction in part iii. $E^{\ominus}_{cell}$ is negative For this reaction to be spontaneous, $E^{\ominus}_{cell}$ must be positive so no reaction occurs.

Chapter 15

1 $Fe^0 = [Ar] 3d^64s^2$ $Fe(II) = [Ar] 3d^64s^0$ $Fe(III) = [Ar] 3d^54s^0$ 2 $Cu = 6$ $Ni = 4$
3 $Cu[(NH_3)_4(H_2O)_2]^{2+}$
4 a   90°   b   109.5°
5 a   6
   b   2+ c neutral d the ligand forms 4 co-ordinate bonds with the central metal ion e
     as shown for the four N atoms in the square planar arrangement, Fig. 9, p. 211.
6 $[Ar] 3d^64s^2$
7 Blue is at the high energy end of the spectrum, so blue is more likely to be absorbed.
8 H_2O because the solution appears pink, so the blue end of the spectrum is being
   absorbed.
9 If the ligand is changed in a complex, the 2:3 split of the d orbitals changes, and the
   energy difference changes so the energy of the light absorbed changes and therefore
   the colour changes.
10 a $[Cu(H_2O)_6]^{2+} = +2$ $[Cu(CN)_4]^{3-} = +1$
   b $[Cu(H_2O)_6]^{2+} = [Ar] 3d^94s^0$ $[Cu(CN)_4]^{3-} = [Ar] 3d^{10}4s^0$
   c   In $[Cu(H_2O)_6]^{2+}$ not all the d orbitals are occupied, so the photons in the visible
     region of the spectrum can be absorbed and colour is produced. $[Cu(CN)_4]^{3-}$ has
     all its d orbitals occupied, so absorption is not in the visible region so the
     substance is not coloured.
11 Absorption of colour is by photons interacting with the ions present in solution, so if
   the solution is diluted there is a lower concentration of ions, therefore fewer interactions
   as the light passes through it. This will result in a less intense (paler) colour but the
   actual colour will be unchanged.
12 Between 610 and 620 nm
Application
1 $[Ar]3d^74s^2$ 2 $[Ar]3d^74s^0$
3 Drawing of an octahedral complex resembling Fig. 13, page 214. Co-ordination
   number 6.
4 The ligands cause a splitting of the 3d energy levels. This produces an energy difference
   and the energy absorbed to move one or more of the d electrons from the lower
   energy levels to the higher ones corresponds to the energy of light in the visible region
   of the spectrum $\Delta E = h\nu$, where ΔE is the energy difference between the groups of d
   orbitals, h = Planck's constant and ν = the frequency of the radiation absorbed. Hence
   some parts of the visible spectrum are absorbed and the remainder is transmitted. The
   light transmitted will not be white light: it will consist of all those colours not absorbed.
   Absorbing the blue end of the spectrum produces transmitted red light.
5 a   The drawn complex is tetrahedral with $SCN\rightarrow$ as ligand; its co-ordination number
     is 4.
   b   The colour depends upon the ligands attached since these will affect the splitting of
     the d orbitals. If the ligand changes the energy difference ΔE, the frequency of the
     radiation absorbed changes, so the transmitted colour will change.
6 The aqua complex is only pale pink, so the intensity is not very great, so not much light
   will be absorbed by the solution. The thiocyano complex gives a very strong
   absorption, so much lower concentrations can be detected easily.
7 Any finger marks or scratches produced by touching the optically flat surfaces will
   scatter some light, reducing the amount transmitted. This will reduce the sensitivity of
   the test.
8 According to the Beer–Lambert law absorbance A depends upon $A = \varepsilon lc$
   ε = molar absorption (extinction) coefficient (a constant for how strongly the
      molecule absorbs as a 1 mol^{-1} dm^3 solution)/mol^{-1} dm^3 cm^{-1}
   l = path length/cm c = concentration of solution/mol dm^{-3}
   If the path length is greater, then more light will be absorbed as it passes through the
   solution.
9 1.15 mg dm^{-3}

Chapter 16

1 a $3Zn + Cr_2O_7^{2-} + 14H^+ \rightarrow 3Zn^{2+} + 2Cr^{3+} + 7H_2O$ b $Zn + 2Cr^{3+} \rightarrow Zn^{2+} + 2Cr^{2+}$
   c $4Cr^{2+} + O_2 + 4H^+ \rightarrow 4Cr^{3+} + 2H_2O$
2 $Cr_2O_7^{2-}$:   Cr $2 \times +6$ O $7 \times 2-$ overall charge -2
   CrO_4^{2-}:   Cr $+6$ O $4 \times 2-$ overall charge -2
3 Moles of MnO_4^- used $21.5 \times 0.018/1000 = 3.87 \times 10^{-4}$ mol
   The balanced half equations give this full equation:
   MnO_4^- (aq) + 5e⁻ + 5Fe^{2+}(aq) + $8H^+$(aq) → Mn^{2+}(aq) + 5Fe^{3+}(aq) + 5e⁻ + $4H_2O$(l)
   1 mol of manganate(VII) reacts with 5 mol iron(II) ions so:
   Moles of Fe(II) = $5 \times 3.87 \times 10^{-4} = 1.935 \times 10^{-3}$ mol
   Mass of $Fe^{2+} = 56 \times 1.935 \times 10^{-3} = 0.108$ g
   % mass of iron in the tablet = $0.108 \times 100/0.780 = 13.9\%$ (to 3 s.f.)
4               $Cr_2O_7^{2-} + 14H^+ + 6e^- \rightarrow 2Cr^{3+} + 7H_2O$
   Oxidation state: +6   +3
                   $6Fe^{2+} \rightarrow 6Fe^{3+} + 6e^-$
   Oxidation state: +2   +3
   There is a transfer of 6 electrons from the Fe^{2+} to the $Cr_2O_7^{2-}$. Adding the two half
   reactions: $Cr_2O_7^{2-} + 14H^+ + 6Fe^{2+} + 6e^- \rightarrow 2Cr^{3+} + 6Fe^{3+} + 6e^- + 7H_2O$
   Cancelling electrons gives the equation.
5 a   Moles of $Cr_2O_7^{2-}$ used $31.0 \times 0.0185/1000 = 5.74 \times 10^{-4}$ mol
   b   The equations for the two half reactions are:
     $Cr_2O_7^{2-} + 14H^+ + 6e^- \rightarrow 2Cr^{3+} + 7H_2O$ $Fe^{2+} \rightarrow Fe^{3+} + e^-$
     Balancing the electron transfer gives: $6Fe^{2+} \rightarrow 6Fe^{3+} + 6e^-$
     and the overall equation is:
     $Cr_2O_7^{2-} + 14H^+ + 6Fe^{2+} \rightarrow 2Cr^{3+} + 6Fe^{3+} + 7H_2O$
     1 mol of dichromate(VI) reacts with 6 mol iron(II) ions.
   c   Moles of Fe(II) = $6 \times 5.74 \times 10^{-4} = 3.45 \times 10^{-3}$ mol
   d   Moles of Fe(II) in the whole solution = 3.45×10^{-2} mol
   e   Mass of $Fe^{2+} = 56 \times 3.45 \times 10^{-2} = 1.93$ g
   f   % mass of iron in the wire = $1.93 \times 100/2.225 = 86.7\%$ (to 3 s.f.)

6 Moles of MnO_4^- used $45.0 \times 0.0200/1000 = 9.0 \times 10^{-4}$ mol
   The balanced half equations give this full equation
   MnO_4^- (aq) + 5e⁻ + $8H^+$(aq) → Mn^{2+}(aq) + $4H_2O$(l)
   SO_3^{2-} (aq) + H_2O(l) → SO_4^{2-} (aq) + 2e⁻ + $2H^+$
   $2MnO_4^-$ (aq) + 10e⁻ + $5SO_3^{2-}$(aq) + $6H^+$(aq) →
                              $2Mn^{2+}$(aq) + $5SO_4^{2-}$ (aq) + 10e⁻ + $3H_2O$(l)
   1 mol of manganate(VII) reacts with 5/2 mol SO_3^{2-} ions so:
   moles of SO_3^{2-} = $5/2 \times 9.0 \times 10^{-4} = 2.25 \times 10^{-3}$ mol (in 25.0 cm^3)
   concentration of SO_3^{2-} = $1000 \times 2.25 \times 10^{-3}/25 = 9.00 \times 10^{-2}$ mol dm^{-3}
7 Ethanedioic acid is $(COOH)_2.2H_2O$ ($M_r = 126$) and in water it behaves as the
   ethanedioate ion $C_2O_4^{2-}$(aq).
   The balanced half-equations give this full equation:
   MnO_4^-(aq) + 5e⁻ + $8H^+$(aq) → Mn^{2+}(aq) + $4H_2O$(l)
   $C_2O_4^{2-}$(aq) → $2CO_2$(g) + 2e⁻
   $2MnO_4^-$(aq) + 10e⁻ + $5C_2O_4^{2-}$(aq) + $16H^+$(aq) →
                          $2Mn^{2+}$(aq) + $10CO_2$(g) + 10e⁻ + $8H_2O$(l)
   1 mol of $C_2O_4^{2-}$(aq) reacts with 2/5 mol manganate(VII) so:
   moles of $(COOH)_2.2H_2O$ (to give $C_2O_4^{2-}$(aq)) = $2.145/126 = 1.702 \times 10^{-2}$ mol (dissolved
   in 250 cm^3)
   moles of $(COOH)_2.2H_2O$ in 25.0 $cm^3 = 1.702 \times 10^{-3}$ mol
   moles of MnO_4^- used = $2 \times 1.702 \times 10^{-3}/5 = 6.81 \times 10^{-3}$ mol
   concentration of $MnO_4^- = 1000 \times 6.81 \times 10^{-3}/35.0 = 0.0195$ mol dm^{-3}
8 A rough surfaced coating has more surface area and more active sites.
9 MnO_4^- (aq) + 5e⁻ + $8H^+$(aq) → Mn^{2+}(aq) + $4H_2O$(l)
   $C_2O_4^{2-}$(aq) → $2CO_2$(g) + 2e⁻
   This is homogeneous catalysis because both reactants are in the same phase.
Application
1 Co, Cr and Cu all have incomplete d shells in either their elements or compounds or
   both. Zinc has the electronic configuration $[Ar]3d^{10}$ $4s^2$ as the element and $[Ar]3d^{10}$ in
   its compounds, giving it a complete d shell in both so it is not normally considered to
   be a transition element.
2 $[Cr(H_2O)_5OH]^{2+}$
3 A catalyst is an element or compound that alters the rate of a reaction and remains
   unchanged in chemical composition at the end.
4 a   The H^+ takes part in one of the reaction stage; it speeds up the reaction, but is still
     present in the same quantities at the end of the reaction.
   b   A homogeneous catalyst is in the same phase as the other reactants, in this case
     the aqueous phase.
5 In the manufacture of chocolate the nickel is a solid and the other reactants are gases
   or liquids – they are in different phases so nickel is a heterogeneous catalyst. The nickel
   is finely divided and indistinguishable from the other reactants in the reaction vessel.
6 The more finely divided the nickel, the greater is the surface area on which the
   hydrogenation reaction can take place, so the reaction will be faster.
7 Tungsten forms very strong links with the molecules that are adsorbed onto it, so the
   products when formed are difficult to desorb. If this happens there are fewer sites left
   for remaining molecules to react. Silver only forms weak links with the molecules
   attached to it, so these molecules are not held in place for as long, so fewer fruitful
   collisions are likely.
8 In a poisoned catalyst, unwanted impurities block the active sites on the catalyst.
   Molecules such as carbon monoxide form very strong links with the catalyst and these
   stick to the active sites which are no longer available for the reactant molecules, so the
   catalysis cannot take place.
9 Enzymes are proteins and the active sites where the reactions take place have a very
   specific stereochemistry (arrangement in space), so that usually only a specific molecule
   or part molecule with an identical stereochemistry can fit into it. This is often referred to
   as a lock and key mechanism. If the protein is heated strongly, the stereochemistry will
   change, the reactant molecule will no longer 'fit' and the reaction cannot take place.

Chapter 17

1 a   Lewis acid = H^+
   b   Lewis base = H_2O
   c   Lewis acid = BF_3
   d   Lewis base = NH_3
2 iv and vi
3 a   octahedral $[V(H_2O)_6]^{2+}$
   b   octahedral $[Cr(H_2O)_6]^{3+}$
   c   octahedral $[Fe(H_2O)_6]^{2+}$
4 Transition metal ions have colour because there is an energy difference between groups of
   the d orbitals. Light is absorbed which corresponds to the visible region of the spectrum
   as electrons move between orbitals. The energy difference between the groups of orbitals
   is affected by the nuclear charge and the nature of the ligands. If the energy difference
   changes then the energy of the light absorbed will be different and the colour of the light
   not absorbed will change.
5 a $[Cr(H_2O)_6]^{3+} + H_2O \rightleftharpoons [Cr(H_2O)_5(OH)]^{2+} + H_3O^+$
   b   pH 3–6
   c   The highly polarising 3+ ion attracts electrons from the metal–oxygen bond and
     therefore weakens the oxygen–hydrogen bond in the ligand. This releases some H^+
     ions forming H_3O^+ ions, producing acidity and lowering the pH value below 7.
6 a   Adding H^+ ions drives the equilibrium to the left, from the brown $[Fe(H_2O)_5(OH)]^{2+}$
     complex to the pale violet $[Fe(H_2O)_6]^{3+}$ complex.
   b   Adding OH- ions means they will react with the H^+ ions (forming H_2O) and removing
     H^+ ions from solution. The equilibrium will shift to replace these H^+ ions, so the
     equilibrium will shift from left to right and produce a brown colour.
   c   Adding excess OH- ions will send the equilibrium as far right as is feasible and a
     brown precipitate of $Fe(H_2O)_3(OH)_3$ forms.
7 a $[Ni(H_2O)_6]^{2+} + CO_3^{2-} \rightleftharpoons NiCO_3 + 6H_2O$, $NiCO_3$ is a green precipitate.
   b $2[Fe(H_2O)_4(Cl)_2]^+ + H_2O + 3CO_3^{2-} \rightleftharpoons 2[Fe(H_2O)_3(OH)_3] + 3CO_2 + 4Cl^-$. A brown
     precipitate of $[Fe(H_2O)_3(OH)_3]$ and an effervescence of carbon dioxide gas would
     be seen.

8   One ammine ligand replaces one aqua ligand at each step.
9   The equilibrium should shift to the right, replacing all the aqua ligands with ammine ligands, but the two aqua ligands are not replaced because they are further away from the central metal ion.
10 a   The co-ordination number is 4 and the oxidation state is +2.
    b $[Co(H_2O)_6]^{2+} + 4Cl^- \rightleftharpoons [CoCl_4]^{2-} + 6H_2O$. The co-ordination number changes from 6 in the aqua complex to 4 in the chloro complex.
11 a $[Co(H_2O)_6]^{2+} + EDTA^{4-} \rightleftharpoons [CoEDTA]^{2-} + 6H_2O$
    b $[Fe(H_2O)_6]^{3+} + 3C_2O_4^{2-} \rightleftharpoons [Fe(C_2O_4)_3]^{3-}$. The structure is the same as Fig. 15, row 2, right structure with Fe replacing M.
12 Energy is released in the process, so the products are at a lower energy level (more stable) than the reactants were.

Application

1   $[Cu(H_2O)_6]^{2+} \rightleftharpoons [Cu(H_2O)_5(OH)]^+ + H^+$; $[Cr(H_2O)_6]^{3+} \rightleftharpoons [Cr(H_2O)_5(OH)]^{2+} + H^+$.
    These are hydrolysis reactions because the ion reacts with water and releases hydrogen ions giving an acidic solution.
2   The Cr^{3+} ion has a higher ionic charge than Cu^{2+} so the attraction for the lone pairs on the water molecule is greater and this weakens the O–H bond more.
3   The equilibrium for Cr^{3+} lies further to the right, the solution is more acidic, and pH is lower for Cr^{3+} than Cu^{2+}.
4   The hydrolysis of the aluminium aqua ion lowers the pH of the water. Copper is more soluble at low pH values, the equilibrium
    $[Cu(H_2O)_6]^{2+} \rightleftharpoons [Cu(H_2O)_5(OH)]^+ + H^+$
    lies further to the left. In tap water the copper comes from the copper pipes.
5   If chloride ions are added, the equilibrium shifts, so the solution will move towards the green complex.
6   $[Cr(H_2O)_6]^{3+} + OH^- \rightleftharpoons [Cr(H_2O)_5(OH)]^{2+} + H_2O$. This reaction proceeds stepwise depending on the amount of OH^- added. Excess OH^- gives
7   $[Cr(H_2O)_6]^{3+} + 6OH^- \rightleftharpoons [Cr(OH)_6]^{3-} + 6H_2O$.
8   $[Cu(H_2O)_6]^{2+} + 4Cl^- \rightleftharpoons [CuCl_4]^{2-} + 6H_2O$.
    a   In $[Cu(H_2O)_6]^{2+}$ each water ligand donates one lone pair to the central copper ion. It is a monodentate ligand and the six positions in the octahedral complex are occupied by six aqua ligands. With $EDTA^{4-}$ a single ligand will donate six lone pairs to the central copper ion so it is described as a multidentate ligand.
    b   There is an increase in entropy when $EDTA^{4-}$ complexes are formed. For example, in the reaction $[Cu(H_2O)_6]^{2+} + EDTA^{4-} \rightleftharpoons [CuEDTA]^{2-} + 6H_2O$, the number of particles on the left-hand side is 2, whereas the number of particles on the right-hand side is 7. As a result, there is an overall increase in entropy. An increase in entropy favours a reaction.
9   a   moles of EDTA = $4.80 \times 10^{-3} \times 0.05 = 2.4 \times 10^{-4}$ mol
    b   amount of Cu^{2+} in 100 cm³ sample = 2.4×10^{-4} mol (because it forms a 1:1 complex with EDTA)
    c   concentration of Cu^{2+} = $2.4 \times 10^{-4}/100 \times 10^{-3} = 2.4 \times 10^{-3}$ mol dm⁻³.
        1 mol Cu^{2+} has a mass of 64 g so concentration = $64 \times 1000 \times 2.4 \times 10^{-3}$
        $= 153.6$ mg dm⁻³

Index